ER

PROPOSED ROUTE

Transcontinental Railroad

MAP OF THE
SOUTHWEST TERRITORIES
OF THE **UNITED STATES**
1855

ROCKY MOUNTAINS

INDIAN
TERRITORY

ERRITORY

TEXAS

O

DOWN THE GREAT
UNKNOWN

DOWN THE GREAT
UNKNOWN

JOHN WESLEY POWELL'S 1869 JOURNEY OF DISCOVERY
AND TRAGEDY THROUGH THE GRAND CANYON

EDWARD DOLNICK

HarperCollins*Publishers*

HarperCollins books may be purchased for educational, business, or sales promotional use. For information, please write: Special Markets Department, HarperCollins Publishers Inc., 10 East 53rd Street, New York, NY 10022.

FIRST EDITION

Designed by The Book Design Group

Maps designed by Warren Cutler

Library of Congress Cataloging-in-Publication Data

Dolnick, Edward
    Down the great unknown : John Wesley Powell's 1869 journey of discovery and tragedy through the Grand Canyon / Edward Dolnick.—1st ed.
        p. cm.
    Includes bibliography references (p.) and index.
    ISBN 0-06-019619-X
        1. Grand Canyon (Ariz.)—Discovery and exploration.   2. Grand Canyon (Ariz.)—Description and travel.   3. Colorado River (Colo.-Mexico)—Discovery and exploration.   4. Powell, John Wesley, 1834–1902—Journeys—Arizona—Grand Canyon.   5. Explorers—United States—Biography.   I. Title

F788.D65 2001
917.91'3044—dc21   2001024819

01 02 03 04 05 ❖/RRD 10 9 8 7 6 5 4 3 2 1

TO RUTH AND LYNN, THE GIRLS IN MY LIFE

CONTENTS

We are now ready to start on our way down the Great Unknown . . . We are three quarters of a mile in the depths of the earth . . . We have an unknown distance yet to run; an unknown river yet to explore. What falls there are, we know not; what rocks beset the channel, we know not; what walls rise over the river, we know not. Ah, well! We may conjecture many things. The men talk as cheerfully as ever; jests are bandied about freely this morning; but to me the cheer is somber and the jests are ghastly.

—John Wesley Powell, August 13, 1869

August 30, 1869, the Colorado River at the foot of the Grand Canyon

The fishermen kept their gaze focused intently on the river, but it was not fish they were looking for. A splintered plank from a broken boat, a torn shirt, perhaps a lifeless body—these were the "fragments or relics" they had been instructed to watch for.

Only a few miles upstream, six exhausted men in two boats pushed themselves into the current. Ragged, half-starved, burnt black by ninety-eight days in the desert sun, the men were in sorry shape. Explorers who dreamed of gold and glory, they had been given up for dead weeks before. "Fearful Disaster," the *Chicago Tribune* had trumpeted on its front page on July 3. The entire party, save one man, had been "Engulfed in a Moment."

But they have *not* drowned. They are still alive, barely, still hoping they have food and strength enough to grope their way to safety.

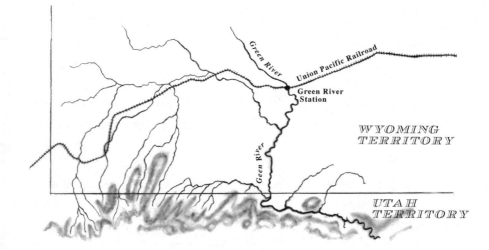

CHAPTER ONE

THE CHALLENGE

Noon, May 24, 1869

The few inhabitants of Green River Station, Wyoming Territory, gather at the riverfront to cheer off a rowdy bunch of adventurers. Ten hardy men in four wooden boats had spent the morning checking their gear and their provisions one last time—bacon, flour, coffee, spare oars, sextants and barometers (their leader, the skinny, one-armed man in the *Emma Dean*, fancied himself a scientist). Their plan could hardly be simpler. They will follow the Green River downstream until it merges with the Grand to become the Colorado, and then they will stay with the Colorado wherever it takes them. They intend in particular to run the river through the fabled chasm variously called Big Canyon or Great Canyon or Grand Canyon, a region scarcely better known than Atlantis. No one has ever done it.

The men hope to make their fortunes; their leader plans to emblazon his name across the heavens. They are brave, they have new boats and supplies to last ten months, they are at home in the outdoors. Most important, they are ready to risk their lives.

At one o'clock, the *Emma Dean*, the *Kitty Clyde's Sister*, the *Maid of the*

Cañon, and the *No Name* push themselves out into the current. A small American flag mounted on the *Emma Dean* flaps proudly in the breeze. Most of the crew are still a bit bleary-eyed. As a farewell to civilization, they have done their best to drink Green River Station's only saloon dry. Now they are suffering what one of them describes as "foggy ideas and snarly hair." The small crowd gives a cheer, the leader doffs his hat, and the four boats disappear around the river's first bend.

John Wesley Powell, the trip leader, was a Civil War veteran who had lost his right arm at Shiloh. Thirty-five years old and unknown, Powell was a tender-foot who barely knew the West, a geology professor at a no-name college, an amateur explorer with so little clout that he had ended up reaching into his own (nearly empty) pocket to finance this makeshift expedition. His appearance was as unimpressive as his résumé—at 5 feet, 6½ inches and 120 pounds, he was small and scrawny even by the standards of the age, a stick of beef jerky adorned with whiskers.

To Powell, a natural leader, all that was unimportant. Overflowing with energy and ambition, he was a man of almost pathological optimism. With a goal in mind, he was impossible to discourage.

He had devised an extraordinary goal. In 1803, with the full and enthusiastic backing of the president of the United States, Lewis and Clark had opened the door to the American West. In 1869, with almost no government support, John Wesley Powell intended to resolve its last great mystery. By this time, the map of the United States had long since been filled in. For two centuries, Boston had been a center of learning and culture. New York and Philadelphia were booming, Nashville and New Orleans struggling to recover from the Civil War. California's gold rush was almost a generation in the past. In May 1869, the pounding of a ceremonial spike at Promontory Summit, Utah, marked the completion of the transcontinental railroad.

The Rockies and the Sierra Nevada and Yosemite and Death Valley were old news. Miners in search of gold, trappers in quest of beavers whose pelts could be transformed into hats for London dandies, a host of government and railroad surveying parties, all had crisscrossed one another's steps in even the most isolated spots of the American continent.

Except one. One mystery remained. In the American Southwest an immense area—an area as large as any state in the Union, as large as any coun-

try in Europe—remained blank. Here mapmakers abandoned the careful nota-
tions that applied elsewhere and wrote simply "unexplored." Venturesome
Westerners knew that the region was desolate and bone-dry; they knew the
Colorado River ran through it; they knew that canyons cleaved the ground
like gouges cut by a titanic axe. Beyond that, rumor would have to do. Men
whispered tales of waterfalls that dwarfed Niagara and of places where the
mighty Colorado vanished underground like an enormous snake suddenly
slithering down a hole.

Powell aimed to fill in that blank in the map. His plan, such as it was, took
audacity to the brink of lunacy. Once they were well under way, he and his
men would have no supplies other than those they could carry. They had
no reliable maps—none existed—and their route stretched across a thousand
miles of high desert. It was Indian territory, and peace had yet to break out.
There were no white settlements (or settlers, for that matter) anywhere along
their river route nor within a hundred miles on either side.

The Grand Canyon itself, Powell knew, was many hundreds of miles
downstream. It was the final canyon the expedition would pass through—and
the longest and the deepest and the least known—but they would have to
confront countless obstacles before they ever drew near it. The first three-
fourths of the route, Powell guessed, led through a series of virtually unexplored canyons. The last one-fourth, if he and the crew were still alive, would
be the Grand Canyon.

Powell's friends feared he was throwing his life away. On May 24, the day
he set out, his hometown newspaper had reported on his plans. "It would be
impossible for a boat constructed of any known material, upon any conceivable plan, to live through the canyon," one supposed expert declared. "We do
not know what kind of boats Professor Powell purposes to descend the Grand
Canyon," the newspaper cautioned, "but we greatly fear that the attempt to
navigate by any means whatever will result fatally to those who undertake it."

Bon voyage!

Despite the dangers, for a man of Powell's character the temptation was irresistible. Perhaps nowhere on earth were science and adventure as intertwined

as in the American Southwest. For Powell, would-be scientist and would-be explorer, it was like a chance to be the first man on the moon. But to achieve his dreams, he would have to survive the Colorado.

The expedition's starting point, Green River Station, Wyoming, sits 6,100 feet above sea level. The destination, any of the small settlements near the mouth of the Virgin River in Arizona, was at about seven hundred feet. Powell and his men, then, were proposing to descend over a mile in the course of their journey. The question was whether the drop was sudden or gradual. Did the river follow a course like an elevator shaft or like a ramp?

No one had a clue. A waterfall as high as Niagara, a mere 170 feet, would be little more than a steep step in comparison with an overall vertical fall of a mile. For all Powell knew, his crew might find themselves trying frantically to pull upstream from a waterfall ten times higher than Niagara. Worse still, they would have almost no warning, for as the river makes its meandering way, it is hemmed in by soaring cliffs that cut off the view downstream. A mega-Niagara could be lying in wait, like some colossal mugger, around any of a thousand river bends. The men would hear it before they saw it, probably, for from water level it would look like nothing more than a sharp, horizontal line, as if the river had vanished into the air.

This was far more than a theoretical hazard. The Grand Canyon of the Yellowstone, for example, is only a fraction the size of the Grand Canyon but has two towering waterfalls. A short distance upstream of the first, the Yellowstone is "peaceful and unbroken by a ripple," in the words of one of the first explorers to describe it. A canoeist who happened on this quiet spot might be tempted to set out on a day trip. Then, suddenly, the river dives over a ledge and plunges one hundred feet. Half a mile downstream, it roars over another rock ledge, this time falling more than three hundred feet, almost twice Niagara's height.

Powell could hope that he would never confront such a sight, but he *knew* there were spots where the Colorado's drop was far from gradual—earlier explorers had tried to follow the river upstream, starting below the Grand Canyon, and had run into unnavigable rapids. Indians contributed their own tales of the Colorado's power. One old man told Powell of a calamity he had seen himself. "The rocks h-e-a-p, h-e-a-p high," he began, or so Powell recorded his words. "The water go h-oo-wooogh, h-oo-woogh; water-pony [canoe] h-e-a-p buck; water catch 'em; no see 'em Injun any more! No see 'em squaw any more! No see 'em papoose any more!" Powell respected the

Indians—in this era of Custer and Sheridan ("The only good Indians I ever saw were dead"), his attitude was rare—but he chose to ignore this warning.

Neither Powell nor any of his men had ever run a rapid. As a young man with a bad case of wanderlust (and two arms), Powell had rowed the length of the Mississippi, rowed the Ohio and the Illinois and the Des Moines. But those were rolling, midwestern rivers that hardly bore comparison with the rambunctious Green and Colorado. One of the crew had perhaps put in some time in fishing boats off the New England coast. In comparison with the others, Powell was an old pro. He had *seen* white water, from a cliff he and his wife had climbed high above the Green River at the Gates of Lodore.

"Reading the river"—identifying a path through the chaos of colliding waves and protruding rocks and sucking whirlpools—is a skill as fundamental to a boatman as reading music is to a musician, but the river was a closed book to Powell and all his men. Powell claimed once that the nine men of his crew were "all experienced in the wild life of the country, and most of them in boating on dangerous streams," but that was a stunning exaggeration. "We were all green at the business," one of the men acknowledged.

There would be no choice but to learn on the fly while careening downstream. They could hardly have picked a more forbidding classroom. (Making matters worse, the ten men had only one life jacket among them, for Powell. The able-bodied men disdained such sissy stuff.) Stretches of the Green are still feared today, and the Colorado is near the top of any list of America's white-water rivers. Powell planned to portage rapids whenever that was possible, on the theory that the heavy labor of carrying the boats (and their tons of supplies) was preferable to drowning in them. But portaging was backbreaking work and dangerous besides, for a false step could mean a broken ankle or a boat impaled on a rock. The alternative was to "line" the boats downstream—to tie ropes to bow and stern and to hang on while clambering up and over the slick rocks along the river's edge. Lining a bucking boat through the rocky margin of a rapid carried all the appeal of dragging a skittish horse through an obstacle course.

The only remaining choice, running the rapids, seemed suicidal. Even today, the big rapids on the Green and the Colorado are the stuff of dry mouths and pounding hearts. For amateurs seeing them for the first time, the rapids must have been a revelation, a rumbling, heaving nightmare. "To get an idea of the scale involved," one modern-day river guide suggests, "think of yourself as sitting in a boat on the floor of your living room. The waves . . . can be as high as the ceiling of a room on the second story! Now think of being on the roof of that two-floor house and looking down twenty feet to the bottom of a dark, churning hole."

Early in the journey, Powell and his men would still have the option of giving up, abandoning the river and hiking overland to safety. Even later on they could hope to find a side canyon that led to freedom. But that was a desperate hope. Without maps, no one could know how many miles it was to the next side canyon or where it led or whether it dead-ended in a sheer, unclimbable wall. For all anyone could know, there *was* no "next one." Once the river had pinned itself between towering rock walls, there would be no chance of escape for days on end. In canyon country, the choice would be to run the river or die.

The problem was that there were only two exits, and both were blocked. First, retreat was impossible. To take a rowboat upstream through mighty rapids was unthinkable, like trying to push a boulder up a cliff. Second, trying to climb up and over the canyon walls, in a kind of outdoor jailbreak, was almost as unlikely. The legends that had grown around the Grand Canyon, for example, were close to the truth. As long ago as 1540, Spanish conquistadors (led by Hopi guides) had crept to the rim and gazed down in slack-jawed stupefaction. The view from water level is even more stunning. From the river, the canyon cliffs soar upward for a *mile*. Writers talk glibly of the "canyons" of Manhattan, but the analogy understates reality. You could stack one of the World Trade Center towers on top of the other, and they would reach only halfway to the rim of the Grand Canyon.

Once the canyon walls closed around them, Powell and his men would be as bound by their decision as sky divers falling through the air. All they could do would be to struggle on, knowing that each mile would carry them deeper into the earth, farther into the unknown, and farther from the possibility of rescue by the outside world.

Between their starting point and safety, though they could not have known it, stretched a thousand miles of river and nearly five hundred rapids. At spots beyond counting, a moment's inattention or the briefest of mistakes could prove fatal. Drowning was only the most obvious hazard. A capsizing that left the food stores soaked or sunk would mean death just as surely, though more slowly. A boat damaged beyond repair could be a calamity. A broken leg could be a death sentence. Skill and will counted only to a degree; luck and caprice were as important. The river could grab a boat and trap it in a "hole," a kind of whirlpool turned on its side, or it could take a drowning man and spit him contemptuously to safety. The river doled out punishment with a kind of casual indifference, as a bored lion might flick a mighty paw.

Today's river runners can only shake their heads in disbelief. "By modern standards," one of them writes, "[Powell's boats] were the technical equivalent of walnut shells." But inadequate boats were only the beginning. Powell and his fellow novices would have been in desperate trouble even with the best of equipment, like beginning drivers trying to take a Ferrari down an icy mountain road.

All this was the penalty for being first. (Nor was there a crowded field of those vying for second place. The Colorado was so fearsome that as late as World War II, seven decades after Powell, only 250 people had ever been through the Grand Canyon in boats.)

In 1869, no one ran white water, and so no one knew what boats were suited to it or how best to maneuver them. Today, boat design and boating technique have had decades to evolve, and 3,500 private boaters a year run the Colorado River through the Grand Canyon. (They wait ten to fifteen years for the privilege.) The National Park Service, which oversees the sign-up process, requires that these boatmen have some degree of skill and experience. Today, Powell and his men would not qualify for a permit to run the Grand Canyon.

The private boaters are far from alone. Every year twenty thousand wet, happy, scared tourists opt for Grand Canyon trips run by commercial outfitters. Every one of those boatmen, whether private or commercial, has a store of information that Powell would have given his *other* arm for. They have maps that detail the river's course, and its hazards, mile by mile. They know from countless books and videos what they will see along the way. They know the boats to use and the food to bring and the gear to pack. They know how the rapids they will confront compare with other rapids they have run. They know that if all goes wrong, the Park Service will swoop down and helicopter them to safety.

Above all else, they have one priceless bit of knowledge whose lack tormented Powell and his companions. They know that what they are trying is possible.

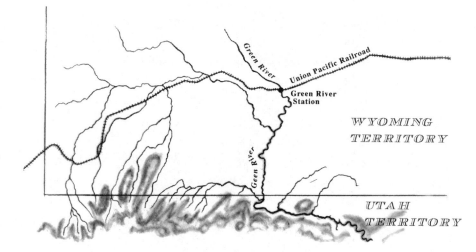

CHAPTER TWO

THE CREW

Inside the palatial railroad car, the mirrors were French and the carpets Belgian, and the menu included such indulgences as raw oysters and lobster salad and omelets made with a splash of rum. A passenger crossing the Green River by train might have been forgiven for neglecting to glance out the window. And, in truth, even if a curious sightseer had turned his eyes from the sparkling chandeliers and the black-walnut woodwork, he might not have deemed the scene below him worthy of a second glance.

He would have seen, in the middle weeks of May 1869, an empty desert and a broad river and Powell's crew of novices struggling to learn how to handle their boats. Powell and his brother Walter, and the expedition's four boats, had arrived in town on May 11, by train from Chicago. By then the rest of the men had been impatiently hanging around Green River Station for three weeks. The town was shabby and tiny—it had only a hundred residents and had not existed a year before—and three weeks was a lengthy sentence.

In its brief heyday, during the single month when it had marked the transcontinental railroad's farthest advance west, Green River Station had bustled with activity, nearly all of it illicit. These end-of-the-line towns were known collectively as Hell on Wheels. Like the others, Green River Station had plenty to tempt hardworking railroad workers and to repel anyone else. "By day disgusting, by night dangerous, almost everybody dirty, many filthy, and with the marks of lowest vice," wrote one newspaperman, "averaging a murder a day; gambling and drinking, hurdy-gurdy dancing and the vilest of sexual commerce the chief business and pastime of the hours."

Those were the good old days. By the time Powell's men arrived in Green River Station, the railroad workers had been gone for six months and the town was almost deserted. The visitor who had sampled Jake Fields's home-made whiskey and Ah Chug's apple pie had largely exhausted Green River's amusements.

Powell had chosen Green River Station as a rendezvous not because of any virtue of the town itself but simply because the train stopped there. This was no ordinary train, but the transcontinental railroad, the marvel of the age. The name alone, in the nineteenth century, conveyed power, glamour, and pizzazz. And it was brand-new. On May 10, 1869, millions of Americans across the nation had waited eagerly for word of the driving of the golden spike that linked East and West. For six years, the Central Pacific's army of laborers had been racing east from Sacramento over the Sierra Nevada, digging their way through forty-foot snowdrifts and blasting tunnels through thousands of feet of solid rock. (Workmen, especially the Chinese, had died in such numbers that the expression "not a Chinaman's chance" came into common use.) In the mean-time, the Union Pacific workforce, at its peak ten thousand strong, had raced westward from Omaha, laying track across the treeless plains and through the scorching desert while fighting off Cheyenne and Sioux raiding parties.

Now, finally, the last link was forged. The spanning of the continent, the *New York Times* exclaimed, marked "the completion of the greatest enterprise ever yet undertaken." At 12:47 P.M. on May 10, 1869, a telegraph operator tapped three dots, signaling "done." The signal triggered a simultaneous nationwide celebration—the first ever. Fire bells rang in every hamlet in the land; cannons thundered in San Francisco and New York and Omaha and Sacramento; ten thousand residents of Chicago took to the streets and formed a parade that stretched seven miles. To an enthralled nation, it was an occasion as momentous as the announcement of victory at the end of a long war.

•   •   •

For John Wesley Powell, the train made feasible the assault on what he liked to call "the Great Unknown." Rather than having to build his own boats on site with whatever materials he could scrounge together, like a desert-bound Robinson Crusoe, Powell could have first-rate, professionally made boats built in Chicago and sent west by train. He had placed his order with Thomas Bagley's boatyard in early spring, 1869. On May 11, Powell stepped off the train at Green River Station, collected his handsome new boats, and joined his restless crew.

For one moment, two of the epic sagas in American history occupied the same stage, though few ventures could have had as little in common as the transcontinental railroad and Powell's expedition. Like the space shuttle or the Concorde a century later, the transcontinental railroad was the racing, roaring embodiment of technological might and engineering elegance; Powell's rowboats were the products of a technology nearly as old as human culture. Newspapers across the country had lavished countless pages on the building of the railroad; John Wesley Powell and his men were all but anonymous. The federal government had backed the railroad with an unending succession of giveaways of cash and land; Powell had knocked on one closed door after another in Washington, looking for funding, and had come up next to empty. Leland Stanford and Collis Huntington and a host of other eager, striving men would make millions from the railroad, piling up fortunes that generations of descendants could barely dent; Powell and his men worried about starving to death. The story of the railroad was an epic with a cast of thousands; Powell and his crew were a force of ten lone men.

Now the two expeditions found themselves at the same lonely spot in the desert, in a juxtaposition that seemed almost to have required time travel. It was as if, on a single parade ground, one could see a battalion of modern soldiers with automatic weapons and night-vision goggles and also a host of knights in battered armor, perched atop gaunt-ribbed stallions.

In their reliance on tiny boats and fragile oars and their own muscle and nerve, Powell's men were a new link in an ancient chain. In common with explorers in every age, they had willingly left the safety and familiarity of home to visit an unknown world beyond the reach of rescue. But unlike modern explorers, unlike astronauts in particular, Powell and the other nine men of the grandly named Colorado River Exploring Expedition were hardly an elite corps, certainly not the survivors of a rigorous selection program.

The crew—not Powell—bore a closer resemblance to a band of rough, experienced camping buddies in pursuit of the ultimate outdoor adventure. The very factors that would have sent more prudent men scurrying toward hearth and home drew these men on. Territory unknown? Death a possibility? Good! When do we start?

The youngest member of the group was barely into his twenties, the oldest thirty-six. Six of the men, in addition to Powell, were Civil War veterans. (All seven had fought on the Union side.) Each man preferred life outdoors with a blanket and a rifle (or, in Powell's case, with a collecting box for fossilized bones and shells) to a sheltered existence of weekly paychecks and regular mealtimes. Only Powell was married.

Five of the crew were "mountain men," hunters and trappers by trade and human tumbleweeds by inclination. The mountain men were as savvy in the ways of the West as anyone on earth and twice as ornery, not much inclined to heed Powell or *any* leader. Their motives for joining the expedition varied, though vague but seductive notions of easy fortune played a key part. In a land almost unknown to white men, perhaps they would find game just waiting to be bagged and gold glistening unclaimed in the sunlight. Certainly they would find adventure, and that was perhaps the greatest temptation of all.

It was not simply that, by temperament, these men craved action. The times played a role as well. The Civil War had been agonizing, thrilling, boring, and terrifying. It was the most important event since the nation's birth and universally recognized as such. For the young men who survived it, it was clearly the most important thing that would *ever* happen to them. Few soldiers could echo the Gettysburg veteran who called that battle the most enjoyable three days of his life and a time of "joyous exultation." But more than a few exsoldiers did admit that sometimes civilian life seemed a bit humdrum, a shade anticlimactic. Most of Powell's men were in their twenties; the war had ended four years before. Against such a backdrop, the prospect of a glorious adventure, tempting at any time, beckoned all the more alluringly.

Powell's previous trips to the West—the first had been only two years before—had been in the company of eager and well-mannered college students. He liked to read aloud to them. Emerson and Scott and Longfellow were particular favorites. The men of this new expedition were a tougher lot.

Campfires were places to drink and brag and swap tall tales. An attempt to read "The Lady of the Lake" aloud was unlikely to go well.

The expedition was strictly low-budget. Powell and six of his men were volunteers. (Powell, who took for granted that everyone felt as he did, noted blithely that the unpaid crew members "give their time, feeling remunerated by the opportunity for study.") The other three men were to receive a small wage, one for taking measurements with the barometer, another for using the sextant, and the third for drawing maps. The wage, as spelled out in a homemade contract, was $25 per man per month. Even in an era when a carpenter or a stonemason earned only $3 a day and a hotel room cost $1 a night, this was far from lavish. In addition, the three men had been promised they would be given five days along the way to prospect for gold and silver and thirty days to hunt and trap. Finally, a long, meticulous list spelled out the prices Powell had agreed to pay for various skins: deer, $1.25 each; elk, $2; grizzly bear, $10; and so on, through two dozen animals.

The contract seemed routine, even optimistic in its presumption that there would be free time and good hunting. Only its last sentence hinted at darker prospects. "Should it be necessary to proceed on the journey without delay on account of disaster to boats or loss of rations," it read, "then the time specified for hunting may not be required by either party, nor shall it be deemed a failure of contract to furnish supplies should such supplies be lost in transit."

Powell's motives were different from those of his men. Hunting, prospecting, and general dare-deviltry held little appeal. For Powell, the expedition was primarily an *intellectual* adventure. "The object," he declared, "is to make collections in geology, natural history, antiquities and ethnology," in the hope of "adding a mite to the great sum of human knowledge." This was sincere, if a bit coy. Powell was a man of vaulting ambition, and he rarely thought in "mites." He intended to fill in a blank space on the map, as he freely declared. What was equally true, though Powell refrained from saying so outright, was that he intended to put *himself* on the map.

This is not to downplay Powell's passion for science, which was heartfelt. In the middle of the Civil War, for example, while his men dug trenches outside the besieged city of Vicksburg, Powell had combed the turned-up ground for fossils. In canyon country, where nature herself had dug trenches a mile deep, Powell's zeal for science would grow all the more fervent. Early on, at least, the men were bemused rather than irritated by the quirky and time-consuming obsessions of "the Professor."

Let us gather the entire crew together for a moment. In the center stands John Wesley Powell, the undisputed leader. Powell was a formidable character—intelligent, impetuous, brave, driven, visionary, a jangle of primary colors with hardly a pastel in the mix. Every gesture was quick, every word emphatic. When he was caught up in an argument, for example, he would wave his maimed right arm, beating the air with his empty sleeve. (Powell made as little concession as possible to his injury, although it made even the most routine task a trial. He could not wash his own hand, for instance, and had to rely on someone else for help.) Magnificently self-assured, Powell had what a later generation would call charisma. He was, in the words of one contemporary, "eminently a magnetic man."

Like Walt Whitman, who contained multitudes, Powell was a grab bag of contradictions. He was a democrat to the marrow of his bones, for instance, but also a firm believer in the privileges of rank. (The "boys" of his crew were "Billy" and "Sumner" and "Frank," but Powell was always "the Major.") He was so reckless that he put together an expedition that struck his contemporaries as more akin to a suicide pact, but so prudent that, to the men's dismay, he insisted on portaging rapid after rapid. He was gregarious, fond of reading aloud and singing, but on the river he often took his meals alone. He was "a Renaissance man," in the judgment of one twentieth-century historian, but "as single-minded as a buzz saw" in the eyes of another.

One contradiction is especially jarring. Powell was honest and straightforward in his financial dealings—and this in the flamboyantly corrupt Gilded Age—but he was light-fingered as a pickpocket when it came to stealing credit from others. Sometimes it was merely a matter of embellishing a story. Sometimes it was bolder than that. Powell was a fine writer, and although his account of his historic expedition has become a classic, it combined reporting and invention in an intricate mix. "He wasn't all saint," Powell's brother-in-law noted. "He could lie on occasion—be generous one minute and contemptible the next."

The fudging began early on. "After three years' study of the matter," Powell told the readers of the *Chicago Tribune* in a May 1869 letter announcing the aims of his expedition, "I think it doubtful whether these canyons have ever been seen by man."

Even without three years' study, Powell knew better than that. For countless generations, Indians had not only seen the canyons of the Southwest but had lived in them. In the sixteenth century, they had guided Spanish explorers to the Grand Canyon's rim. In the eighteenth century, the Spanish had come back for a closer look. In 1857, the Army Corps of Topographical Engineers had surveyed a stretch of the Colorado beginning near its mouth and had descended into the Grand Canyon at two different spots. In 1861, in an official document entitled the *Report upon the Colorado River of the West*, the Corps had described its findings.

The early visitors had not been tempted to linger. "The region last explored is, of course, altogether valueless," wrote Lieutenant Joseph Christmas Ives in concluding his 1861 *Report* on the Grand Canyon. "It can be approached only from the south, and after entering it there is nothing to do but leave. Ours has been the first, and will doubtless be the last, party of whites to visit this profitless locality. It seems intended by nature that the Colorado River, along the greater portion of its lonely and majestic way, shall be forever unvisited and undisturbed."

All this was common knowledge, and Powell had no reason to gloss over it. To have *seen* the Grand Canyon was one thing; to become the first to ride the wild Colorado through it was something else entirely. Powell's predecessors had gazed upon the dragon. They left to him the task of taking up a sword and battling it.

The second-ranking member of Powell's expedition was his younger brother Walter. A Civil War veteran himself and a onetime prisoner of the Confederates, Walter was as much a war casualty as his brother. Half crazy with all he had seen and endured, Walter was surly, sullen, alternately melancholy and bad-tempered. Even to the indulgent eye of his older brother, he appeared "silent, moody, and sarcastic." One of the mountain men was even more direct. Walter Powell, he griped, was "about as worthless a piece of furniture as could be found in a day's journey."

Among the others, Jack Sumner was first among equals. Sumner, another ex-soldier, was a guide and outfitter who ran a trading post at Hot Sulphur Springs in the Colorado Rockies. ("Trout fishing can now be indulged in to the fullest extent. Game is abundant. Mr. John Sumner has supplies of all kinds at the Springs. Come ye sweltering denizens of the plains to the mountains and enjoy life.") Sumner was fearless, quietly competent, short-fused, sharp-eyed, and sharp-witted. He kept an acerbic journal that serves as a counterbalance to Powell's more polished and more politic account. Where Powell inclined to starry-eyed excess, Sumner favored a sly wink. He stood a fraction under five feet, six and was fair-haired and deceptively delicate in appearance.★

Powell had met Sumner in the summer of 1867 when he ventured to the West for the first time, leading a group of eleven that included his wife, his brother-in-law, and several of his Illinois State Normal students on what amounted to an extended field trip. This was an era when the young nation still tilted eastward. The population of New York City had already neared one million and was climbing fast, but the West was barely settled. San Francisco, with a population of nearly 150,000 thanks to the gold rush, stood almost alone. Salt Lake City had fewer than 13,000 residents, Denver fewer than 5,000, Los Angeles fewer than 6,000. Las Vegas was an almost empty oasis in the desert. Brattleboro, Vermont, was a bigger city than Denver, and Pawtucket, Rhode Island, more of a metropolis than Los Angeles.

Powell made no claim to know the West—he freely acknowledged that his 1867 party consisted entirely of "amateurs like myself"—but he was never one to tiptoe into a new project if he could fling himself into it headlong. "Mountains, hills, rocks, plains, valleys, streams," Powell exclaimed in delight, "all were new." Let loose in the Rockies, Powell and his fellow amateurs began a frenzy of collecting that soon yielded box after box of birds and butterflies and plants and insects and rocks and minerals and fossils.

Sumner served as Powell's guide. Like many a guide, he was proud of his skills, and he liked to moan about the ignorance and naiveté of the dudes he was obliged to chaperone. "In our evening talks around the campfire," he wrote, "I gave the Major some new ideas in regard to the habits of animals, as he had gotten his information from books, and I from personal observation of the animals themselves."

★By the standards of the day, Sumner and Powell (at 5 feet, 6½ inches) were a bit under average size. The typical soldier in the Civil War stood between 5 feet, 7 inches and five feet, eight inches and weighed 142 pounds.

A less prickly character than Sumner might have been more generous. Sumner had been raised not on a cloud-shrouded mountain or in a desert gulch but on an Iowa farm, and he had come to Colorado only in June 1866, just a year before Powell's first visit. Even so, Sumner's know-how was genuine, and his nerve was beyond question. Like life in the army, life in the West brought new experiences at faster than the accustomed rate. Last year's raw recruit was this year's grizzled veteran.

In the course of his first few years in the West, Sumner had nearly managed to lose his life a multitude of times. He was generally on good terms with the Utes, for instance—they called him "Jack Rabbit"—but once, while sleeping near the Fraser River, he found himself under ambush. Woken by the sound of his horse snorting, Sumner lifted his head, drew his gun, and took an arrow in the arm. He shot his attacker and escaped. As Sumner told the story later—he liked to show the scar the arrow had made—twenty Utes in war paint had followed him to his cabin. Sumner let three of them inside and argued that he had fired in self-defense. He sat perched on a keg of powder, revolver cocked, poised to fire into the powder and annihilate both himself and his guests if the need arose. It didn't.

The highlight of Powell's first summer in the West was a climb to the top of Pikes Peak. Powell's wife, Emma, resplendent in long dress, felt hat, and green veil, became only the second woman to have reached the summit. (The party believed mistakenly that Emma had been the first.) Decades later, Emma's hometown newspaper would recall her as "a very beautiful young woman with golden hair and blue eyes." The skinny, bulb-nosed Powell was less attractive—"Wes," his sister once exclaimed, "I think you're the homeliest man God ever made"—but he was immensely proud of his lovely and adventurous wife. She "could ride all day on horseback like a veteran," he boasted, and he liked to point out that in months of travel through the West they had not seen another white woman.

The next year, in the summer of 1868, Powell had traveled to the Rockies again. This time he had a new and larger group in tow, as well as Emma, and an even more ambitious agenda. Again Sumner served as guide.

So when Powell set out on his dazzlingly ambitious Colorado River expedition, in 1869, it was natural that he turned first to Sumner. In response,

Sumner helped enlist several of his own acquaintances. He recruited Oramel Howland, a native Vermonter who had transplanted himself to the West in a search for adventure. A hunter by choice and a printer (and sometime editor) by trade, Oramel was, at thirty-six, the oldest member of the expedition. With his long hair and longer beard, he reminded Powell of mad King Lear roaming the heath. Oramel enlisted his brother Seneca, younger by ten years. A former soldier who had been wounded at Gettysburg, Seneca was, Powell tells us, "a quiet, pensive young man." Sumner also signed up yet another ex-soldier, a silent, mysterious character who gave his name variously as Billy Hawkins or Billy Rhodes. Rumor had it that Hawkins/Rhodes was only a step or two ahead of the law. The fugitive became camp cook at a salary of $1.50 a day.

Bill Dunn, another acquaintance of Sumner's, was a mountain man who dressed in greasy buckskin and wore a wild beard and long, black hair that fell down his back. Like Sumner, Hawkins, and Oramel Howland, he had helped guide Powell on his Rocky Mountain explorations.

Powell himself recruited one last Civil War veteran, an unhappy career soldier named George Bradley. At 5 feet, 9 inches and 150 pounds, Bradley was not especially big, but in Sumner's admiring words he was "as tough as a badger." Powell had lured him with a promise to get him out of the army. With that incentive, Bradley said, he "would be willing to explore the River Styx." (It was an age when such allusions abounded. Bradley had dropped out of school in the sixth grade.)

At his best when times were worst, Bradley was sulky and conspicuously long-suffering whenever time permitted. "The Major as usual has chosen the worst camping-ground possible," he complained to his diary at the end of one long day. "If I had a dog that would lie where my bed is made tonight I would kill him and burn his collar and swear I never owned him."

Two men rounded out the crew. Andy Hall, the youngest expedition member and a last-minute addition, was a twenty-year-old who had been idly fooling in a rowboat at Green River Station when he caught Powell's eye. Ready for adventure and cheery under even the worst circumstances, Hall would become a sort of expedition mascot. One last man recruited himself. Frank Goodman, a round, red-faced, eager Englishman, had been roaming the West in pursuit of "experience." He had come to the right place.

•   •   •

The newspaper editor Samuel Bowles had met Powell in 1868, at Hot Sulphur Springs, Colorado. Bowles was influential and worldly, with interests that extended beyond the usual newsroom bounds—among other things, his *Springfield Republican* had been the first to publish Emily Dickinson's poetry. He had already written one book on the exotic sights to be found in the American West and was gathering material for a second. Powell, who in 1868 was nearly as new to the West as Bowles himself, nonetheless dazzled the editor with his bold schemes, chief among them his plan to resolve the mysteries of the Colorado's myth-shrouded canyon.

Bowles warned his readers that, most likely, "whoever dares venture into this canyon will never come out alive," but he believed that Powell had a chance if anyone did. "Professor Powell is well-educated, an enthusiast, resolute, a gallant leader . . . ," Bowles wrote, "seemingly well-endowed physically and mentally for the arduous work of both body and brains that he has undertaken." That task, Bowles went on, as if infected with Powell's own strain of virulent impatience, was "more interesting and important than any [other project] which lies before our men of science." The blank space on the map was an embarrassment, a rebuke that shamed a country trying to leave behind its gawky adolescence. "Is any other nation so ignorant of itself?" Bowles demanded reproachfully.

The prize was there for the taking. If America's men of science could fill that blank, they would make an invaluable contribution to knowledge and find themselves showered with honors besides. "The wonder," Bowles wrote, "is they have neglected it so long."

They would neglect it no longer.

Still, only a gambler with a taste for long shots would have bet on Powell and his men to make history. One of the leaders was maimed in body, another in mind, and most of the crew had as their outstanding credential a willingness to risk their lives for no very clear reason.

Even a casual observer would have noted two looming problems. One was straightforward, the other harder to pin down, but both were crucial. The straightforward problem was simply put—the men were about to trust their lives to their boats, but there was not a single white-water boatman among them. The other problem had to do with psychology. The trip that Powell had planned—down an angry river through an empty desert, and alone—

demanded on penalty of death that everyone work together like a well-rehearsed military team. But the Colorado River Exploring Expedition had nothing to bind it together but a name.

The men were a ragtag band of friends and strangers, haphazardly assembled and essentially untrained. Worse still, with respect to the need for harmony, the leader was an ex-officer (as was his brother) who was accustomed to giving orders and having them unquestioningly obeyed. That expectation had not been much eroded by civilian life, for it had been only four years since the end of the war. In contrast, the other five ex-soldiers had all been enlisted men. They had not been delighted to take orders when they were obliged to, and they were damned if they were going to be ordered about now that the war was over.

That went double for the mountain men, who were temperamentally unsuited to discipline in any form. (Walter Powell, complained Billy Hawkins, had "a bull-dozing way that was not then practiced in the West.") Independence and self-reliance were a mountain man's defining traits, as strength defined a blacksmith. Dunn's contract with Powell specified that he was "to make baro-metrical observations night and morning of each day," and, what was more dif-ficult, that he was "to have the air of an assistant" while doing so.

Contrary even in good times, these short-tempered outdoorsmen found themselves caught in especially irritating circumstances. Only the summer before, they had been the guides and Powell the wide-eyed rube. Now *he* pre-sumed to command *them*.

One last irritant was undeniable but unmentionable, protected by a strict taboo—Powell's injury made his imperious manner that much harder to take. With only a stump of a right arm, he could give orders but could not row, could not bail, could not line or portage, could not perform his share of any of a thousand daily tasks.

Powell could issue whatever orders he liked. Whether his cantankerous and essentially unpaid crew would leap to obey the commands of the college pro-fessor turned conquistador was another matter.

But with the excitement of their impending departure, no one had time to fret about hazards that might lie ahead. In late May 1869, Andy Hall stole a moment to write a brief letter home. "Dear Mother," he began, "It is a long time since I wrote you but I want you to know I am still alive and well and hope you are the same." He pointed out, in the manner of all sons writing obligatory letters, that he only had time to scribble a few words and then explained that he would not be able to write again for some time.

"I am going down the Colorado River to explore that river in boats with Major Powell, the professor of the Normal college in Illinois. You need not expect to hear from me for some time ten or twelve months at least. You can write to me at Collvile, Arizona give my love to all."*

"Yours till death," Hall concluded, and then he scrawled his signature. "Yours till death" was nineteenth-century rhetoric, not a premonitory shiver. But Hall's warning that he was about to drop out of sight was no exaggeration. Both literally and figuratively, the ten men of the Powell expedition were about to disappear off the face of the earth.

---

*In this quotation and in all others that follow, the punctuation and spelling are as in the original.

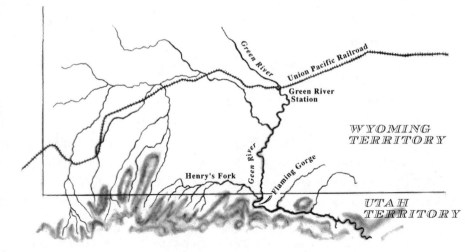

CHAPTER THREE
THE LAUNCH

The trip began well. Relieved to be under way at last and thrilled to find the river fast but manageable, the men raced along in high spirits. Most were new to boats and all were new to high-spirited rivers, but novelty was a good thing and novelty spiked with adrenaline better still. Sumner exulted in the "swift, glossy river," and Bradley noted with unaccustomed good cheer that they swept along "almost without effort."

Powell was so pleased to be on the move that he saw his own exuberance mirrored in the fast-flowing river. The Green, he wrote, was "swollen, mad and seeming eager to bear us down through its mysterious cañons." But, in truth, the first mysteries (and the first real danger) lay several days ahead. At this earliest stage of the journey, the Green cut through familiar territory. Powell and his guides had tramped over some of it on his natural-history excursions during the previous two years. They had seen the snowy Uinta Mountains, for example, and scouted stretches of the Yampa and the White Rivers, two tributaries of the Green, and peeked down into canyons from cliffside perches. On one of these excursions, Powell had even hidden a stash of barometers and other supplies for the boats to pick up later.

This was bleak country, abundant in rock and deficient in nearly everything else. The badlands were sculpted in sandstone and shale, in layers of gray and green and brown and black unsoftened by soil or vegetation but for the occasional grove of cottonwood trees along the river's edge. Erosion had carved buttes here and there, but the first canyons—the first grand and inspiring vistas—were miles downstream. The weather seemed a match for the scenery. "Raining hard," reads one of the earliest journal entries, followed soon after by "Rained all day and most of the night," and "Rained most of the day."

The men would quickly come to think of the river as moody and fickle. Dependent on its every whim, like fleas clinging to a psychotic dog, they would watch it with a wary and obsessive eye. But for now, they were delighted to find it in a forgiving mood. This was a good thing, for as Hawkins freely noted, "We knew nothing about a boat."

They set out as if to prove it. Hall and Hawkins, in the *Kitty Clyde's Sister*, made it only a mile or two from the launch point before running aground on a sandbar in the middle of the river. (The Green was not quite so "swollen" as Powell had thought.) The two men jumped overboard, managed to push the boat clear, and relaunched themselves. They promptly ran aground again, this time on the riverbank. Eventually, to the whooping and teasing of the rest of the crew, Hall and Hawkins managed to catch up with the others and pull to shore. Hall came into camp sputtering indignantly about his boat's unmanageability. He was an experienced mule driver—"*Kitty's* crew have been using the whip more of late years than the oars," Sumner noted gleefully—but even a mule seemed cooperative in comparison with a boat. The *Kitty* would "neither gee nor haw nor whoa worth a damn," Hall cursed, and in fact it "wasn't broke at all."

The others were having problems of their own. Even Powell, more inclined than his men to try to put a presentable face on chaos, described the journey's start in a way that calls to mind Laurel and Hardy more than Lewis and Clark. "In trying to avoid a rock," he wrote in his account of that first afternoon, "an oar is broken on one of the boats, and, thus crippled, she strikes. The current is swift, and she is sent reeling and rocking into the eddy. In the confusion, two others are lost overboard and the men seem quite discomfited, much to the amusement of the other members of the party."

These were mishaps, not crises, and the crew seemed to revel in the horseplay. But, as Hawkins noted, they were still on "very good water." The men were having trouble staying afloat though the Green had yet to flex its muscles. For some seventy miles downstream from Green River Station, it wandered benignly this way and that across the barren landscape. Only when it left the badlands and smacked into the towering Uinta Mountains would the river truly rouse itself. Powell and his men, floundering already, were in the position of a boxer who is knocked to the mat by an opponent still shrugging off his robe.

Even so, they were confident, the mood a happy mix of can-do optimism and "school's out for summer" boisterousness after the dreary weeks at Green River Station. Ignorance was not quite bliss, but it veered in that direction. Powell, for one, took for granted that outdoor expertise in general would translate into river expertise in particular. "The hunters managed the pack train last year, and will largely man the boats this," he wrote matter-of-factly. The tone implied that the switch from horses and mules to boats was trifling, akin perhaps to a switch from guitar to banjo.

The men, especially those who had worked as guides, were less inclined to gloss over their inexperience. Fully aware of the gulf that separates amateurs from professionals, they knew they were no boatmen. But they subscribed to the logic that impels people to buy lottery tickets—they had as good a chance as anyone else. They did not claim to have knowledge they lacked, but they focused on the strengths they did have. Of the four ingredients crucial to success in exploration—skill, experience, equipment, and courage—they had two. Those seemed like decent odds. "If we fail," Sumner noted cheerily, "it will not be for the want of a complete outfit of material and men used to hardships."

They were, indeed, hardy souls. "After supper," Powell wrote in an account of one of the trip's first nights, "we sit by our camp fire, made of drift wood caught by the rocks, and tell stories of wild life, for the men have seen such in the mountains, or on the plains, and on the battle fields of the South."

The stories they told vanished with the campfire smoke. Did Powell describe the carnage at Shiloh? Did Seneca Howland break his usual silence to tell the others what he had seen at Gettysburg? Did Sumner spin a yarn about the grizzlies he had killed two winters before? "It is late before we spread our blankets on the beach," Powell wrote. (The men slept in pairs, sharing a blanket. The strangest bedfellows were Powell and Sumner, the college

professor and the feisty frontiersman, one a mini-autocrat and the other a natural rebel allergic to authority in any form.)

The men were tough and their gear was good, or so they felt. The boats,
brand-new and well-made wooden rowboats, met with everyone's approval.
They were Whitehalls, a classic design that by most accounts originated in
New York City in the 1820s and then spread across the nation. Speed was the
Whitehall's great virtue—indeed, an article in 1995 hailing a resurgence of
interest in the design was entitled "Survival of the Fastest." Wherever speed was
crucial, there were Whitehalls. In New York Harbor, police relied on them; so
did the thieves trying to make off with cargo stolen from ships lying at anchor;
so did the "runners" for various waterfront businesses who raced one another
across the water to bag new customers.

In boat design, every choice represents a trade-off. A sea kayak, for example, is intended for long-distance travel in a straight line and is therefore long
and skinny, like an arrow pointing toward a distant destination. It has relatively
little "rocker," or curvature from bow to stern, so that it touches the water
almost all the way along its length. A white-water kayak, which is designed to
pivot away from danger at the touch of a paddle, has a different design.
Comparatively short and stubby, it has a good deal of rocker, meaning that
bow and stern are lifted up out of the water, so the boat is easy to turn. (So
easy, in fact, that beginners in a white-water kayak are usually unable to paddle in a straight line, even on a lake.) In the case of Powell's boats, there were
two main trade-offs. Whitehalls were heavy and sturdy, which made them
rugged but hard to maneuver. They were round-bottomed, which made them
fast but tippy in rough water.

From the front, the Whitehall looked like a conventional rowboat. Viewed
from the rear, the boat resembled a wineglass in cross section. The wineglass
look reflected the design of the keel. From about the midpoint of the
Whitehall to the stern, the keel—the "breastbone" that runs the length of a
boat—thickened into a rigid fin that acted like a rudder fixed in the "straight
ahead" position. It made for an elegant look, but the design was eminently
practical. "Maneuverability was not a high priority," notes Robert Stephens, a
modern-day boat builder and Whitehall aficionado, "since two or more strong
oarsmen could manhandle her in close quarters by backing and filling, but
steady tracking was very desirable in the open waters of a large harbor. Above
all, speed was the overriding criterion—especially speed under oars [the boat

could also be rigged with a sail], speed which was essential for procuring business, delivering passengers, capturing criminals, or eluding police."

This was crucial. The Whitehall had two distinguishing features—it was fast, and it was hard to knock off course. For nearly every purpose, that made it an ideal choice. Suppose, though, that you planned to bring your rowboat not to a lake or harbor but to a surging river. Suppose, further, that you needed desperately to keep from being smashed against house-sized boulders and plunged into boat-hungry whirlpools. Then the combination of speed and the inability to change direction would be the last things you wanted, and maneuverability would be beyond price. Your best hope would be a boat designed to inch its way through the minefield, poised at every instant to pirouette away from danger.

In hindsight, Powell may have had better options. Grand Banks dories, for instance, were sturdy, flat-bottomed fishing boats that were perhaps more maneuverable than Whitehalls. But their home base was the ocean off New England (the dories were stored on a schooner's deck and lowered into the sea) and no one had ever thought to try them on white water. In any case, hindsight was one of countless luxuries unavailable to Powell. In search of the best rowboat available, he had picked a Whitehall. In 1869, anyone might have done the same. Powell had made the natural choice, and it was all wrong. Now he was headed into a fire wearing a gasoline suit.

Powell's 1869 boats no longer exist, not even in drawings or photographs. (Nor, for that matter, do we have a group photo of Powell and his men.) There were four boats altogether, three of them built to identical specifications. Each of the three "freight boats," as the men called them, was twenty-one feet long and four feet wide and could hold about four thousand pounds of cargo, though they were never loaded that heavily. Even empty, these long, narrow boats were a burden—it took four men to carry one empty boat. And as time passed and the boats grew waterlogged, they would grow heavier still.

They were far from empty. The gear and food together weighed seven thousand pounds. The plan was to carry supplies to last ten months, so that if the expedition was still under way in winter and the river froze, they could "winter over" until the spring thaw. The food was standard army issue—rice, flour, beans, coffee, sugar, bacon, dried apples. (Powell's trip to Washington to find government financing had not panned out, but in lieu of money he had been granted army rations for himself and his men.) They would supplement that dreary fare with fish and whatever fresh meat the hunters brought in.

The gear was as straightforward as the food. The men carried tents, ponchos, bedrolls, extra clothing, hundreds of feet of rope, knives, rifles, guns, traps, gold-panning equipment, axes, hammers, saws, nails, screws, and a miscellany of other tools, as well as sextants, chronometers, barometers, thermometers, and compasses. It was a lengthy and careful list, as it had to be for a caravan through the desert.

Vital as the tons of supplies were, they were dangerous as well. To upend a hardware store into a boat that was hard to maneuver even when empty was asking for trouble. The boats rode so low, Powell noted, that even without rapids to make life complicated, it required "the utmost care to float in the rough river without shipping water."

Powell had anticipated that the boats would take a beating, so he had ordered them built of oak. They were "stanch and strong," he noted proudly, "double-ribbed, with double stem and stern-posts." At each end of the boat a decked-over bulkhead provided storage space. This was essentially an off-the-shelf design. The few modifications, notably the choice of oak, provided extra strength but made the boats heavier and more ungainly than they would have been otherwise.

Built of pine and only sixteen feet long, the fourth boat looked like the others except that it was smaller and lighter. This was Powell's boat, the *Emma Dean*. The food and supplies were divided into three identical parts and distributed among the three large boats, as a precaution in case a boat was lost. The *Emma Dean* carried only a few of the scientific instruments, three guns, and three bundles of clothing.

Not knowing what the river had in store for him and his men, Powell had devised a river-running plan intended to keep surprises to a minimum. His boat was faster than the others, because it was smaller and lighter and less heavily loaded. Powell's idea was to proceed downriver ahead of his clumsier companions and scout a safe course for them to follow. If no safe course presented itself, he would give a signal to pull to shore, and the men would begin the dangerous, exhausting business of wrestling the boats and their thousands of pounds of cargo around the foaming, mocking rapid in their path.

In theory, the system was simple and sound. "The boats were ordered to keep one hundred yards apart," Sumner explained. "Flag signals were arranged as follows, always to be given by Major Powell from the pilot boat: flag waved right and left, then down, 'Land at once'; waved to right, 'Keep to right'; and waved to left, 'Keep to left of pilot boat.'" Everyone was well-pleased with the boats and the planning (and boaters running unknown rapids today follow a similar system). "We feel quite proud of our little fleet," Powell acknowledged, and even Sumner, temperamentally allergic to gush, conceded that "we make a pretty show."

Since the boats were so heavy, each needed two men at the oars, one seated in front of the other. Powell was in the *Emma Dean*, with Sumner and Dunn rowing. Walter Powell and Bradley crewed the *Maid of the Cañon*, Hall and Hawkins the *Kitty Clyde's Sister*, and the Howland brothers and Frank Goodman the *No Name*. The men rowed in the age-old fashion, facing upstream, their backs to the action. To row "backward" might sound odd to someone who has never tried it, but it has the great advantage of permitting the boatman to use the big muscles of his back and legs. (Rowing "forward," facing downstream, places heavy demands on the boatman's arms.) And since a rowboat on flat water moves slowly, it is easy to keep on course by stealing an occasional peek over one shoulder or the other. For countless years, boatmen on harbors and lakes and lazy rivers have rowed in just this way.

In rapids, though, Powell and his men would not be moving slowly, and they would be headed—blind—into desperate danger. So Powell would be the eyes for the entire group. Not rowing, he was free to face forward and fix all his attention on perils downstream. There he would stand while Sumner and Dunn strained blindly at the oars, his left hand clutching a strap that ran across the boat, balancing like a circus rider on the back of a cantering pony, looking for trouble.

On the expedition's first day, the river was kind, as if allowing its fledgling challengers to get their bearings. After a shaky start, Powell and his men managed to fish their dropped oars back out of the river and point themselves downstream. They made their way more or less uneventfully for another seven or eight miles, where they made camp for the night. Hall and Hawkins, in the last boat, reacted too slowly. "I saw they were all landing," Hawkins recalled, "and I told Andy they were camping at this point. The river was straight and the water smooth and Powell signalled to me and we tried to land, and did finally get to shore some four hundred yards below." No one was much inclined to preach a sermon on a four-football-field miscalculation—the upstream boats joined their wayward colleagues and, in Hawkins's grumpy summary, "the rest of the boys had the laugh on us"—but everyone knew the story had a moral. It was easily put: The river was fast and strong, and they were all novices; they had better hope they were quick learners.

Their problem, Hall and Hawkins decided, was that they were overloaded. Each of the boats was supposed to be carrying the same weight, but the *Kitty Clyde's Sister* seemed to be riding several inches lower than the others. Even in calm water, the river was within four inches of spilling into the boat. The two men set to work removing supplies, figuring that "we better unload some of the bacon and take chances of replacing it with venison and mountain sheep later on. So we unloaded five hundred pounds of bacon in the river." At the time, throwing food overboard seemed like a good idea.

They camped that first night at the foot of an overhanging cliff, perhaps ten miles downstream from Green River Station. The two Powells and Bradley set out for "a couple of hours geologising." While they searched for fossils, Oramel Howland and Dunn set out to hunt dinner. The two men returned at dark with one small rabbit. It made, Sumner noted dryly, "rather slim rations for ten hungry men."

Camp was cheery, though it was wet and raw. Unfazed, the men did their best to keep dry and to outboast one another. We "exchanged tough stories at a fearful rate," Sumner recalled, but the crew was still suffering the effects of the nights of hard drinking at Green River Station. Everyone turned in early.

The first day of the trip was complete. So far, so good.

While the men snore in their bedrolls, let us take a moment to talk about their journals. Powell, Sumner, and Bradley all kept diaries. (Somehow the taciturn Bradley managed to keep a detailed daily record of the trip without any of the others catching on.) Several of the others chimed in briefly, adding still more voices to the unruly chorus. We have two short accounts of the trip from Billy Hawkins, a few brief letters from Andy Hall, a long newspaper article by Walter Powell, two long newspaper stories by Oramel Howland.

The men had two favorite modes of speech, wild exaggeration and ludicrous understatement. Ideally, both were delivered deadpan. Time and again, the accounts overflow with an offhand vitality that reminds us that we are listening to Mark Twain's contemporaries. One remote spot was "desolate enough to suit a lovesick poet." An eddy snagged a boat and "whirled it around quick enough to take the kinks out of a ram's horn." In one especially wild rapid, "we broke many oars and most of the Ten Commandments."

Some of the handwritten originals have been lost. But Powell's notes (or, more precisely, notes that cover just over half the expedition) survived a long journey, from the depths of the Grand Canyon to a silent, dusty archive at the National Museum of Natural History in Washington, D.C. There, at a wooden table under flickering fluorescent lights, a visitor can hold the pages that Powell held and study the notes he scrawled on long, thin pages marked with water spots and splashes of coffee that fell more than a century ago. The handwriting is large and looping, and the words sit awkwardly on the page.

Bradley's journal has come to rest only a mile or two from Powell's, in the Library of Congress, a second message-in-a-bottle from a single shipwreck. Meticulously neat, it looks nothing like Powell's. Bradley wrote on small pages in impeccable but infinitesimal script, as if he were one of those people who can inscribe the Twenty-third Psalm on a grain of rice. From its appearance alone—with no bold underlinings, no exclamation points, no crossed-out words, no stars or arrows in the margin to hint at excitement—one might take Bradley's journal to be a record of experiments in a none-too-promising chemistry lab. No one would guess that it records one of the great American adventures.

It is tempting to see Powell's bold, sprawling handwriting as reflecting his taste for splash and melodrama, especially when his writing is compared with Bradley's tiny, finicky penmanship, but the true explanation is simpler. Powell had lost his right arm at Shiloh seven years before, and he had not quite mastered the art of writing with his left hand. Writing outdoors, in the wind, perhaps on a rock serving as a makeshift desk, made matters worse. A friend who came to know Powell later observed "the difficulty of writing with his left hand and keeping the paper from blowing away by trying to keep it in place with the stump. I have often seen him struggling this way."

Powell's description of the 1869 expedition is the most compelling but the least straightforward of the firsthand accounts. He kept two journals, first of all, and the two are quite different. Powell's river diary contains short, spare entries, written in free moments in camp and not published in his lifetime. One day's entire entry, for example, reads: "Wrote until 10:00 A.M., and then came to camp with Walter." (This is the diary at the National Museum of Natural History.) Powell's published account, which is the one nearly always cited when people quote him, was lovingly and painstakingly composed years after the expedition had ended.

Powell purposely blurred the distinction between the diary he composed on the river and the account he published in 1875 as an official government report. The published journal is written in the present tense, in diary format, as if each entry had been composed by firelight at the end of a hard day on the river. In fact, it was dictated half a dozen years later to a secretary as Powell paced back and forth in his office, waving his cigar and gesturing as though he were addressing a vast lecture hall and not a lone listener.

For the most part, the reader is grateful for Powell's literary sleight of hand. His river diary is as dry as the Southwest it described, but the published account continues to draw new readers even after a century and a quarter. What other government publication can say as much? At a glance, few books look less inviting. The title, usually abbreviated to *Exploration of the Colorado River*, runs on for more than two dozen words; the publisher is the Government Printing Office; the text is introduced by a formal note carrying the signatures of the secretary of the Smithsonian Institution and the Speaker of the House of Representatives. But even in such an austere setting, Powell's personality bursts forth like a dancer from a birthday cake.

The government imprimatur gave Powell's version of events a credibility it did not always deserve.* Powell described his formal account as simply a carefully worked-out version of the telegraphic river diary. "I decided to publish this journal, with only such emendations and corrections as its hasty writing in camp necessitated," he declared. This seems a stretch—the river diary contains only three thousand words, for example, in comparison with the published journal's nearly one hundred thousand—but diaries, after all, are not written for outsiders.

---

*Even the illustrations in the *Exploration* are less than straightforward. A drawing captioned "The Start from Green River Station," for example, shows three boats rather than the actual four. The illustration on the cover of this book, taken from Powell's Figure 28, "Running a Rapid," depicts boats with a decked-over compartment in the middle, which is incorrect, and shows a sweep oar at the stern (for reasons discussed in Chapter Thirteen, historians disagree about whether the boats had sweep oars). More important, the drawing shows a one-armed man at the sweep, though both Powell and his men agreed that Powell never touched an oar.

And perhaps those three thousand words spoke to Powell even years later, summoning complete memories, as a scrap of melody can call forth a symphony. But at some points in this official report, Powell soared beyond his own adventure, magnificent as it was, and touched up stories or actually invented them.

May 25, the first full day on the river, was another day of bad weather and good spirits. The men were under way by six in the morning and made it until about nine-thirty before running into trouble. Then Powell ran aground on a sandbar and, before he had time to signal, so did the next boat and the one behind that. Bradley, trailing the others, just managed to steer to the right and sneak by. Two of the men jumped out of their boats and pushed everyone free.

They continued downstream for another hour. The rain continued to pelt down, and the men pulled ashore to try to wait it out. By this point, everyone was "wet, chilled, and tired to exhaustion," Powell wrote, but with the help of a roaring fire and many cups of coffee they were soon "refreshed and quite merry." (They also cooked up some "villainous bacon," as Sumner put it, but that was less satisfactory.) When the sky looked as if it might clear, they set out again. After five or six miles, they saw some bighorn sheep on a cliffside and stopped to give chase. Two or three hours later, the hunters straggled home empty-handed. Only Hawkins had not struck out completely. He had found a sleeping lamb, which he had caught by the heels and thrown off the cliff, toward camp. The hunters consoled themselves for their failure by teasing Hawkins—they pretended to believe that the lamb was dead when he found it—but they all agreed it made a fine lunch.

It was now about four in the afternoon, time to move on. All the boats except *Kitty Clyde's Sister* soon ran aground on another sandbar and found themselves unable to budge. Bradley and Walter Powell managed to plant the *Maid* so firmly they had to pry her off with oars. It took "a great deal of tall lifting and tugging" to get free, Bradley wrote, which was especially irksome because Powell had given a danger signal and Bradley had decided to ignore it.

Their third day was another fairly easy one, noteworthy only for an entry in Bradley's journal. They had encountered, he wrote, "the largest and most difficult rapid yet seen." From here on, Bradley's diary would be dotted with similar observations, as if to convince himself that *this time* they must have taken the river's worst blow. Sumner, no more inclined to bluster than the hard-to-faze Bradley, was just as impressed by the rapid. "It cannot be navigated by any boat with safety, in the main channel," he wrote, though it was possible to hug the bank and scoot by safely.

Three of the boats made it through in fine fashion, but Hawkins and Hall, in the *Sister* (the shortened name they used for their boat) found themselves pinned on a rock. Hawkins climbed overboard and managed to pry the boat free. "No injury done except *one man took a bath*," Bradley noted unsympathetically. The rain, which had continued throughout the day, kept up at night, but no one paid it much heed. The hunters, for once, had something to show for their efforts, and everyone tucked happily into an excellent dinner of duck and goose.

There was no great significance to running aground. Heavy boats moving fast on low water might almost be expected to beach themselves. But in pinning the *Sister* to a rock, even if only briefly, the river had provided a far more telling warning of the havoc it could unleash at any moment. It happens in an instant—one minute a boat is racing along and then, suddenly, it is sideways to the current, wrapped against a rock or another obstacle, and helpless. The river holds the boat in place with overwhelming force, like a sumo wrestler smothering a kitten. Worse still, it perpetually replaces itself as it flows, so that there is no wriggling out from under. A kitten might claw or bite a wrestler and sneak away in the ensuing confusion, but a river never "shifts its weight." It simply persists in its assault, unceasingly and unforgivingly, until the obstacle in its way is an obstacle no longer.

"Wrapping is, in the estimation of many, the worst fate that can befall a riverboat," writes the historian and river runner Roderick Nash. "In an upset, at least, the boat washes downstream where it can usually be recovered and righted. But a wrapped boat is bent around a rock and pinned there by the force of moving water. Some boats can be freed using lines from shore, but often they remain wrapped until the river shreds them into rubber ribbons or wooden or metal splinters."

Nash tells the story of one recent wrap, on a rock in Crystal Rapid, one of the Grand Canyon's notorious danger spots. The boat was a thirty-foot-long rubber raft, motorized and carrying a dozen people. Many boatmen disdain such behemoths. They prefer small, oar-powered rafts, or elegant craftlike dories and kayaks that flit across the water like dragonflies. (The trade-off is that dories and kayaks are harder to patch after a wreck.) In comparison with its small, maneuverable rivals, a giant inner tube has all the grace of a brontosaurus. But though they are ungainly, these wallowing rubber beasts are as close to invulnerable as anything on the Colorado.

Close to invulnerable, but not all the way there. The rubber raft at Crystal turned sideways against a boulder near the top of an obstacle course called the Rock Garden. The river, rushing downstream, wrapped the raft's two ends tight to the rock. The raft sat glued in place, while passengers and crew scrambled up and out onto their new island home. They sat on the rock through the night, trapped in mid-rapid, listening to the water rise around them (the Colorado, dammed upstream of the Grand Canyon in the 1960s, rises and falls depending on electricity demand downstream). Dawn found the scared, chilled passengers still with a bit of rock to cling to, and the boat itself freed by the rising water and straining against the lines the boatmen had used to tie it in place. Boatmen and passengers climbed back into the boat, which was still tied down but now being yanked violently downstream. Three knives came out; on a signal, all three touched the mooring lines. A touch was all it took. With a crack like a whiplash, the taut ropes snapped, and the boat and its passengers shot free.

The Green was not yet up to such malevolent tricks. May 27, the fourth day on the river, marked the end of the first and easiest leg of the journey. After a late, leisurely start and a quiet day on what Powell called a "placid stream," the men reached the junction of the Green and Henry's Fork, a beaver stream the mountain men knew well. Here they found the barometers, chronometers, and sextants Powell had stashed earlier in the spring, still safe beneath the overhanging rock where he had hidden them. Powell and his crew settled in for a few days, making scientific observations, Powell displaying more enthusiasm than the men.

The rain continued (it had hardly let up since they set out), but the scenery had improved and spirits were high. The men had yearned for canyons. Now they had them. "The river winds like a serpent through between nearly perpendicular cliffs 1200 ft. high but instead of rapids it is deep and calm as a lake," Bradley wrote. "It is the most safe of any part we have yet seen for navigation. Found some marine focils [fossils] in hard limestone—first yet found."

The exuberant tone was new. At camp on the first night, Powell had climbed a cliff and struggled manfully to admire the view. "Barren desolation is stretched before me," he had noted, "and yet there is a beauty in the scene." Still looking with eyes accustomed to the rolling terrain of the green midwest, he had found himself bewildered. "The fantastic carving . . . with the bright and varied colors of the rocks conspire to make a scene such as the dweller in verdure-clad hills can scarcely appreciate."

Even so, he had done his best to wax rhapsodic. "Dark shadows are settling in the valleys and gulches," Powell wrote in an account of that first afternoon, "and the heights are made higher and the depths deeper by the glamour and witchery of light and shade." Sumner was more succinct. "Country worthless," he scrawled in his journal.

Now things were picking up, and there was no need to manufacture enthusiasm. "It is the grandest scenery I have found in the mountains and I am delighted with it," Bradley exulted. "I went out to see the country this morning and found it grand beyond conception."

Powell was just as excited. The Green at this point ran south, and the Uinta Mountains, running east-west, lay smack in its path, and "yet it glides on in a quiet way as if it thought a mountain range no formidable obstruction to its course. It enters the range by a flaring, brilliant, red gorge, that may be seen from the north a score of miles away."

That gorge would be their first canyon. "We name it Flaming Gorge," Powell wrote proudly. Like Adam in the Garden of Eden, where everything was new, Powell had the opportunity to bestow names as he chose.

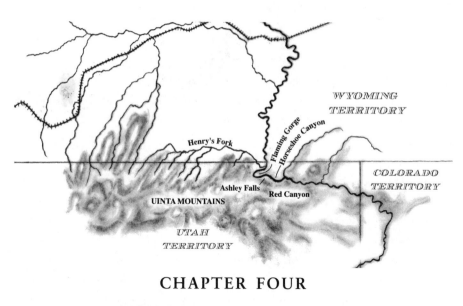

CHAPTER FOUR
ASHLEY FALLS

For the next several days, life was easy. This was the trip as Powell had envisioned it, a scientific expedition rather than a mad dash. There were cliffs to explore and measure, repairs to make, fossils to find. Bradley and Powell climbed one day to a vantage point some one thousand feet above the river and surveyed the panorama—the long, sinuous curve of the Green, the valley of Henry's Fork stretching to the west, desert and hills and buttes to the north, the Uinta Mountains to the south, the peaks of the Wasatch just barely visible in the distance to the west.

Bradley went off exploring on his own early one morning, hoping for fossils, but came up empty. Worse still, he managed to spend the day lost in a driving rainstorm. After eleven hours, he finally made it back to camp, "tired and hungry and mad as a bear." Powell was nearly immune from such frustration, in part because he had a gift for imbuing even the most mundane chore with drama. He and Sumner had spent most of a rainy day in camp repairing one of the barometers. The barometers were crucial—they were the tool used to determine altitude—but they were finicky and fragile. The repair was a matter of taking a long glass tube open at one end, adding mercury to it a few inches at a time, heating the tube (without cracking it) in order to create a vacuum, and then repeating the procedure again and again until the tube was filled to the proper volume. When the tedious repair was finally completed succesfully, Powell beamed with pleasure. "[We] are ready," he cheered, "to measure mountains once more."

The men were harder to excite. "Tramped around most of the day in the mud and rain to get a few fossils," Sumner wrote crabbily on May 27, and his journal entry the next day began with a weary "Still in camp." But the crew was not being paid for their opinions (or for anything else), and Powell seemed unaware of the grumbling.

By May 30, after a morning hike to survey the local geology, even Powell was prepared to move on. "We are ready to enter the mysterious cañon, and start with some anxiety," he wrote. They entered Flaming Gorge on a fast current and emerged into a little park and then, when the river swung sharply left, headed toward another canyon cut into the mountain. "We enter the narrow passage," Powell continued. "On either side, the walls rapidly increase in altitude. On the left are overhanging ledges and cliffs five hundred—a thousand—fifteen hundred feet high. On the right, the rocks are broken and ragged, and the water fills the channel from cliff to cliff."

Then, as if unsure of what to make of the mountains, the river made a sharp turn back to the right. In the lead boat, Powell strained to make sense of the chaos ahead of him. Bradley braced for trouble. "We took off boots and coats and prepared for a swim," he wrote. Powell struggled to stay cool. "Here we have our first experience with cañon rapids," he wrote. They had seen other rapids, in fact, but in comparison with what he now confronted, Powell seemed to consider them not worth mentioning. "I stand up on the deck of my boat to seek a way among the wave-beaten rocks. All untried as we are with such waters, the moments are filled with intense anxiety. Soon our boats reach the swift current; a stroke or two, now on this side, now on that, and we thread the narrow passage with exhilarating velocity, mounting the high waves, whose foaming crests dash over us, and plunging into the troughs, until we reach the quiet water below; and then comes a feeling of great relief. Our first rapid is run."

This was Horseshoe Canyon, the name chosen to indicate the U-shaped course of the river. Then came another valley and soon another canyon, this one fed by a beautiful creek. They named both creek and canyon Kingfisher, for a bird they saw on the branch of a dead willow. It stood on its perch above the river, Sumner wrote, "watching the finny tribe with the determination of purpose that we often see exhibited by politicians while watching for the spoils of office."

The river curved again, bending around a dome-shaped rock where hundreds of swallows had built their nests. The rock, a thousand feet of gray-white sandstone, looked like an enormous beehive, with swallows playing the role of bees. The men named it Beehive Point, and camped on the opposite bank. The hunters set out before dinner, returning with the usual lack of success. "Goodman saw one elk, but missed it," Sumner noted with dismay, and no one else came even that close.

Powell was in an expansive mood nonetheless. The first serious rapid lay behind them, the river had become "broad, deep, and quiet," and the view from camp was inspiring. On the far shore, behind the beehive, rose a kind of natural amphitheater perhaps fifteen hundred feet high, formed of sandstone cliffs and terraces of pine and cedar. "The amphitheater seems banded red and green," Powell wrote, "and the evening sun is playing with roseate flashes on the rocks, with shimmering green on the cedars' spray, and iridescent gleams on the dancing waves. The landscape revels in the sunshine."

The next day brought everyone back down to earth. It began well enough. Powell, Bradley, and Dunn had set out "to examine some rocks," in Sumner's none too enthusiastic words, and Oramel Howland and Goodman had climbed high above the river to survey the landscape for Howland's map. By ten o'clock, the men were on the river and into two miles of barely interrupted rapids. Then came half an hour of flat water and after that, in Sumner's words, "a bad rapid through which no boat can run; full of sunken rocks, and having a fall of about ten feet in two hundred yards." They lined it in two hard hours, one boat at a time, most of the men stumbling along the shore clutching ropes tied to bow and stern while two of the group grabbed oars and stationed themselves on the rocks along the riverbank to fend off the boats as needed.

At five o'clock, Sumner continued, "we came to the worst place we had seen yet; a narrow gorge full of sunken rocks, for 300 yards, through which the water runs with a speed that threatened to smash everything to pieces that would get into it." The men pulled to shore to make a plan and quickly saw that they had landed on the wrong side of the river. Now the problem was to get across the rushing river without being swept into the rapid. "Dunn and the trapper"—Sumner referred to himself in the third person—"finally decided to take the small boat across or smash her to pieces." They made it across, emptied the boat, and crossed back to help their companions. In the meantime, the three heavy boats had each unloaded about half their cargo so that they would not be quite as clumsy as usual. Five crossings later, the *Emma Dean* had ferried the excess cargo to the far side. Now, one at a time, the big boats set out. They made it, barely, the men pulling desperately while the current rushed them toward the rapid. Tomorrow they would line their way past it. Now they were too exhausted to do more than collapse. "Had supper," Sumner wrote, "turned in, and in two minutes all were in dreamland."

Not quite all. "As the twilight deepens," Powell wrote, "the rocks grow dark and somber; the threatening roar of the water is loud and constant, and I lie awake with thoughts of the morrow and the cañons to come."

Away from its rivers, the desert at night can be eerily silent. "Was there ever such a stillness as that which rests upon the desert at night!" the writer John Van Dyke asked a century ago. "Was there ever such a hush as that which steals from star to star across the firmament!" We can be sure that Van Dyke was not camped by a rapid when he wrote those lines. Many people find the sound of running water soothing. They may have in mind a babbling brook or a gurgling fountain; they are not thinking of rapids, which do not murmur. They rumble. They roar. They crash. The sound evokes a thunderstorm just overhead, a jet skimming the ground, a runaway train. The noise echoes all the louder when it is amplified by stone cliffs that soar upwards of a thousand feet. And, in nerve-racking contrast with the other ground-shaking sounds it calls to mind, this river thunder never stops.

The message is worse than the sound itself—the roar of a rapid is a proclamation of danger as clear as a giant's bellowed curse in a fairy tale.

Small wonder that Powell and his men all rose early the next morning. The first order of business was to make it past the rapid they had sneaked in front of the day before.

It took three tense hours of lining, but then, as if to reward the men for their hard labors, the day took a sudden turn for the better. In the boats at last, they found the river fast and free of obstacles. The contrast between the misery of lining rapids and the delight of "shooting" them leaps off the journal pages. Even the dyspeptic Bradley seemed excited, writing happily that the boats raced along "like lightning for a very long distance." Powell, characteristically, was the most high-spirited of all. "To-day we have an exciting ride," he wrote. "The river rolls down the cañon at a wonderful rate, and, with no rocks in the way, we make almost railroad speed. Here and there the water rushes into a narrow gorge; the rocks on the side roll it into the center in great waves, and the boats go leaping and bounding over these like things of life." The swooping, soaring motion of the boats over the waves reminded Powell of "herds of startled deer bounding through forests beset with fallen timber." For the time being, the rapids were thrilling rather than threatening.

Even on as fine a run as this, it was not all play. Occasionally a wave had the bad manners to jump uninvited into one of the boats. This "necessitates much bailing, and obliges us to stop occasionally for that purpose," Powell complained. (Powell had made the wise decision to cover bulkheads in the bow and stern of each boat, but he had left the center section of each boat open. Had he known the size of the waves he would face, he might have chosen to cover this center section, too, except for open cockpits for the men at the oars.)

The expedition's introduction to the fine art of bailing was comparatively gentle. A boat that has swallowed a wave is instantly heavier and clumsier by more than a thousand pounds. For a boatman, already struggling to find a route that will take him safely through a rapid, it is as if some prankster god has dropped a piano into the boat at the worst possible time. The only good news is that a boat carrying an extra half ton of ballast is hard to flip. But passengers and pilot may end up swimming anyway, because the newly heavy boat may plow directly into the next giant wave rather than float over it, and the impact of that collision can overpower the most white-knuckled grip.

About two decades ago designers devised "self-bailing" rafts with an inflated floor that rides considerably higher than the river surface. Water that spills into the boat drains out through strategically placed holes along the floor. Self-bailing boats have obvious advantages but disadvantages as well—they are harder to repair, slower in flat water, and, because they drain their watery ballast so quickly, comparatively easy to flip.

So the old design is still widely used, and it is still common for white-water trips to be punctuated by shouted orders to bail. On commercial trips on the

Colorado today, passengers quickly learn what is at stake. At the foot of every rapid, the boatman yells, "Bail!," and three-hundred-dollar-an-hour lawyers and thirty-year-old millionaires grab their bailing buckets and start slinging water overboard.

Despite the pauses for bailing, Powell and his men were making good time. In a single hour, they reckoned, they had covered twelve miles and run perhaps a dozen rapids. (The mileage figures were rough guesses, obtained by stringing together estimates of the "half a mile to that black rock" sort. The guesses tended to be high, especially early on.) Then came trouble. "As the roaring of the rapids dies away above us," wrote Sumner, "a new cause of alarm breaks in upon us from below . . . when, turning an abrupt corner, we came in sight of the first fall, about three hundred yards below us." Powell, in the lead boat, signaled the other boats to land. The *Emma Dean* approached the rapid within about twenty feet and pulled to shore to reconnoiter. No one liked what he saw. "[We] found a fall of about ten feet in twenty-five," Sumner reported. "There is a nearly square rock in the middle of the stream about twenty-five by thirty feet, the top fifteen above the water. There are many smaller ones all the way across, placed in such a manner that the fall is broken into steps, two on the east side, three on the west."

They were in a chasm they had named Red Canyon, in honor of the red sandstone walls that soared anywhere from a thousand to two thousand feet above the river. Powell had been warned about these rapids; they were the ones he had been told about the previous spring by the Indian who had painted a word picture of roaring waves and bucking canoes and rocks that were "heap, heap high." The rapid in front of them confirmed that description. As miserable as it would be to carry the supplies and line the boats here, where there was no decent land route, no one even suggested they try to run the rapid.

The lining strategy was born of desperation, as if house movers who needed to transport a safe ended up tying ropes to it and pushing it down the stairs. (Today, lining is still dreaded, although techniques have improved.) First, the boats were unloaded. Then, one boat at a time, a long rope was attached to the boat's stern. Five or six men on the shore grabbed that stern line. A second line was attached to the bow and tied off downstream, well below the fall. Next came the critical move—the half dozen men clutching the stern line with all their strength did their best to ease the boat over the fall, while others in the crew stood on the sharp rocks along the river's edge, armed with oars, poised to ward off the boat if it careened into the rocks. When the river overpowered the men on the stern line and tore the rope from their hands, the boat shot over the fall. Now it was up to the men stationed downstream to reel in the runaway craft by taking up the slack in the bowline, as if our housemovers had to snag the safe tumbling down the stairs before it crashed through the front door.

All four boats, and at least three tons of cargo, had to be manhandled past the rapid. The men began with the *Emma Dean*, the lightest boat, and lowered her to safety in fifteen minutes. Then they unloaded the *Kitty Clyde's Sister*, but it had grown late and they put off any further work until the following morning. Everyone woke early the next day and soon had all the boats safely below the fall. "Then came the real hard work," wrote Sumner, "carrying the freight a hundred yards or more over a mass of loose rocks, tumbled together like the ruins of some old fortress. Not a very good road to pack seven thousand pounds of freight."

It was brutal work. The cargo came to nearly eight hundred pounds per able-bodied man, to be carried on their backs a distance of a couple of city blocks (and with those blocks torn up by a construction project). The trail, Bradley wrote, led over and around an array of "huge bowlders recently fallen from the mountains." With all their surroundings so outsized—house-sized boulders, thousand-foot cliffs—the men staggering under their enormous loads seemed like ants making off with crumbs from a picnic.

At one point, someone noticed an inscription cut into a rock. "Ashley," it said, and then, less clearly, a date. Some of the men thought it said 1825, some 1835 or 1855. They named the rapid Ashley Falls in honor of their mysterious predecessor.

No one knew who Ashley was, though Powell thought that he had heard the name. He had been told a story about a group of men who had started down the Green. Their boat had swamped, and some of the men drowned. Powell believed (mistakenly) that Ashley was one of that group. The whole episode was unsettling, the faded name and date evocative of an inscription on a gravestone. "The word 'Ashley' is a warning to us," Powell wrote, "and we resolve on great caution."

William H. Ashley was an entrepreneur and a major figure in the fur trade, though he himself seldom ventured west of St. Louis. In February 1822, he had placed a newspaper advertisement seeking "one hundred men, to ascend the river Missouri to its source, there to be employed for one, two or three years." A host of men who would become explorers of almost legendary stature—Jim Bridger, Jim Beckwourth, and Jedediah Smith among them—took Ashley's offer and marched into history.

These mountain men were not only surpassingly brave but surpassingly accomplished as well, vastly more at home in the unsettled regions of the West than any other white men of the day. "They went about the blank spaces of the map like men going to the barn," in the words of the historian Bernard DeVoto.

Illiterate though many of them were, they knew how to read the landscape. "It is hardly too much to say that a mountain man's life was skill," De Voto wrote. "He not only worked in the wilderness, he also lived there and he did so from sun to sun by the exercise of total skill. . . . The mountains, the aridity, the distances, and the climates imposed severities far greater than those laid on forest-runners, rivermen, or any other of our symbolic pioneers. . . . Why do you follow the ridges into or out of unfamiliar country? What do you do for a companion who has collapsed from want of water while crossing a desert? How do you get meat when you find yourself without gunpowder in a country barren of game? What tribe of Indians made this trail, how many were in the band, what errand were they on?" Such questions were all in a day's work for the mountain men. And Ashley's men were among the best.

Jim Beckwourth, who wrote about Ashley in a swashbuckling autobiography, lived a life with all the plot twists of a boy's adventure book. The son of a slave mother and a plantation-owner father, Beckwourth was born in Virginia and raised in Missouri but became a celebrated hunter and explorer in the West. Along the way, he was adopted by the Crow Nation, took a Crow wife, and became a war chief. A man of extremes, he stood out even among his fellow mountain men, one historian notes, as "a tough hombre, a daredevil, a thug, and a liar."

Ashley placed his newspaper ad in 1822. By 1824, the mountain men had brought him the welcome news that the tributaries of the upper Green teemed with beaver. Determined to see for himself, Ashley headed West in 1825. He divided his men into four scouting parties and assigned three to explore particular valleys and streams. The fourth group, Ashley and seven mountain men, prepared to set out down the Green itself. Beckwourth was, at least sometimes, part of Ashley's party.

First Ashley's crew had to make their boats. These fragile craft, called bull-

boats, were formed by draping buffalo skins across a framework made of willow branches. The first boat was sixteen feet long and seven wide, made from
half a dozen buffalo hides. It bore a none-too-encouraging resemblance to an
upside-down umbrella. Soon the men found it necessary to build a second,
smaller boat as well. Awkward at the best of times, these bullboats grew even
more unwieldy as they became waterlogged.

Beckwourth claimed that the boats proved unreliable from the moment they
were built. "One of our boats being finished and launched," he wrote, "[Ashley]
sprang into it to test its capacity." Immediately the mooring line snapped, and
Ashley was pulled from the river's edge by what Beckwourth called the "Green
River Suck." The ominous name referred to a "fall [that] continued for six or
eight miles" while descending "upward of two hundred and fifty feet." Ashley
made it to the far shore but capsized, and Beckwourth swam to his rescue.
But both men found themselves caught in the Suck and "began slowly to
recede from the shore toward inevitable death." At the last possible moment,
Beckwourth managed to catch a line flung from shore, and the other mountain
men dragged their two half-drowned companions to safety.

From that dismal beginning, the trip quickly grew even worse. The men had
not brought enough food, and they could not get more because they had lost
their guns in various capsizings. After six days without food, they had decided
to draw lots to see who should be sacrificed to serve as dinner for his companions. Ashley prevailed on the men to hold out one more day. The next day,
the canyon opened up and they found a party of trappers waiting with food.

So Beckwourth told it, at any rate. It is worth emphasizing that, just as good
cooks pride themselves on being able to conjure up a feast from a wedge of
cheese, a tomato, and a few other bits and pieces, so the mountain men prided
themselves on crafting an elaborate tale from a handful of half-truths. "To be a
liar was as much a part of mountain honor," one historian notes, "as hard drinking or straight shooting. Embroider your adventures, convert to use any handy
odyssey, and spin it all out in the firelight. The only sin is the sin of being dull."

By this standard, Beckwourth was a man of unimpeachable honor.
Remarkably, his account was taken as gospel for decades. Only in 1918, when
Ashley's journal was posthumously published, did the true story of the 1825
trip emerge. The danger had been real enough, even without exaggeration,

and both men and boats had taken a beating. The portages were frequent and difficult, and sometimes the men had deemed it better to take a chance on capsizing in a rapid than to volunteer for the certain misery of portaging. "Exosted with the fatiegue of portages and the tediousness of our progress . . . we crossed many dangerous places This day without examining them previously," Ashley wrote. The men *had* run low on food (although never to the point of considering cannibalism), and once Ashley *had* been rescued from a rapid "just as from all appearance [the boat] was about making her exit and me with her for I cannot swim."

All in all, though, the trip was less colorful in reality than in Beckwourth's dramatization. But even the unadorned facts might have made Powell hesitate, if he had known them. Consider what Ashley and his mountain men had done. They had seen the Green, struggled through some of its fearsome canyons and rapids, and gladly left the river to rendezvous with their colleagues. It is true that Ashley's goals were not Powell's—Ashley's concern was business, not exploration. But even allowing for their different motives, it is worth noting how much more ambitious Powell's plans were than Ashley's. Ashley covered only a fraction of Powell's route; he saw the Green only before its junction with the Colorado, before that second mighty river added its power to what was already a life-threatening torrent; and he never came near the Grand Canyon.

Ashley's men were bold, and proud of their boldness. But enough was enough. They were hungry, and the river was growing steadily more threatening. Farther downstream, Ashley was told, the Green was not only "destitute of game" but also "verry dangerous." Powell would eventually hear the same warnings. It was a mark of his daring—some would say of his recklessness— that, within two years of seeing the West for the first time, he crossed a line that the mountain men had crept up to and then shied away from.

The first to find Ashley's inscription, as far as we know, was an ill-fated young man named William Manly. He had himself proudly written his own name near the river, a few miles upstream of Ashley Falls, but Powell and his men had missed it. "Capt. W. L. Manly, U.S.A," it read, in large letters painted on a rock. Manly's "paint" was a combination of gunpowder and grease, his brush a bit of cloth tied around a stick. He had signed his name twenty years before Powell set off down the Green, and he, too, had found himself overmatched at Ashley Falls.

Manly was a restless young man, a hardworking but not notably successful hunter and trapper who, in 1849, caught a bad case of gold fever. California was strewn with gold, everyone said, just waiting for someone to come and get it. No story was too outlandish to believe. For $2.50 a bottle, one newspaper advertisement declared, you could buy a special salve that would guarantee your fortune. All you had to do was climb a hillside in gold country, rub the salve all over yourself, and roll downhill. By the time you reached the bottom and scraped off the gold dust, you would have riches that would last the rest of your days.

Manly, then living in Wisconsin, found a wagon train headed for the gold fields and hired on to drive an ox team. They had started too late in the year, he soon learned; the Sierra Nevada would be deep in snow and impassable. The only choice was to wait out the winter in Salt Lake City. By this time, the party had reached the Green River. Manly and a small group of impatient companions came up with their own plan. "We put a great many 'ifs' together and they amounted to about this: If this stream were large enough; if we had a boat; if we knew the way; if there were no falls or bad places; if we had plenty of provisions; if we were bold enough to set out on such a trip, we might come out at some point or other on the Pacific Ocean."

Then came a sign from heaven—someone found a small ferry boat abandoned on a sandbar. It was beat-up and filled with sand, but the men soon dug it free and declared it usable. Better still, they found a pair of oars buried in the sand beneath the boat. Manly and six companions bade farewell to the wagon train, gulped hard when they realized that "all our worldly goods were piled up on the bank, and we were alone," and prepared to head downriver. The boat had awkward proportions—it was about twelve feet long and six wide—but it held all the men, six guns, and their gear. "It was not a heavy load for the craft," Manly noted, "and it looked as if we were taking the most sensible way to get to the Pacific, and we almost wondered that everybody was so blind as not to see it as we did."

The first inklings of doubt crept in soon enough. One man was at the oars, the others at the sides of the boat armed with long sticks they intended to use to keep away from rocks in the channel. The river was fast and shallow. At one point, Manly, who had been chosen captain, pushed his pole against the bottom to dodge a boulder. Instead, the stick caught between two rocks and flung Manly out of the boat and into the river, pole vaulter style. He managed to swim to shore to the sound of his companions' cheers.

They soon reached a stretch of river choked by "huge rocks as large as cabins." Here Manly and his companions tried a version of lining that makes Powell's approach seem stodgy. "While the boys held the stern line," Manly explained, "I took off my clothes and pushed the boat out into the torrent which ran around the rocks." Then Manly scrambled into the boat, and the men slowly paid out the stern line. Manly cried out, "Let go!" when he thought the time was right, and the boat leapt ahead—with Manly still in it—and over the fall. "I grasped the bow line, and at the first chance jumped overboard and got to shore, when I held the boat and brought it in below the obstructions."

At about this point, Manly chanced to look up and see the rock where Ashley had painted his name. He didn't have much time for sightseeing. Three hundred yards downstream an enormous rock nearly blocked the channel. "The current was so strong that when the boat struck the rock we could not stop it," Manly wrote, "and the gunwale next to us rose, and the other went down, so that in a second the boat stood edgewise against the big rock, and the strong current pinned it there so tight that we could no more move it than we could move the rock itself."

Checkmate! Or so it seemed, for the boat was irretrievably lost and the men were "afoot and alone." Manly, though, saw two pine trees that he believed could be shaped into canoes. The men set to work with their axes and soon had hacked out two dugout canoes, rough but perhaps workable, each about fifteen feet long and two feet wide. They lashed the two together for stability but found that the makeshift craft was too small. They struggled back to shore, chopped down more trees, and set to work building a third, larger canoe. The new boat, about twenty-five or thirty feet long, was soon finished and loaded. Manly and a companion set out in the lead, in the big boat.

They rattled their way along, sometimes dodging rocks, sometimes crashing into them, riding over some waves and being flipped by others. In one stretch where the channel ran straight and unobstructed, "the stream was so swift that it caused great, rolling waves in the center, of a kind I have never seen anywhere else," Manly wrote. "The boys were not skillful enough to navigate this stream, and the suction drew them to the center where the great waves rolled them over and over, bottom side up and every way."

Some of the men struggled to shore, but one, Alfred Walton, could not swim and clung to his canoe with a death grip. "Sometimes we could see the man and sometimes not, and he and the canoe took turns in disappearing. Walton had very black hair, and as he clung fast to his canoe his black head looked like a crow on the end of a log. Sometimes he would be under so long that we thought he must be lost, when up he would come again still clinging manfully."

Walton lived, and somehow the men made it downstream several more miles. Exactly how many is a matter of historical dispute; Manly and Walton and the rest of their crew seem to have covered roughly the easiest one-fifth of the distance Powell intended to travel. Somewhere in the Uinta Basin they had the good fortune to encounter a Ute Indian, Wakara, whose name Manly rendered as "Walker." Wakara asked, in sign language, where Manly thought he was going. Manly did his best to explain that he and his party were taking a shortcut to the Pacific Ocean and the land of gold. "When I told Chief Walker this he seemed very much astonished," Manly reported, deadpan.

Wakara then led Manly to a sandbar by the river, took up a stick, and began to draw a map in the sand. Here was the Green, here were its upstream tributaries, here (a trail of pebbles and much waving of the stick, as if Wakara were driving oxen) was the white man's overland route. It was all, Manly saw, "exactly correct." When Manly nodded his understanding, Wakara began to map the section of river that Manly would encounter next. At last, after depicting a series of mountains, valleys, and canyons, Wakara began gathering stones and piling them up one on top of the other, to represent the deepest canyon yet. "Then he stood with one foot on each side of his river and put his hands on the stones and then raised them as high as he could, making a continued e-e-e-e-e-e as long as his breath would last, pointed to the canoe and made signs with his hands how it would roll and pitch in the rapids and finally capsize and throw us all out. He then made signs of death to show us that it was a fatal place. I understood perfectly plain from this that below the valley where we now were was a terrible cañon, much higher than any we had passed, and the rapids were not navigable with safety."

Manly and four of his companions decided that it was time to take the overland route.★ (Two of the men chose to stick with the river, although they, too, soon decided that they had had enough.)

★Few travelers have endured journeys as harrowing as Manly's. He ended up taking yet another shortcut; this one turned out to lead across Death Valley. (The name was given later, in testimony to the suffering of Manly's party.) In comparison with the trek across Death Valley, the trip down the Green was a pleasure jaunt. Manly eventually reached California, although he never did strike it rich. He told his story in his autobiography, *Death Valley in '49*.

•   •   •

Powell did not know these harrowing stories, and he might have proceeded anyway had he known them. But even John Frémont, "the Pathfinder," perhaps the most renowned Western explorer of them all, had rejected an expedition like Powell's as too reckless. That this man had looked at the Colorado and blinked should have been worrisome indeed, for Frémont risked his own life (and the lives of his companions) with blithe indifference.

Frémont was a larger-than-life character, part hero and part comic-opera buffoon. He was the Republican presidential candidate in 1856 and, for a time during the Civil War, commander of all Union forces in the West. (He surrounded himself with a guard of thirty men, all of them five feet, eleven or taller so that the effect would be suitably impressive.) Frémont was "the damndest scoundrel that ever lived," Abraham Lincoln once observed, "but in the infinite mercy of Providence . . . also the damndest fool."

He may also have been the first, in 1842, to try to run rapids in an inflatable rubber boat. The attempt began well. Frémont and his crew dashed their way down the Platte River, singing happily. "[We] were, I believe, in the midst of a chorus when the boat struck a concealed rock immediately at the foot of the fall," Frémont wrote later, "which whirled her over in an instant. Three of my men could not swim, and my first feeling was to assist them, and save some of our effects; but a sharp concussion or two convinced me that I had not yet saved myself."

No one drowned, though a few men came close, but, in one historian's summary, "the party was left stripped of every morsel of food, ammunition, and arms, at the mercy of savages and in danger of starvation." The experiment could hardly have been less successful, but Frémont's zeal for "firsts" and for adventure was undiminished.

In 1848, he set out to cross the Rockies in winter, against the fervent advice of everyone who heard his plans. His goal was to demonstrate the feasibility of a transcontinental train route that followed the thirty-eighth parallel. Such a continent-spanning route, joining New York and San Francisco, had long been a pipe dream of Missouri senator Thomas Hart Benton, Frémont's patron and father-in-law. (Not coincidentally, the route would have passed through St. Louis, where Benton lived.) The problem was that the thirty-eighth parallel ran directly through some of the most forbidding terrain in the

Rocky Mountains. Frémont's reasoning, apparently, was that the worse the conditions he survived, the more plausible the case for a rail line. At twelve thousand feet, with his men starving and their mules frozen to death, the storms that had bedeviled the trip all along suddenly grew even worse. The snow lay in drifts as deep as thirty feet. The men were so desperate for food that they ate their candles and the soles of their moccasins. Some of the survivors cannibalized their dead companions. In the end, ten members of the thirty-three-man party died; another fifteen were rescued at the point of death by the heroic scout Alexis Godey.

None of this fazed Frémont, who declared the mission's outcome "entirely satisfactory." And yet even he paled at the dangers of the Green and the Colorado. It was undoubtedly true that those untamed rivers presented "many scenes of wild grandeur" and "many temptations" generally, Frémont wrote, but he conceded that neither he nor anyone else wanted any part of them. "No trappers have been found bold enough," he wrote, "to undertake a voyage which has so certain a prospect of a fatal termination."

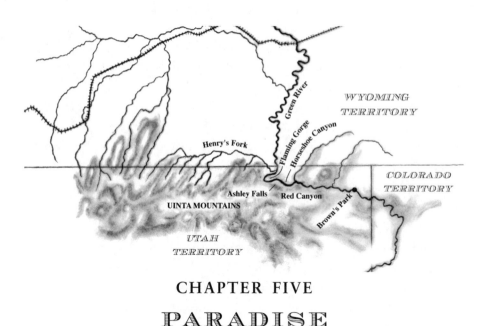

CHAPTER FIVE

PARADISE

If Frémont had seen Powell and his men on June 2, 1869, he might have taken back his words of warning. That glorious spring day marked perhaps the high point of the entire Powell expedition, and the hours between eleven o'clock and three marked the best part of the best day. The men had finished lining the boats and wrestling the cargo past Ashley Falls by late morning—"had dinner and smoked all round," Sumner reported contentedly—and then returned to the river. They were safe and on the move and all was bliss.

"Beautiful river, that increases its speed as we leave the fall . . . but clear of sunken rocks," Sumner noted with glee. "So we run through the waves at express speed; made seventeen miles through Red Stone canyon in less than an hour running time, the boats bounding through the waves like a school of porpoise." Sumner's boat, smaller and lighter than the three "freight" boats, bounded even more than the others. "The *Emma* being very light is tossed about in a way that threatens to shake her to pieces, and is nearly as hard to ride as a Mexican pony. We plunge along singing, yelling, like drunken sailors, all feeling that such rides do not come every day."

Sumner ran through his memories in search of a similar thrill and finally hit on a comparison. "It was," he wrote, "like sparking a black-eyed girl—just dangerous enough to be exciting."

Suddenly the sandstone canyon opened out into a beautiful little valley. This was the best land they had seen yet, the low, grass-covered hills a welcome change from the towering cliffs and bare rocks they had grown to expect. The men made camp at about three in the afternoon, spread the bedding out to dry, and set out to shoot dinner or flung themselves on the ground to rest, depending on how ambitious they felt. The hunters eventually returned empty-handed, except for Powell, who could wield a pistol but not a rifle and had managed to shoot two grouse. Sumner was in too good a mood to be dismayed by the meager haul. "Spread our blankets on the clean, green grass," he wrote, "with no roof but the old pines above us, through which we could see the sentinel stars shining from the deep blue pure sky, like happy spirits looking out through the blue eyes of a pure hearted woman."

Bradley, never much inclined to lyricism, went so far as to acknowledge that "the sky is clear—weather warm and delightful." Still, he had little trouble keeping his euphoria in check. "The men are out hunting and have not yet come in," he wrote, "but if they have their usual luck *we shall have bacon for breakfast.*"

Powell, in contrast, was in full Wordsworthian mode. He rose early the next morning and went for a five- or six-mile hike "up to a pine grove park, its grassy carpet bedecked with crimson, velvet flowers, set in groups on the stems of pear shaped cactus plants; patches of painted cups are seen here and there, with yellow blossoms protruding through scarlet bracts; little blue-eyed flowers are peeping through the grass; and the air is filled with fragrance from the white blossoms of a Spiraea." To complete the picture, a mountain brook ran through the meadow. "It is a quiet place for retirement from the raging waters of the cañon," Powell noted happily.

While Powell was basking in nature's glories, the rest of the party were out hunting and fishing. The hunters came back to camp disappointed again— "they didn't bring in enough game to make a grease spot," Bradley grumped— but the fishermen had succeeded, after a fashion. The riverbank was lined with strange-looking specimens. Some were a "queer mongrel of mackerel, sucker, and whitefish," Sumner wrote, and the rest "an afflicted cross of white fish and lake trout." Their taste was a match for their appearance. "Take a piece of raw pork and paper of pins, and make a sandwich, and you have the mongrels,"

Sumner wrote. "Take out the pork and you have a fair sample of the edible qualities of the other kinds."

The next day was nearly up to the standard of the previous two. Rested and content, the men had a "splendid ride" on a fast river with "only few rappids and mostly deep water." At one point all the boats ran aground yet again, but this counted as little more than a nuisance. Indeed, the day was marred only by the dismaying appearance of "a most disgusting looking stream" they encountered about six or eight miles after setting out. "It is about ten feet wide, red as blood, smells horrible, and tastes worse," Sumner noted. But they quickly left the muddy stream behind and came to a large and handsome valley about twenty miles long by five miles wide. They made camp beneath a giant cottonwood tree that Sumner claimed was big enough "to furnish shade and shelter for a camp of two hundred men." The weather was good, the river was easy, and the men had killed some ducks (and, less enticingly, caught more fish). Life was sweet.

The valley was Brown's Park, a hideaway nestled in the mountains and, in winter, sheltered from the worst snow and wind. Though the river itself was nearly unknown, at this early stage of the trip the canyons occasionally opened up into well-explored valleys like Brown's Park. Because of mild winters and abundant grass, this was a good place for cattle—Sumner had passed through the previous winter and had seen "4000 head of oxen pastured in it without an ounce of hay"—and it was remote as well. That combination would prove enticing. Brown's Park later became a favorite spot of outlaws and cattle rustlers and other entrepreneurs not much hemmed in by conscience. Butch Cassidy and the Wild Bunch would come to know Brown's Park well.

Powell's men seemed more inclined to sprawl on the grass than to make trouble. "Everything [was] as lovely as a poet's dream," Sumner declared, and he painted a picture of his own vision of celestial bliss, which ran less to harps and angels than to a harem full of beckoning beauties. He lay half asleep the morning of June 5, entranced by "the sweet songs of birds, the fragrant odor of wild roses, the low, sweet rippling of the ever murmuring river at sunrise in the wilderness . . . I was just wandering into paradise; could see the dim shadow of the dark-eyed houris . . ." The fantasy was interrupted by a bellowed call to breakfast—hot coffee and fried fish. Perhaps not worth leaving paradise for, but a step up from bacon.

The rest of the day continued almost as pleasantly, the men exploring or mapping or loafing. Powell and Dunn scrambled their way up a cliff; Oramel Howland worked on his map; and Bradley, Seneca Howland, and Hall crossed the river to survey the far side. On the way back to camp, Bradley shot a rattlesnake and tried (unsuccessfully) to *shoot* some trout he saw in a brook.

The following day brought a return to work. The river was broad and sluggish, and the men were rowing directly into a strong head wind. They fought it for twenty-five miles—"no easy task," Sumner conceded—and dragged themselves into camp dog-tired. "I feel quite weary tonight," Bradley wrote. "Would rather have rappids than still water but think I shall be accommodated."

Accommodated soon, he meant, and he was right. The view downstream was unsettling. It looked, in the words of one early settler, like "a mountain drinking a river." Bradley provided a more detailed picture. "We have now reached the cañon at the lower end of Brown's Hole and have camped tonight at the mouth of the Cañon," he wrote. "It looks like a rough one for the walls are very high and straight and the sides are of sand-stone much broken with seams but at the mouth nearly perpendicular; in such the worst bowlders have been found and I expect them below here."

Powell took a long look the next day, and, unlike Bradley, he liked what he saw. "When we return to camp, at noon, the sun shines in splendor on vermilion walls, shaded into green and gray, where the rocks are lichened over; the river fills the channel from wall to wall, and the cañon opens, like a beautiful portal, to a region of glory."

Then he made the mistake of reexamining the view later in the day. The sun had begun to set, and the reds and greens had given way to long, black, foreboding shadows. Now the canyon that had beckoned so enticingly only hours before appeared as "a dark portal to a region of gloom—the gateway through which we are to enter on our voyage of exploration to-morrow. What shall we find?"

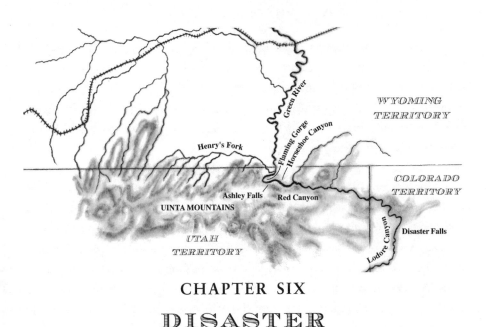

CHAPTER SIX

DISASTER

Melodrama came naturally to Powell, but his dread here was sincere. At the same point on the Green, Ashley, in 1825, had written just as fearfully. A businessman and a politician, Ashley was not given to flights of fancy. Nor were his mountain men. But when they left the sanctuary of Brown's Park and approached Powell's "dark portal to a region of gloom," they seemed as jumpy as schoolchildren trembling at the door of a haunted house. "As we passed along between these massy walls, which in a great degree excluded from us the rays of heaven ... I was forcibly struck with the gloom which spread over the countenances of my men," Ashley wrote. "They seemed to anticipate (and not far distant, too) a dreadful termination of our voyage, and I must confess that I partook in some degree of what I supposed to be their feelings, for things around us had truly an awful appearance."

The entrance to Lodore Canyon, as Powell named it, retains its dismaying appearance to this day. "Each canyon of the Green has its own distinctive presence but none is as dramatic as Lodore," writes Ann Zwinger, an acclaimed and notably levelheaded natural historian. "The cliffs rise two thousand feet, immediate, all the more striking because of the pale landscape from which they spring, almost without transition. The Gates of Lodore hinge inward, cruelly joined, hard rock, ominous, and when mists skulk low between the cliffs, they become an engraving by Gustave Doré for one of Dante's lower levels of hell." And this, mind you, from a writer far more inclined to celebrate the river than to cower before it.

The Gates of Lodore, the entranceway to the canyon, are forbidding partly because they tower so high. Castle builders in the Middle Ages knew well that a doorway that loomed twelve or even twenty feet high would make a visitor feel timid and puny. But castle doors, even colossal ones, bear some relation to human proportions; this narrow stone gateway stands nearly half a mile high. In addition, Lodore Canyon seems foreboding partly because it is dark and shadowy. Red Canyon, just upstream, has lower walls and runs east-west, so that it is often bathed in sunlight. Lodore, hemmed in by taller cliffs and running north-south, is gloomy for all but a few hours a day.

But it was not high cliffs and weak sunlight that played the largest part in making Ashley and Powell so fretful. Lodore is truly fearsome because it is here, in a series of roaring, seemingly interminable rapids, that the Green finally reveals its power. The rapids and falls and fast water they had seen so far, Powell and his crew now realized, amounted to no more than a warm-up routine on the river's part, the equivalent of stretching exercises. Now the preliminaries were complete.

Powell's men were still novices when it came to running rapids, but they had figured out some of the basics. They knew that dark, smooth water was probably deep (and therefore safe). Small, apparently random ripples meant rocks just below the surface. A smooth, rounded mound of water usually signaled a big, barely submerged rock. Powell and the men watched eagerly for V-shaped stretches of water, wide at the top and narrowing downstream, where the river squeezed itself between nearby obstacles. Their river-running strategy at this point consisted essentially of two options—aim for one of these V-shaped "tongues" and pray, or portage.

They knew enough to be wary, but not much more than that. Often, for example, the river pooled up quietly just above a rapid. That calm zone was safe, Powell noted, but "sometimes the water descends . . . from the broad, quiet spread above into the narrow, angry channel below, by a semicircular sag. Great care must be taken not to pass over the brink into this deceptive pit."

The problem with Powell's rudimentary strategy—"If I can see a clear chute between the rocks," he wrote, "away we go"—is that the chute rarely led all the way through a rapid. More often, it provided a glorious mini-ride that served as a shortcut to trouble, as if a magic carpet passed above a line of angry boulders only to drop its passengers into a whirlpool. And once into the heart of a rapid, all bets were off. "On the ocean, where waves are rhythmic, predictable, and usually from the same direction as the wind, there is time to anticipate and prepare . . . ," observes Martin Litton, one of the towering figures in modern river running. "Not so in a rapid where—if you're in really big water—things are coming at you from all directions at once."*

But rapids were riddles formed of water, and any tentative conclusions the men reached in one rapid washed away in the next. Often the waves *did* come from every direction at once, for example, but at other times they lined up in perfect order, seemingly eternal. "The water of an ocean wave merely rises and falls," Powell noted, so that the wave *form* moved toward shore but the water itself bobbed and sank in place. "But here," he noted, transfixed, "the water of the wave passes on, while the form remains." At Hermit Rapid in the Grand Canyon, for example, the waves stand in perfect rank, the first one some seven feet tall, then the second, third, and fourth each a bit taller in turn, and the fifth tallest of all, followed by a set of waves that diminish just as uniformly. Since each wave stays in place, a boatman surfing a river wave could theoretically stay poised in the same spot forever, freeze-framed in ecstasy as if on a Grecian urn. When the Colorado is running chocolate brown, these standing waves look so much like abstract sculptures that it is hard to remember they are made of water. Touching one, it seems, would feel like patting a dolphin.

The men had learned at the outset that it was crucial to take waves head-on rather than sideways, but how to convince the waves to wait their turn? Let yourself turn sideways even for a moment, so that the river has the full length of the boat to attack, and it can flip you before you have time to curse.

A homemade experiment helps make the problem clear. Hold your hand out the window of a fast-moving car, palm toward your cheek, and notice that your hand slices through the wind effortlessly. Now abruptly rotate your hand so that the palm faces forward, as if you were a policeman signaling "Stop!" Suddenly the wind has something to hit, and it pummels your hand.

---

*Litton, now over eighty and still running the Colorado and other rivers, is the oldest person ever to have rowed the length of the Grand Canyon.

Try the same experiment while standing in shallow water in a fast river. Place a hand or a canoe paddle in line with the current, and the water flows smoothly around the obstacle. But turn the paddle so that it catches the current broadside, and the force can jolt you. Now try the same experiment with the canoe itself rather than with a mere paddle. Fasten your life jacket first.

Whether they wanted to or not, Powell and his men would become students of rapids. Lesson one was that rapids formed wherever the channel shrank in size. That happened where colossal chunks of rock broke off a cliff side and tumbled into the river or where giant loads of rocks and boulders swept down a side canyon and finally came to rest in the main channel. Suddenly squeezed into a small space but still retaining all its power, the pent-up river was transformed and newly dangerous, a tornado in a box.

All the trouble in rapids stems from one simple fact—unlike a wad of cotton, for instance, water cannot be compressed. Arnold Schwarzenegger could squeeze a water balloon until he was red-faced and gasping, and although he could change the balloon's shape—it might bulge *here* instead of *there*—he could not shrink its overall volume. Now think of a river. It flows at the same steady rate whether it is traveling through a broad valley or a skinny bottleneck. Suppose that rate is, say, one thousand cubic feet per second. All that means is that if someone could divert the river into a giant bathtub, it would fill a one-thousand-cubic-foot tub every second. But there *is* a tricky point. Whether the river channel is broad and deep or narrow and shallow, the river would fill the same one thousand-cubic-foot bathtub in a second. How can that be?

Since water cannot squeeze in on itself, the only possibility is that the river speeds up whenever the channel contracts. This is the discovery that every delighted five-year-old makes when he learns to put his thumb partway across the end of a hose and spray his little brother. More to the point, it is the reason a placid, gently flowing, almost lakelike river transforms itself into a chaotic, churning nightmare when it squeezes through a narrow, rock-choked channel.

Like the five-year-old with a hose, Powell understood perfectly well that a narrow channel meant fast, tumultuous water. The difference is that Powell was on the *receiving* end of the waterpower, like a ladybug caught in a hose's blast.

The problem for Powell and his crew of novices was that their experience in other domains, such as driving wagons, provided little guidance in river

running. In similar fashion, beginners today often run into trouble because
they assume that a boat on a river will behave like a car on a highway. But in
crucial ways, rivers and highways are opposites.

On man-made roads, traffic speeds up on the straightest, broadest stretches
and slows down where the road suddenly shrinks from six lanes to two.
With rivers, as we have seen, the opposite applies. The more a river narrows,
the faster it flows. To picture the plight of novices in white water, imagine a
highway where drivers have only the most rudimentary control of their car's
steering wheel, gas pedal, and brakes. Suppose, further, that each car's speed
changes automatically and inevitably, depending on the road conditions.
Picture, in particular, that the *worse* the road—the more suddenly it squeezes
itself into a single lane with a cliff on one side and a sheer drop on the other—
the *faster* each car hurtles along. On such a diabolical highway, we might begin
to appreciate the special, unnerving qualities of white water.

In fact, it is worse than that. The speed of the current is not the greatest
hazard in a rapid. Waves in a rapid ricochet off rocks and cliffs and collide
with one another; water rushes over rocks and dives down into holes and
moves *upstream* to fill in "empty" spaces behind obstacles. The problem is not
that the current is moving so fast but that it is flowing in so many different
directions at so many different speeds, downstream and upstream and even
straight down toward the bottom of the river. Think of our diabolical highway again, and this time throw in not only a bottleneck but a hairpin curve
and some potholes and patches of ice and broken-down cars abandoned in
the middle of the road. If this highway behaved like a river, it would not only
speed up as it narrowed but would form itself into a complex series of mini-
roads, some heading straight into a ditch, others speeding toward junked cars
or ice slicks or—perhaps—safety.

The rocks in the river, it should be noted, provide a double dose of danger. They *make* trouble, first of all, by choking the channel and providing the
structures that create waves and falls and boat-sucking holes. But even in their
passive role, as obstacles rather than as creators of chaos, rocks can be formidable. The risk, as Powell's men had already found, is in getting hung up sideways
against a boulder, pinned against an unyielding obstacle by the concentrated
force of a surging river. Here highway analogies fall short. Think instead of a
python's prey, immobile in the giant snake's relentless coils, struggling fiercely
but futilely against a vastly stronger opponent.

•   •   •

On June 7, Powell and some of the men climbed the cliffs to survey the new canyon. It was not an easy climb, for the rocks were split with dark, threatening fissures. The cliff top, which proved to be 2,085 feet above the river, provided a river view of some six or seven miles. From this vantage point, the Green looked small and harmless, almost inviting. On returning to the river, the men quickly found that rapids that looked easy from half a mile above were not so easy when seen from a bucking, tossing rowboat. Andy Hall, undaunted, dredged up from his memory a bit of English verse by Robert Southey about a waterfall's "Rising and leaping, / Sinking and creeping, / Swelling and sweeping," and so on. The poem, which continued on in singsong fashion for another hundred-plus lines, was called "The Cataract of Lodore," and Hall proposed the name Lodore for the canyon they were passing through. Sumner was not pleased—"the idea of diving into musty trash to find names for new discoveries on a new continent is un-American, to say the least," he grumbled—but Powell liked the name, and Lodore it is to this day.

On the following day, no one was reciting poetry. It brought, Sumner wrote, "as hard a day's work as I ever wish to see," and it was as dangerous as it was difficult. The morning alone saw a dozen bad rapids to line and portage, each one seemingly worse than the ones before. The scenery, for those in the mood for sightseeing, was spectacular. The men stopped for lunch at the foot of a perpendicular, rose-colored wall some fifteen hundred feet high. At one o'clock, they started up again.

In half a mile, they came to a maelstrom that earned Bradley's customary description as "the wildest rapid yet seen." This time Sumner echoed him. They had reached "a terrible rapid," he wrote, a place "where we could see nothing but spray and foam." The lead boat, the *Emma Dean*, pulled safely to shore above the rapid.

So did the *Maid of the Cañon*. Powell began climbing the rocks along the shore to size up their predicament. Then he heard a shout—the Howland brothers and Frank Goodman, in the *No Name*, were speeding down the river in midstream, out of control and headed for the rapid. In the meantime, where was the *Sister*? Unable to help the *No Name*, Powell ran back upstream to try to warn the *Sister* to land. Racing along the rocks, shouting, waving, he saw nothing. Then, rounding a bend and pulling hard to shore and safety, there she was. Powell turned around yet again, chasing desperately back downstream in search of the *No Name*.

Where was she? The first part of the rapid was a drop of ten or twelve feet, which was bad but perhaps manageable, and then came a steep, boulder-strewn stretch of forty or fifty feet, beaten into foam and churned by whirlpools. Powell scrambled over the rocks and finally caught sight of the *No Name*, straining to pull toward shore. Suddenly she hit a rock, tipped alarmingly, and filled with water. The men lost their oars. Helpless, the three men sat while the boat raced sideways several yards, crashed into another rock, and broke in two. The three crew members—none of them in life jackets—struggled frantically toward the broken boat and grabbed on to a chunk of its bow in the surging waters. Down they drifted a few hundred yards into another rapid, this one, too, filled with huge boulders. Twice the men lost their grip on the fractured bow and sank into the water; twice they managed to struggle back again.

At one point, the river carried them near a sandbar, almost a mini-island, in midstream. Oramel Howland made a leap, found himself momentarily protected from the current by a rock, and dragged himself ashore. Frank Goodman tried the same move but vanished into the river. One hundred feet downstream, Seneca Howland leaped and pulled himself onto the same sandbar.

Goodman reappeared a few seconds later, clinging to a barrel-sized rock in the middle of the torrent, gagging on river water, and calling for help. Oramel Howland found a branch that had washed up onto the sandbar. Wading into the river as near Goodman as he dared, he extended the branch toward him. Goodman let go of his rock, dove for the branch, and Howland pulled him to safety. "And now the three men are on an island," Powell wrote, "with a swift, dangerous river on either side, and a fall below."

Worse yet, the river was rising. "Our position on the bar soon began to look serious," noted Oramel Howland, a hard man to rattle. It fell to Sumner to rescue Howland and his fellow castaways. The men unloaded the *Emma Dean*, so that Sumner would be able to maneuver her more easily, and then lined her past the upper rapid. Then it was up to Sumner. He managed to cut a diagonal path to the island. "Right skillfully he plies the oars," Powell wrote, "and a few strokes set him on the island at the proper point." Now the trick would be to get back across a river that was running "with the speed of a racehorse" while carrying three extra passengers and without getting swept into the rapid.

Sumner and the other men dragged the boat upstream and then waded out with it as far as they could into the river. Three of them clambered aboard while the fourth stood perched on a rock, holding *Emma* ready. Then he pushed the boat's nose into the current and flung himself aboard. Fearful that a false stroke meant "certain destruction," Sumner instructed his passengers to lie flat on the bottom of the boat while he alone manned the oars. At one point, he struck a rock and the boat tipped up at a forty-five-degree angle, but it slid safely off. Pulled downstream by the current, Sumner and his three beat-up passengers made it back to shore a scant twenty-five yards above a madhouse of waves and foam. "We are as glad to shake hands with them," Powell wrote, "as though they had been on a voyage around the world, and wrecked on a distant coast."

It had nearly been a calamity. Sumner was rarely one to exaggerate. (He disposed of his role in the rescue in a single sentence, in the third-person voice he used whenever he had pulled off anything especially difficult. "The trapper," he wrote, "crossed over and brought them safely to shore on the east side.") But not even he could deny the narrowness of the three men's escape. Seneca Howland, Oramel's nearly silent younger brother, had been the last to leap to the safety of the sandbar from the wreckage the men had ridden through the waves. "Had he stayed aboard another second," Sumner wrote, "we would have lost as good and true a man as can be found in any place."

In another thirty feet, Sumner continued, "nothing could have saved them, as the river was turned into a perfect hell of waters that nothing could enter and live." Certainly not the *No Name*. "The boat drifted into it and was instantly smashed to pieces," Sumner wrote. "In half a second there was nothing but a dense foam, with a cloud of spray above it, to mark the spot."

After the rescue came the recriminations. What had gone wrong? First of all, the boatmen of the *No Name* had failed to allow for the speed and power of the current. On the first day of the trip, by this point almost an ancient memory, Hall and Hawkins had tried pulling to shore but started too late and missed the campsite by four hundred yards. It had been an embarrassing but harmless mistake, as if a skier on a slope too difficult for him had tried to pull to the side to stop but had badly overshot his destination. Now the Howlands and Goodman had made the same mistake, this time with a hungry rapid waiting to gnaw their bones.

On a big river, things can go bad in a hurry; to react, rather than to anticipate, is almost always to respond too late. And the river was wild as well as swift. The rapids had come in such quick succession, Oramel Howland wrote, that there was no time to bail. At precisely the moment that quick responses were vital, the *No Name* had "so much water aboard," Howland recalled, "as to make her nearly or quite unmanageable." Howland would have been in grave difficulty even without a wave or two in his boat. With that extra weight, he had no chance. For a driver skidding across our demonic highway, it would be as if the power steering chose that moment to quit working.

But why was the *No Name* caught by surprise? Powell's plan, after all, called for him to precede the slower boats and signal them about what lay ahead. This was a touchy subject at the time, and it has stayed that way. For nearly a century, Powell partisans and critics have fought over those signals. Were they sent? Were they seen? We have several descriptions of the first moments after the *Emma Dean* had pulled to shore to scout the new rapid. "I walk along the bank to examine the ground," Powell wrote, "leaving one of my men with a flag to guide the other boats to the landing-place." Sumner, the lead boatman in the *Emma Dean*, provided a similar description. "The scouting boat came to a place where we . . . pulled ashore on the east side and the freight boats [were] instantly signaled to land with us."

Oramel Howland, in the *No Name*, was the target of those signals. Howland's account is not quite clear, although he seems to have seen some kind of signal. "About one o'clock," he wrote, "the signal boat signaled at the foot of a very bad rapid to go ashore; boats nearly full of water—two were made fast, but owing to not understanding the signal, the crew of the *No Name* failed very effectually, owing in the main, to having so much water aboard as to make her nearly or quite unmanageable; otherwise, the mistake was seen by us in time to save her."

Later, though, Sumner would provide a far different account. "As soon as Howland got out of the boat after the rescue Major Powell angrily demanded of him why he did not land," he wrote. "Howland told him he saw no signals to do anything, and could not see the other boats that had landed until he was drawn into the rapid, when it was too late. I asked Hawkins and Bradley in charge of the other boats if they saw signals to land, and they said no signals were given, but as they saw me turn in they suspected something wrong and followed suit at once."

It may be that these accounts can be reconciled. Powell may indeed have given a signal to land, and Howland, fighting for his life on the river, may have missed or misunderstood it. The story of the angry confrontation is more troubling. Powell and his men were a long way from help and a longer way still from their destination. Bad blood between leader and crew boded ill.

In the meantime, there were more pressing problems to confront. No one had drowned, but the *No Name* was gone. The men climbed along the shore to search for the wreckage. Half a mile downstream, they spotted what remained of the boat in the middle of the river, wedged into some rocks. There was nothing but a few battered boards and the splintered stern bulkhead. Powell decided that no one should risk his life to see if it contained anything worth salvaging.

Gone was about a ton of cargo, one-third of the total—great stores of food, three rifles and a revolver, ammunition, all Oramel Howland's maps, half the mess kit, and many of the scientific instruments. The Howland brothers and Goodman lost all their gear, their bedding, and all their clothes except the ones on their backs. (That didn't leave much. Because clothing dragged a swimmer down, the men had taken to running the river clad in only their underwear.)

Powell was despondent. As a precaution against accidents like this one, he had made sure that the three freight boats all carried the same cargo. But somehow all three barometers had been placed in the *No Name*. The fundamental purpose of the entire expedition, in Powell's eyes, was to provide a scientific description of the territory they were passing through. The barometers were crucial to that work, and now they were gone.

The barometers had a practical role fully as important as their role in mapmaking. Although they could not reveal how much farther the men had to go, the barometers *did* show the river's altitude. By comparing that figure with the known elevation at their downstream destination, the men could know how much farther the river had to fall. That was hardly a full picture, for there was no way of knowing whether the drop was sudden or gradual, but it was better than nothing. A relatively flat river, even if it stretched a long way, was a far more docile beast than a sharply dropping (and therefore rapid-infested) river.

Without barometers, no one could know if they had already survived the worst or if their troubles so far only hinted at the ordeal still to come.

Powell spent a sleepless night. The barometers had been in the *No Name*'s stern bulkhead, which the men had seen caught on the rocks in mid-river. Was there a chance that the fragile glass tubes were still in the wreckage? Could they possibly have survived the smashup? "But, then," Powell asked himself miserably, "how to reach them! The river is rising. Will they be there to-morrow?" Perhaps it would be better, he thought, to abort the trip and hike to Salt Lake City, where he could order new barometers from New York City.

The men were scarcely cheerier. "We are rather low spirited tonight," Bradley acknowledged, "for we must camp right at the head of a roaring rapid more than a mile in length and in which we have already lost one of our boats and nearly lost three of our number."

Powell later named the rapid Disaster Falls.

Everyone was up at sunrise the next day, June 9. The men began the weary work of unloading the remaining boats and hauling the cargo up and over the rocks along the river's edge. Then it was time for more lining. In the meantime, Powell set out to examine the wreckage of the *No Name*. As he had feared, the river had dislodged it during the night. As he had not dared hope, it had drifted only another fifty or sixty feet downstream and had run aground on a sandbar. In the wreck's new position, Powell decided, it might be reachable.

Sumner and Andy Hall volunteered to try. (Powell gave the credit, mistakenly, to Sumner and Dunn. The biggest of big-picture thinkers, Powell was not much for dates and other details. In his published journal, for example, he inadvertently put the crew of the *Maid of the Cañon* in the *Kitty Clyde's Sister* and vice versa. *This* mistake is worth noting, though, because we will want to keep an eye on Powell's dealings with Dunn. At this juncture, Powell evidently thought well enough of Dunn to praise him even where no praise was due.)

Sumner and Hall made it to the *No Name*'s remains more or less unscathed. "Away they went and got to it safely, after a few thumps on the rocks," Sumner reported with his customary third-person brevity. (Bradley was a bit more effusive. The two boatmen, he noted admiringly, had overcome "great risk.") They found all three barometers, unbroken, as well as some spare barometer tubes, two thermometers, one pair of old boots, some sole leather, and an untapped ten-gallon cask of whiskey that Oramel Howland had smuggled aboard at Green River Station. Everything else had vanished.

Then came the problem of getting back to shore. It took "an hour's floundering" and several dashes through "pretty rough passes," Sumner recalled, but they made it back to their colleagues' eager welcome. "The Professor was so much pleased about the recovery of the barometers," said Sumner, "that he looked as happy as a young girl with her first beau."

The men were happy, too. They had all watched Sumner and Hall whooping in glee in mid-river as they unpacked the *No Name*'s treasures, but no one could see what the two had found. Powell had been delighted by the one-for-all-and-all-for-one good fellowship. "The boys set up a shout, and I join them," he wrote, "pleased that they should be as glad to save the instruments as myself."

He soon learned about the whiskey keg, "which is what the men were shouting about." A good leader is adaptable, though, and on this night, at least, no one could accuse Powell of being a martinet. "They had taken it aboard, unknown to me," Powell admitted, "and now I am glad they did, for they think it will do them good, as they are drenched every day by the melting snow, which runs down from the summits of the Rocky Mountains."

Drenched every day and cold and bone-weary and three men nearly drowned, and the trip just begun. Disaster Falls changed everything. A few days before, life had seemed almost carefree. Then, when one-third of the supplies vanished in an instant, the trip that Powell had envisioned as a contribution to science threatened to become a test of survival instead. Would the food last? Would the rapids worsen? Could the boats hold up? They had lost one boat already. The loss of another would be a calamity, for two boats could not possibly carry ten men. Already it was impossible to deny that the river was gaining power, and so far they had seen only the Green. What would happen downstream when the Green and the Grand merged and Powell and his men at last confronted the full might of the Colorado?

They had been under way two weeks.

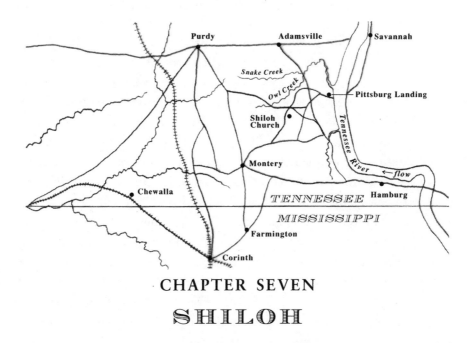

CHAPTER SEVEN

SHILOH

The success of the expedition—the *survival* of the expedition—would depend on Powell's skills as a leader and his ability to persevere in the face of calamity. Powell had been tested before. For the men of his generation, the Civil War was the great, bloody swath that cut their lives in two. "War was a proving ground," notes Stacy Allen, the historian at Shiloh National Military Park, "and in that first year and a half, you saw some meteoric rises and you saw some disappearances." In May 1861, Powell had entered the army as a private. When he left, in January 1865, he was a major and a changed man—wounded, perpetually in pain but burning with ambition, fond of command, and confident that he could keep his head in the worst circumstances that man or fate could throw his way.

Like many of his contemporaries, Powell had known responsibility from an early age. His father, Joseph Powell, was a halfhearted farmer and a dedicated preacher. Nearly always on the road, Joseph left the great bulk of the farmwork to his oldest son. By the age of twelve, John Wesley, known to the family as Wes, had charge of clearing and planting the family's sixty acres and then hauling the crops to market and selling them. That left little time for formal schooling. Though he would eventually spend part of a year at Oberlin College, Powell's education was always a patchwork affair.

Many nineteenth-century farm boys had similar backgrounds. But in two respects, Wes Powell stood apart from most of his peers. He had grown up in a fiercely abolitionist household at a time when few whites gave more than a passing thought to slavery. (Part of Oberlin's appeal was its strong antislavery stance.) And he was plagued by an itch, compounded of intellectual restlessness and wanderlust, that seemed not to afflict many others. "People in those days mostly stuck close to home," the historian Bruce Catton observed. "They had to, because it was so hard to move about . . . A young man stayed at home, and his fatherland was what he could see from his bedroom window, along with the few square miles he might tramp about in the area near his home. Everything else he took on faith."

Powell, the least complacent of men, never took much on faith. As we have seen, he had set out to explore the great rivers of the nation's heartland, on his own, as soon as he could free himself from the family farm. Then, when war came, Powell found that there was more to him than he could have known. He was familiar with hard work and long days, but he did not know command, did not know what it was to make life-and-death decisions, did not know if he could keep his head when death loomed.

In later years, a geologist colleague of Powell's would summarize the lessons that wartime service had taught his old friend. Powell had learned the arts of "reaching prompt decision, giving authoritative command, delegating work to others, and securing loyal obedience from his subordinates," William M. Davis noted. Just as important, he had learned to act quickly even when hampered by incomplete information. "It does not follow that the decisions reached were always the wisest possible," Davis noted, "still they were the best available, and action had to be taken on them without hesitating deliberation."

In ways both obvious and subtle, the war would stay with Powell for the rest of his life. His stump of an arm tortured him without letup, and, though he grew to detest war, the sights and sounds of battle would ever after come unbidden to his mind. Even when he discussed subjects as far removed from the battlefield as geology, Powell favored warlike imagery. Snow and rain were "missiles" in the "storm of war," and crumbling rocks broken down by that attack "fled to the sea" like panicky soldiers running from the front lines. Geology properly understood revealed a tale as violent as the *Iliad*, "a history of the war of the elements." Similarly, the buttes that marked Western land-scapes looked to Powell "as if a thousand battles had been fought on the plains below," with warriors of titanic size in deadly combat, "and on every field the giant heroes had built a monument."

Even late in his life, when Powell turned to philosophy, the squalid years in the trenches were never far away. A discussion on the nature of truth and falsehood might begin with abstractions but would quickly return to solid ground and the mystery of how it could be that a soldier struck by a splinter of wood "interprets it as a fragment of shell, has the illusion of being wounded, and feels the pain and expresses all the agony which a real wound may actually produce."

Powell felt a lifelong bond with those who had seen and endured what he had. When he met a Mississippi congressman, C. E. Hooker, who had lost an arm fighting for the Confederacy at the Battle of Vicksburg, Powell made a pact with his new friend and wartime enemy. Powell had lost his right arm, Hooker his left. Whenever either of the two bought a new pair of gloves, Powell proposed, he should send his friend the extra, useless glove. For thirty years, the two men kept to their bargain, shipping spare gloves back and forth.

In the nineteenth century, Americans saw "duty" and "honor" and "courage" as the highest values and fervently believed that war was a test of manhood. A good soldier was a good man, and a dazed and overwhelmed one a coward and a failure. Combat was a diagnostic test that revealed what you were made of. Powell passed the test, in the world's eyes and in his own. Small wonder that, when the war finally ended, he found himself in search of another outsized challenge.

Powell's first taste of battle came at a sleepy hamlet called Shiloh, in western Tennessee and just north of the Mississippi border. Shiloh would be the site of one of the Civil War's first major battles. At the war's outset, almost no one had a clue about what lay ahead. Leander Stillwell, a teenage soldier from Illinois, wrote that "what I didn't know about war, at that stage of the proceedings, was broad and comprehensive, and covered the whole field." He might have been speaking for Americans everywhere, on both sides.

Americans who came of age at the time of the Civil War had no vivid picture of war. The War of 1812 was a fading memory, notable mainly for having inspired "The Star-Spangled Banner." The Mexican War, though more recent, had been a far-off skirmish against an outmanned foe. In 1861, young men in both the North and the South greeted the news of war with jubilation. Swept up in a vague but romantic vision of the brief, glorious struggle just ahead, they volunteered in droves. "They were convinced that they were committed to something which was larger and grander than life itself, perhaps even a kind of purification, a release from the pettiness of things," a historian wrote of English recruits in World War I, and the description applies equally well to Americans in 1861. The young soldiers marched off to war to the cheers of happy crowds, the music of brass bands, and the waves of pretty girls in their best dresses.

At Shiloh, 110,000 soldiers met in battle. Three out of four had never heard a gun fired in anger. "Do you think any of these men knew what they were getting into?" Stacy Allen, the Shiloh historian, angrily demands. "They'd never seen violence of this magnitude before—no one had. These were a deeply religious people. The only thing they could approximate this with was Armaggedon. It just seemed to them like the whole world had exploded, and they were caught up in the fire."

Shiloh was "the most confused battle of the Civil War," in the judgment of the military historian S.L.A. Marshall. It was "just a disorganized, murderous fistfight, a hundred thousand men slamming away at each other," the historian Shelby Foote observed. "The generals didn't know their jobs, the soldiers didn't know their jobs. It was just pure determination to stand and fight and not retreat."

In 1862, Shiloh was a backwater, a remote spot a long way from Memphis and a longer way still from Nashville, the nearest cities of any size. The name comes from the long-vanished Shiloh Meeting House, a whitewashed square about thirty or forty feet on a side, with cracks large enough to take a man's hand and places for windows but no glass. The name, from the Hebrew, is traditionally translated as "Place of Peace."

War came to this lonesome spot because it was a stopping-off point for the Union army. Shiloh was also called Pittsburg Landing, after a spot on the western bank of the Tennessee River where steamboats could deliver men and supplies. Major General Ulysses S. Grant, fresh off a victory at Fort Donelson near Tennessee's northern border, intended to drive even deeper into Southern territory. The plan was for Grant's 50,000 men and Major General Don Carlos Buell's 35,000 to join forces at Pittsburg Landing, and then to march on Corinth, Mississippi, twenty-two miles farther south.

Corinth was a target because it was a railroad junction, the crossing point of a north-south rail line and a vital east-west one. The Memphis & Charleston Railroad, which meandered more or less horizontally across much of the Confederacy, was known as "the vertebrae of the South." The Union aimed to smash it at Corinth, as if it were killing a snake with a shovel.

Grant had been assembling troops at Pittsburg Landing since mid-March while waiting impatiently for Buell, on his way from Nashville. Anticipating Grant's move, the Confederacy had massed 44,000 soldiers in and around Corinth. Their plan was not to sit and wait to be attacked, but to march on Pittsburg Landing and attack Grant before Buell arrived to reinforce him. Grant, supremely confident, brushed off any talk of a surprise attack.

He had established temporary headquarters in a borrowed town house at Savannah, Tennessee, nine miles downstream from his troops at Pittsburg Landing. (This stretch of the Tennessee River flows from south to north. Grant's headquarters were north of Shiloh and on the opposite side of the river.) Union forces had occupied the small town. One newcomer to town was Emma Powell, who had married Powell four months before, and had taken up temporary residence in Savannah in order to be near him. Early in the war especially, it was not uncommon for wives to accompany their husbands in this way. Grant and Sherman, among others, sometimes went so far as to bring their *children* to war with them.

The Southern plan was to march all day on April 3 and to attack Pittsburg Landing on April 4. Even in good weather, novice soldiers had little chance of covering the twenty miles from Corinth in a day. But rain poured down, and horses and cannons sank in the mud, and entire divisions wandered off, lost. It was late afternoon on April 5, too late to attack, before the Confederates were in position.

By then, it was unclear whether a Confederate attack would be a surprise or an enormous blunder. The Southern commanders, and in particular General Pierre G. T. Beauregard, "the Napoleon of the South," made an urgent argument for canceling the attack. The element of surprise was gone, Beauregard insisted. Confederate soldiers, worried that the rain had wet their powder and rendered their weapons useless, had fired off test shots that might have given the game away. Beyond that, the delay had provided Buell two extra days to join Grant. In addition, Grant's men were surely dug in behind earthworks by this time.

Albert Sidney Johnston, the supreme Confederate commander in the West, insisted on going ahead with the attack regardless. "I would fight them if they were a million," he proclaimed. His troops seemed just as eager. Perhaps they had been inspired by Johnston's rousing call to "offer battle to the invaders of your country" and "march to a decisive victory over the agrarian mercenaries sent to subjugate you and to despoil you of your liberties, your property, and your honor. Remember the precious stake involved; remember the dependence of your mothers, your wives, your sisters, and your children, on the result; remember the fair, broad, abounding land, and the happy homes that would be desolated by your defeat."

Grant had no idea that the Confederate army had spent the night only a mile from the nearest Union troops. It was not true, as the newspapers later reported, that Northern soldiers were bayoneted in their tents as they slept, but they had been badly surprised. At dawn on April 6, 1862, a lazy Sunday morning, many Union soldiers were just rousing themselves. Bored with waiting, or sick— dysentery was so common that the men had nicknamed it "the Tennessee two-step"—they were scarcely poised for battle. Camp was filled with the usual morning sounds of bacon sizzling in skillets and rifle butts grinding coffee beans into powder (each man ground his own beans because no supplier could be trusted to supply unadulterated coffee). Buell's troops had, in fact, not yet been deployed.

Suddenly a high-pitched wail pierced the morning quiet as thousands of Confederate soldiers in full rebel yell rushed from the woods. "My God," cried General William T. Sherman, "we are attacked." Grant, whose breakfast in Savannah was interrupted by the roar of cannons, hurried to the sounds of fighting.

Powell, in the meantime, was seething with frustration. He was captain of an artillery battery, in charge of six cannons and about 110 men, along with roughly 100 horses that were needed to pull the guns. (The horses were considered more valuable than the men, because without them the guns were stuck in place and vulnerable to capture. It was standard practice for each side to train its weapons on the enemy's horses.) Powell was "unassigned." He had, in other words, arrived at Pittsburg Landing so recently that he had not yet been assigned to a particular division. Early on the morning of April 6, 1862, Powell recalled later, he was "awakened by the rattle of musketry and the roar of artillery," and hurried to order the men of Battery F, Second Illinois Light Artillery to prepare for action. He paced about, waiting for orders. None came.

Powell's men, standing by their guns, gulped down a few bites of breakfast. Still no orders came. The sounds of battle grew louder. Wounded men, dragging themselves along or carried by their companions, moved slowly past Powell and away from the fighting. Terrified soldiers fleeing the front lines ran by as fast as they could move. Powell continued to fret. Finally, fearful that in the confusion of battle he might *never* receive orders, Powell moved forward on his own initiative. It was, he wrote later, about eight in the morning. With characteristic boldness and impetuosity—and utterly contrary to regulations—Powell set out to find a place to pitch in, as if the biggest battle America had ever seen were a snowball fight between rival schoolboys.★

Shiloh was slaughter on an industrial scale. What the factory was to the lone craftsman in his cottage, the Civil War battlefield was to the village green at Lexington. The cannonball, that familiar battlefield icon, was one of the *least* horrifying projectiles flying around a Civil War battlefield. Cannons also fired canister, in effect a large tin can filled with some two dozen golfball-sized lumps of iron. The can itself would disintegrate in midair, sending a deadly metal hail in all directions as if from a colossal sawed-off shotgun. A single cannon firing a single canister could destroy everything within a sixteen-by-sixteen-foot space at a range of fifty yards. Trees and bushes, not to mention humans, were uprooted and ripped apart as though a tornado had touched down. "There is some places," a dazed soldier wrote at Shiloh, "that it Looks as if a mouse could not get threw alive."

In the sense that a guillotine is a humane killing machine, a modern steel-jacketed bullet is humane. It travels through the air so quickly that it may be hot enough to cauterize the wound it makes. It may pass through the body

---

★Official records listed Powell as absent without leave at Shiloh. It took him considerable trouble to correct the record, although his amputated arm served as hard-to-refute evidence that he had indeed been present on the battlefield.

entirely and is more likely to clip a bone neatly than to mangle it. It wounds cleanly and kills efficiently. Death on a Civil War battlefield was an uglier affair.

By far the greatest number of wounds—94 percent—were inflicted by the innocuous-looking minié ball. Cone-shaped (though confusingly called a "ball"), a minié ball was a rifle bullet made of soft lead and weighing about an ounce. It was, says Shiloh's Stacy Allen, "a beautiful, inhuman piece of killing machinery. The minute that piece of soft lead hits anything harder than it is— and your bones are harder—it's going to mushroom to the size of your fist as it plows along. When it hits flesh and blood, it does amazing things. It rips muscles to shreds, it destroys arteries and veins, and if it hits bone, it shatters and pulverizes it. You might be hit by shattered bits of bone or teeth from the man standing next to you. And anything that soft lead engages as it enters a human body—dirt, filthy bits of clothing, wood, metal—it's going to carry with it right into the wound, so we're talking instantaneous infection."

The minié ball was shot from a rifle. Before the Civil War, Americans had fought with smooth-bore muskets, which were essentially powerful peashooters that sent a round ball rattling and ricocheting its way down the barrel and then wobbling through the air like a knuckleball. Muskets were notoriously inaccurate, with a deadly range of no more than eighty yards. A man with a musket standing a few hundred yards away, General Grant once observed, "might fire at you all day without you finding it out."

Rifles were not new—Daniel Boone and other sharpshooters had used them—but they had been uncommon until the Civil War, when "rifled" barrels could finally be mass produced. The word refers to a long, shallow groove that spirals along the barrel's inside surface like a stripe on a barber pole. The groove imparted spin to the minié ball as it shot through the barrel, so that it cut through the air and held its line: not a knuckleball but a fastball.

The result was a *threefold* increase in the distance from which a soldier could kill an enemy. A marksman with a rifle could reliably hit a ten-inch circle at two hundred fifty yards.* At five hundred yards, he could hit a square six feet on a side. A line of men advancing in formation toward enemy lines across an open plain might as well have been heading off a cliff.

---

*At Spotsylvania, in 1864, Major General John Sedgwick of the Union Army's Sixth Corps rebuked a soldier for cowering under fire. Sedgwick's last words were, "Why, what are you dodging for? They could not hit an elephant at this distance."

•   •   •

But weapons changed more quickly than tactics. Because musket fire was so inaccurate, soldiers in previous wars had marched against the enemy in close formation, elbow to elbow, and fired in volleys. Offense was a match for defense, for an attacker was relatively safe until he approached within a couple of hundred yards of the enemy lines (thus the famous Bunker Hill command, "Don't fire until you see the whites of their eyes!"). Then it was a desperate race—the attackers, firing in unison and then dashing across the last bit of ground separating them from the enemy, the defenders tearing open paper packets of gunpowder with their teeth, pouring the powder into the shotgun muzzle, ramming the ball in place, frantic to complete the finicky nine-step loading process and fire on the onrushing enemy before he could drive his bayonet home.

With rifles deadly at several hundred yards, such massed assaults were suicidal. In the Civil War, a defense lying in wait in a trench or behind a wall had an overwhelming advantage over any attackers. In July 1862, at Malvern Hill, Confederate officers ordered *fourteen* assaults on an entrenched Northern position. The Southerners fell in waves, like scythed wheat. "It was not war," one stunned Northern officer declared. "It was murder."

The great, romantic emblem of warfare, the cavalry charge, was suddenly obsolete, and for the same reason. The onrushing, saber-brandishing cavalry was exposed to withering fire hundreds of yards before it could threaten the enemy. At Chancellorsville, a force of Northern cavalry launched a gallant but doomed charge against Southern troops. "We struck the Confederate infantry as a wave strikes a stately ship," one survivor recalled. "The ship is staggered, maybe thrown on her beam ends, but the wave is dashed into spray, and the ship sails on as before." Union surgeons later found thirteen bullets in the body of the major who had led the Northern charge.

Even with rifles, infantry had to be massed, because the new weapons took as long to load as the old ones. "It doesn't take any great reasoning to figure out why these guys were still fighting shoulder to shoulder," Stacy Allen observes. "The individual soldier had great accuracy, but he had limited firepower. At best he's getting off one to two rounds a minute. In time, repeating weapons would change all that, but they didn't play much role in the Civil War."

So huge numbers of men fought side by side. And if they were to act in unison—bearing in mind that, in the absence of radios and telephones, officers had to rely on orders carried by messenger or signals conveyed by drums and trumpets and flags—there was no alternative to having soldiers grouped in great, vulnerable mass formations.

At Shiloh, inevitably, the result was chaos, a toxic mix of panic and inexperience. The air was thick with the roar of cannons and the moans of wounded men and the shrieks of mangled horses. The uneven terrain disoriented the men, and the woods and bushes and low-hanging clouds of gun smoke obscured their view. Unsure whether nearby fire came from friend or foe, both sides fought in a frenzy of fear and confusion. "This was the first great modern battle," Shelby Foote wrote. "It was Wilson's Creek and Manassas rolled together, quadrupled, and compressed into an area smaller than either . . . a cauldron of pure hell."

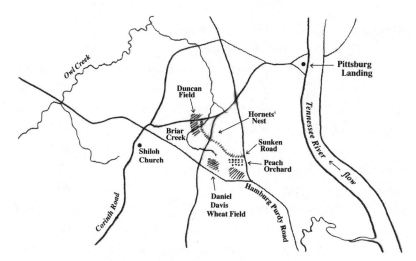

CHAPTER EIGHT
THE HORNETS' NEST[9]

Desperate for a place to join the battle, Powell found the Hornets' Nest. (The name, coined by the Confederates, referred to the fierceness of the gunfire.) Today one of the Civil War's most resonant place names, the Hornets' Nest was a nondescript patch of woodland, a tangle of trees and underbrush that ran half a mile along a rutted dirt road. The road separated the Hornets' Nest on one side from a large pasture called Duncan Field on the other. At the Hornets' Nest, 6,200 Northern troops tried to stem the Confederate attack. General William Wallace commanded most of the Union forces fronting Duncan Field, and Powell and his battery joined the fighting.

Southern troops made at least seven assaults on the Hornets' Nest during the course of the day's fighting. A total of some ten thousand Confederates rushed the Union lines, though never all at the same time. On both sides, planning quickly gave way to mayhem. Blinded by smoke and struggling through woods and thick brush, soldiers scarcely knew what was happening around them. "For God's sake cease firing," Colonel James Fagan of the First Arkansas sent word to a fellow Confederate officer when he finally realized that "we were killing his men and he was killing ours."

The horrors of the Hornets' Nest lay almost beyond description. After two Confederate attacks had been beaten back, Colonel Henry Allen of the Fourth Louisiana exhorted his men to uphold the reputation of their state and attack again. The Confederate troops charged into another hail of fire. Allen took a bullet in the face, but lived. The Southern troops began to fall back, and Confederate Major General Braxton Bragg sent a man to take the Fourth Louisiana's regimental flag and carry it forward. To carry the colors was a tremendous honor and a terrible risk, for the flag drew heavy fire. Allen saw Bragg's man and rode up to him, a bullet hole in each cheek and blood gushing out of his mouth. "If any man but my color bearer carries these colors," Allen declared, "I am the man." Then he turned to his troops. "Here boys, is as good a place as any on this battlefield to meet death!"

The Confederates charged again and were repelled again. Allen told Bragg that the Union position was too strong to carry from the front; they would have to make a flank attack. Bragg, bad-tempered and autocratic, curtly rejected the advice. The dismissal was characteristic. "When he has formed his own opinion of what he proposed to do," observed Bragg's colleague General Simon Bolivar Buckner, "no advice of all his officers put together can shake him; but when he meets the unexpected, it overwhelms him because he has not been able to foresee, and then he will lean upon the advice of a drummer boy."

"Colonel Allen," Bragg snapped, "I want no faltering now." The Confederates charged yet again. Fifty feet from the Union lines, one of the men of the Fourth Louisiana had half his head shot off by a cannonball. While his stunned companions looked on, he staggered on for a step or two. As the barrage continued, the Southern troops fell back again.

Survivors from both North and South would remember scenes from the Hornets' Nest as long as they lived, with a kind of flashbulb vividness. "I am lying so close to Capt. Bob Littler that I could touch him by putting out my hand," a soldier from Iowa recalled, "when a shell bursts directly in our front and a jagged piece of iron tears his arm so nearly off that it hangs by a slender bit of flesh and muscle as he jumps to his feet, and crazy with the shock and pain, shouts, 'Here, boys! Here!' and drops to the ground insensible." An awestruck Confederate soldier searched for an analogy to convey the scale of the bloodshed. "Men fell around us," he wrote, "as leaves from the trees." The roar of the guns grew so loud, one Union soldier noted, that a rabbit, trembling with fear, rushed out of the brush and snuggled up to him.

Powell fought in the Hornets' Nest through the morning and into the late afternoon. Hour after hour, against assault after assault, Battery F, under Powell's command, carried out its complex choreography. With five cannons to tend (a sixth had been lost earlier in the day), each man had a precise role—delivering the round, inserting it, ramming it home, aiming, firing, swabbing out the cannon barrel with a sponge mounted on a pole—while the enemy's shot and shell poured down. At about four o'clock, Powell raised his right arm yet again to give the signal to fire. A minié ball plowed into his wrist and careened toward his elbow. Powell's brother Walter, serving with him, ran over to examine the wound. (It was Walter who would eventually emerge from the war half crazed and join Powell on the Colorado.) Powell slumped against a tree, losing blood but not yet in tremendous pain. General Wallace appeared and took note of Powell's injury. Wallace, a big man, lifted Powell onto a horse and ordered him and a sergeant to get to a doctor.*

Riding with a mangled arm through the battle zone, Powell managed to make it to the boat landing on the Tennessee River. ("Certainly hundreds, perhaps thousands, of shots were fired at us," he wrote later.) Someone helped him onto a steamboat for the short trip to Savannah, Tennessee, where the town hall had been converted into a hospital. A wounded soldier on the steamer *Continental* described one such trip. "The scene upon the boat was heart-rending—men wounded and mangled in every conceivable way. The dead and dying lying in masses, some with arms, legs, and even their jaws shot off, bleeding to death, and no one to wait upon them or dress their wounds—no surgeon to attend us."

As Emma Powell would later tell it, she happened to be on hand at the boat landing in Savannah when two soldiers came walking down the gangplank carrying her husband on a stretcher. "Now, now," Powell whispered to her, "everything is going to be all right."

---

*Powell liked to tell a slightly more flamboyant variant of the tale. "We're going to be captured in a few minutes," Wallace told the wounded Powell, in this version. "Get onto my horse and go back to the landing at once." Wallace himself was mortally wounded later the same afternoon.

•   •   •

Some men were overwhelmed by what they saw at Shiloh. Five thousand soldiers—fifteen thousand by some accounts—fled the battlefield in terror. Union cavalrymen rode among the Union deserters, sabers drawn, threatening to cut the heads off anyone who refused to fight. Even so, deserters fled to the banks of the Tennessee, as far as they could get from the shooting, and hid beneath the bluffs. "Most of them would have been shot where they lay, without resistance," General Grant observed with dismay, "before they would have taken muskets and marched to the front."

Punctuating the scenes of heroism and cowardice were surreal vignettes that in other settings would have been almost comic. At one point, thirty or forty Union soldiers cowered behind a single thick tree, each holding the belt of the man in front while the officer supposedly in charge ran back and forth in panic. One Confederate soldier on a captured mule valiantly charged the enemy lines while his comrades cheered his bravery. Finally he wheeled around, yelling, "It aren't me, boys. It's this blarsted mule. Whoa! Whoa!"

Most scenes were unadulterated horror. One Northern soldier saw a Southerner fleeing on all fours when "a cannonball struck him, tearing him in pieces, and scattering his limbs in different directions." Another found himself haunted by the sight of a dead man "leaning back against a tree as if asleep, but his intestines were all over his legs." In many places, the underbrush had caught fire, and wounded men, unable to flee, cried in anguish from within the flames.

Powell had been wounded near the end of the first day's fighting. By day's end, the Union forces had fallen back two miles. The South had won "a complete victory," the Confederate commander telegraphed Jefferson Davis, and all that remained for the next day was a bit of mopping-up. Darkness put a halt to the fighting, but the night of April 6 was in its way as grotesque as the day had been. At this early point in the war, there was no real system for carrying the wounded from the field, and men too badly hurt to move lay where they had fallen. Tens of thousands of weary soldiers tried to ignore the moans of dying men and the thunder of shells from the Union gunboats and get some sleep. (The shelling, which continued every fifteen minutes through the night, was intended to harass the Confederates, but it tormented both sides.)

At ten o'clock, it began to rain; by midnight, the drizzle had become a downpour, with cold, swirling winds and frequent thunderclaps. Corpses littered the battlefield, and lightning flashes revealed hogs feeding on the bodies.

General Grant spent the night in the field with his troops rather than in a steamship bedroom. He tried to sleep under a large oak tree but was driven out by the rain. A log cabin nearby seemed a likely refuge, but surgeons had found it first and a stack of amputated arms and legs was already rising. "The sight was more unendurable than encountering the enemy's fire," Grant wrote later, "and I returned to my tree in the rain."

By this time, Buell's troops had joined Grant's. The second day's fighting reversed the first day's result. By midafternoon, the South had lost all the ground it had gained the day before. General Beauregard's chief of staff noted with dismay that the Confederate troops were akin to "a lump of sugar thoroughly soaked with water, but yet preserving its original shape, though ready to dissolve." Beauregard conceded the point, and the Southern forces turned back toward Corinth. The exhausted Union soldiers sank down in their own recaptured camp.

The aftermath of battle, when the men could no longer rely on adrenaline to distract them, was perhaps worse than the battle itself had been. In one account after another, as if in a communal nightmare, the same phrase echoes: "I could have crossed the battlefield stepping only on dead bodies and never touched the ground." Even the hardest men found themselves appalled. Sherman described "piles of dead soldiers' mangled bodies . . . without heads and legs" and noted that "the scenes on this field would have cured anybody of war." The landscape itself seemed wounded. "Scarcely a tree or bush had escaped the musket balls," one soldier observed. ". . . Trees had been shivered into splinters, while the ground was covered with brush and downed timber. In many places could be seen where the huge shells from the gunboats had ploughed great pits in the ground."

It fell to the victorious Union side to bury the dead. Over a thousand men were assigned to the ghastly work, slinging dead friends on wagons or dragging them with ropes and then flinging them into long, open pits. (Soldiers did not yet wear dogtags, and many of the bodies were unidentified.) In one trench, the dead lay seven men deep. The work was rushed, to get it over with as quickly as possible and to avoid an epidemic, but even so it took nearly a week. Rain poured down and washed the dirt away, exposing graves that had not been dug deep enough. "*Skulls* and *toes* are sticking from beneath the clay all around," one horrified soldier wrote, "and the heavy wagons crush the bodies turning up the bodies of the buried, making this one vast Golgotha."

Even days afterward, there was no escape. One soldier fled into the woods in search of refuge. "A tramp of four or five miles would not take one out of the silent and polluted woods, the rotting debris of an army, or the all-pervading stench of decay," he wrote. Instead, he found a board nailed to a tree and inscribed "137 DeD rEBeLs BuriED heRE." Looking down, he saw a piece of skull covered with hair and, not far away, a skeleton hand protruding from the ground. A bit of gray uniform showed at the wrist.

The battle, though apparently a stalemate, was in fact a crucial lost opportunity for the South and a strategic victory for the North. Having come close to routing the Northern troops, the Southerners found themselves forced to retreat. But Shiloh had a broader significance. For the nation as a whole, and for tens of thousands of farm boys and factory workers and office clerks, the fighting at Shiloh marked the end of innocence. Before that battle, in the spring of 1862, novice soldiers had fretted that the war might end before they had had a chance to join in. Two days of slaughter on the banks of the Tennessee River put an end to all such talk.

Some soldiers chose not to speak of the battle at all, in the belief that only an eyewitness could grasp what had gone on, while others struggled to find words for events that were beyond description. "There's been a lot written about the gallant charges and the flags fluttering and all that kind of stuff," says Stacy Allen, the Shiloh historian, "but there is an unwritten war, and the unwritten war is the most terrible thing imaginable, and the most severe thing imaginable, and the most tragic thing imaginable."

Shiloh opened the nation's eyes and forced it to confront sights it had never imagined. At the time the largest battle ever fought on American soil, Shiloh was not a duel writ large but a collision of two immense, relentless, man-devouring machines. In two spring days, some twenty thousand men were killed, wounded, or reported missing. Nothing on that scale had ever been seen on the American continent.

There was worse to come. The Civil War was unfathomably bloody. With a death toll of well over half a million, it remains America's deadliest war by far, claiming more American lives than World War I, World War II, and Korea combined. The figures are startling enough when considered as raw numbers; they loom even larger when we remember how small the population of the United States was in the 1860s. If a war today claimed 2 percent of the population, as did the Civil War, well over five million Americans would lie dead.

Powell's wound did not kill him, though it came close. But his ordeal had only begun. Soldiers who had nerved themselves to stand their ground while men fell all around them quailed when they ventured too near a hospital tent. A surgeon from one's own side was as feared as an enemy's cannon. "The sorriest sights . . . are in those dreadful field hospitals," one Southern officer wrote, for the ear was assailed with "the screams and groans of the poor fellows undergoing amputation" and the eye was confronted with "the sight of arms and legs surrounding these places [and] . . . thrown into great piles."

Surgeons were quick to amputate; three out of four operations in the Civil War were amputations. It is not by chance that nearly every Civil War memoir contains a reference to the grim stacks of arms and legs. With the number of wounded men at a battle like Shiloh surpassing fifteen thousand, and with only a comparative handful of surgeons to tend to them all, assembly-line treatment was inevitable. A battlefield surgeon stayed on his feet as long as he could keep his eyes open, perhaps twenty-four hours at a stretch, operating on one maimed man after another. A surgeon could remove a limb in six minutes.

Even so, there were not enough surgeons. The man who amputated Powell's arm, William H. Medcalfe, had been a druggist before the war, and such cases were common. Medcalfe was in over his head, but almost no one was prepared for the demands of wartime surgery. Few civilian surgeons had much experience with gunshot wounds, and even those few were daunted by the horrendous injuries inflicted by the newfangled minié balls.

But a lack of surgical know-how was not the heart of the problem. Many surgeons *did* wield saw and scalpel with dexterity. The Civil War was a medical catastrophe not because of surgical misadventures but because of rampant and uncontrollable infections. A nearly certain follow-up to surgery, infection was the central fact, and the unsolved mystery, of Civil War medicine. Infection rates were so high that Confederate surgeons suspected that Northern soldiers had somehow poisoned their bullets.

Amputation was a desperate attempt to solve the problem by lopping off infected tissue before it could spread its poisons. "Hospital gangrene," for example, began as a dime-shaped black spot. In days, it could transform a healthy arm or leg into a seeping, putrid hunk of dying flesh. No one knew where it came from or how it could be stopped.

Poignantly, the great advance that could have revolutionized the care of the hundreds of thousands of soldiers wounded in the Civil War—the insight that germs cause disease—came a few years after the war's end. That revolution in medicine was the work of Lister in England, Pasteur in France, and Koch in Germany, among others. It is worth noting that there are no American names on that list. The mid-nineteenth century was a low point in the history of American medicine. In medicine and science, Europe led the way and the United States trailed far behind. It was 1868 before Harvard Medical School acquired its first stethoscope, for example, 1869 before it obtained its first microscope. Even if there had been no war, the 1860s were not a good time to look for medical care in America. But war did come, at the worst time, and with it came hundreds of thousands of sick and wounded men clamoring for help.

The Civil War still holds us in its grip, in part because the soldiers who fought in it seem so close to us. From the newly somber young man who wrote home, after Shiloh, that "we were playing Soldiers but now we know what it is to Soldier" to plainspoken, rumpled, cigar-smoking Ulysses S. Grant, who became a general only after failing as a farmer, failing as a debt collector, failing as a clerk in his father's business, these are people we can imagine knowing. But, in crucial ways, they lived in a world that is utterly foreign to us. In its medical aspect, above all, a battlefield in the 1860s presented scenes of almost medieval grimness.

From the instant a man entered the army, he was in trouble. In North and South alike, a soldier was twice as likely to die from disease as from a battlefield injury. Many men had been sick even before they arrived—physical examinations of new recruits were so cursory that *four hundred* women passed as men and went on to fight in the war—and sick soldiers infected others. Crowded into huge and filthy camps where a single latrine might serve for hundreds and garbage lay strewn everywhere and the water supply was almost certainly contaminated, men fell ill in droves. Poor nutrition made matters worse. A soldier's diet consisted largely of beans, bacon, and the rock-hard crackers called hardtack. Meal after meal, complained one Union soldier at Shiloh, was "mouldy crackers and sowbelly with hair on it."

Measles, mumps, and smallpox struck the camps first. Country boys fell harder than city dwellers, who were more likely to have been exposed to such childhood illnesses before. Then came a second wave of illness, headed by typhoid, malaria, and dysentery. It was common for a regiment to lose half its men to death and disease before it ever saw battle. "A man risked his life simply by being in the army," Bruce Catton observed, "even if he never got near a battlefield."

For those who did see combat, the casualty rate at Shiloh and other major battles was about one in five. Worse still, as we have seen, the wounded were largely left to their own devices, especially at early battles like Shiloh. (In theory, a regiment's musicians were responsible for carrying its wounded men safely away from the line of fire, but they performed this nonmusical mission with about the degree of enthusiasm and skill one might expect.) Injured men found someone to drag them off the field, or dragged themselves, or lay where they had fallen. At Fort Donelson, in February 1862, two months before Shiloh, one wounded man lay on the ground for two days before help arrived. He had to be chopped free from the frozen ground before he could be put on a stretcher.

Nurses and orderlies were in short supply. At Shiloh, a few volunteer nurses roamed the edges of the battlefield, doing their best to help the wounded men by tearing strips from their dresses to make bandages. The Union army had established some primitive field hospitals, and surgeons treated the living while surrounded by dead men no one had yet had time to bury. "You may imagine the scene," one exhausted and overworked surgeon wrote, "of from two to three thousand wounded men at one point calling to have their wounds dressed."

Once the wounded arrived in a hospital, further dangers lurked. "There stood the surgeons," wrote a Union officer at Gettysburg, "their sleeves rolled up to the elbows, their bare arms as well as their linen aprons smeared with blood, their knives not seldom held between their teeth, while they were help-ing a patient on or off the table or had their hands otherwise occupied."

Knowing nothing of bacteria, surgeons of the day believed that a finger was the ideal tool for probing wounds. They poked inside the torn flesh of arms and legs with unwashed, ungloved hands. (A man who had been shot in the gut, in contrast, was virtually beyond help. Photographs of Civil War battlefields often show dead men with their clothing ripped apart, as if some thief in a great hurry had ransacked their bodies. In fact, the men had clawed their clothes open themselves, frantic to see whether or not they had taken a bullet in the belly.)

To the harm done by the bullet itself was added the risk of bone infection and blood poisoning caused by the surgeon's attempts at treatment. Time after time, surgeons noted that an operation had gone well and the patient seemed to be recovering, only to find a few days later that he had sickened and died. "However bad the wound may be," cynics observed, "art can make it worse."

For the operation itself, the patient was given chloroform or ether. Despite Hollywood images to the contrary, the era of "biting the bullet" predated the Civil War. "The surgeon snatched his knife from between his teeth . . . ," the officer at Gettysburg continued, "wiped it rapidly once or twice across his bloodstained apron, and the cutting began. The operation accomplished, the surgeon would look around with a deep sigh, and then—'Next!'"

A single sponge and a single unspeakable basin of water might be used to wash every wound in a ward. Looking back on his Civil War experience in his old age, the eminent surgeon W. W. Keen could hardly believe what he had participated in. "We operated in old blood-stained and often pus-stained coats, the veterans of a hundred fights," he recalled. "We operated with clean hands in the social sense, but they were undisinfected hands. . . . We used undisinfected instruments from undisinfected plush-lined cases, and still worse used marine sponges which had been used in prior pus cases and had been only washed in tap water. If a sponge or an instrument fell on the floor it was washed and squeezed in a basin of tap water and used as if it were clean."

Powell had been shot late in the afternoon of April 6 and brought to a hospital that evening. Emma nursed him through the next day while Medcalfe, the druggist turned surgeon, occasionally checked the wound. On April 8, Medcalfe decided to operate. The surgery was quick but not especially skillful—Powell's stump of an arm would torment him throughout his life, and he would eventually undergo two more operations in the hope of easing the pain—but the true danger still lay ahead.

Powell came to after the operation long enough to acknowledge Emma but then fell into a feverish, drugged sleep. At this point a surgeon had no cards left to play. The patient, newly transported from the battlefield, himself became a battlefield in a vast but silent war between disease-producing microorganisms and the body's natural defenses. Powell slept a great deal in the next several days, and recovered enough to ask for news of the battle and of his men, and fell back into a feverish sleep, and complained of mysterious, phantom pains in the forearm he no longer had. Lean and hard at the best of times, Powell grew positively scrawny as the fever hammered at him. But he lived.

On June 30, he rejoined the men of Battery F, by now in occupied Corinth. They gave him three rousing cheers—Powell had to wipe the tears away—and then he shook hands, left-handed, with each man in turn. The reunion was brief. Powell was far from fit, and he was assigned to a desk job for six months.

He had decided, though, that even with only one arm he would return to active duty. His sole condition, quickly granted, was that Emma receive a "perpetual pass" entitling her to visit him wherever the war might bring him. Over the next three years, notably at Vicksburg and at Nashville, Powell returned to his place with the men of Battery F.

In the summer of 1864, he endured a second operation intended to ease the pain in his arm. It failed. Finally, in 1865, with peace imminent, Powell resigned from the army. It was time to decide what to do next. Powell was thirty years old, with a wife and responsibilities and a homemade education and no particular prospects. "Wes, you are a maimed man," his father advised him. "Settle down at teaching. It is a noble profession. Get this nonsense of science and adventure out of your mind."

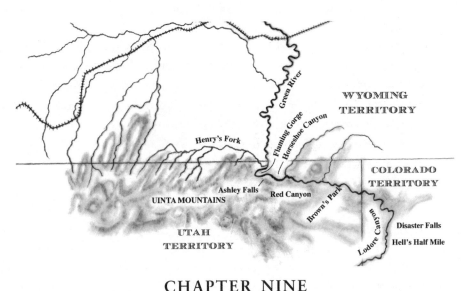

CHAPTER NINE
HELL'S HALF MILE

We have seen already how well Powell heeded his father's advice. Powell was not one for second thoughts—indeed, he had such faith in his spur-of-the-moment decisions that sometimes he neglected first thoughts as well—but at Disaster Falls, on the evening of June 9, 1869, we can imagine that he found himself mulling over his father's words. He had bet his life on "science and adventure," and at Disaster Falls it had begun to look as if the bet had gone bad.

Disaster Falls had been a near miss. Everyone had made it through alive, but the expedition had lost a boat and more than a ton of supplies. Now the exhausted and jumpy men needed to regroup. Bradley managed to spare a thought for the scenery, which was "sublime" but hardly soothing. "The red sand-stone rises on either side more than 2000 ft.," Bradley wrote, "shutting out the sun for much of the day while at our feet the river, lashed to foam, rushes on with indescribable fury. O how great is He who holds it in the hollow of His hand, and what pygmies we who strive against it."

Religious sentiments came easily to Bradley, who was the most devout of the party, but this was more than boilerplate. The scale of the canyons seemed designed to make *any* human being feel puny and insignificant, regardless of his creed. Worn-out and disheartened after their first truly life-threatening adventure, the men were perhaps more vulnerable than usual to such discouraging thoughts. Where Bradley turned to thoughts of divine providence, Sumner sought more practical consolation. "After taking a good drink of whisky all around," he wrote, "we concluded to spend the rest of the day as best suited."

The men loafed or puttered around camp according to their taste. "Some packed freight for future use," Sumner wrote, "[and] the rest slept under the shade of the scrubby cedars." Powell was neither lugging freight nor napping. "Major and brother," wrote Bradley, "are gone ahead *to see what comes next.*"

Much as he hoped for good news, Bradley knew perfectly well that any reassurance would be fleeting at best. Scouting the river more than a short distance ahead was impossible because the terrain was so difficult. "Can't rely on anything but actual tryal with the boats," Bradley noted, "for a man can't travel so far in a whole day in these cañons as we go in a single hour." And so the men were constantly in the position of motorists speeding down an unknown road—and possibly headed off a cliff—while never able to see more than a car length ahead.

Never knowing what surprise the river would spring next was the great hardship Powell and his men were forced to endure, far outweighing the dunkings and near-drownings and injuries and exhaustion and inadequate food. They could all have braced themselves to cope with nearly any privation—Powell wrote a three-hundred-page, first-person account of his expedition and never found reason to mention that he was missing an arm—but this was worse. Perpetual uncertainty was a greater burden than any physical ordeal.

"To travel hopefully is a better thing than to arrive," Robert Louis Stevenson remarked, but the observation is also true if it is turned on its head. To travel in dread is worse than to face even the grimmest reality. Brave as they were, Powell and his crew were forever in fear that they would round the next bend and find themselves flung against knife-edged boulders or tumbling over a waterfall. At the outset, not knowing what lay ahead had been a comfort. Had they known, Powell and the men might never have ventured downstream. Now, having committed themselves to the expedition, they found themselves nearly crushed by the weight of living with endlessly dashed hopes and endlessly renewed fears that the next ordeal would be the worst one yet. They lived each day, Powell wrote, with "the shadow of a pang of dread ever present to the mind."

All explorers confront that burden, which is the cost of being first into the unknown. It can also be a feature of ordinary life. Unknown terrors often loom larger than specific, named ones because the mind can project whatever torments it fancies on a blank screen. (Perhaps this is part of the reason for our fear of the dark.) "There's a spot on the X ray I want to look at," the doctor says, and the weeks of waiting, which one might think would be mollified by hope, are in fact exquisite torture. "I'm afraid it's cancer," she finally announces, and, faced with an ultimatum that cannot be appealed, people somehow manage to endure.

The same is true of children who live with abusive parents. Beatings are bad in themselves, but the capriciousness of life with a parent who might deal out blows one day and hugs another is far worse. This was, in a sense, the load that weighed so heavily on Powell and his men. Their challenge was not to endure an almost insupportable burden but to live perpetually in suspense, like a servant subject to a tyrant's whims.

"The sensation on the first expedition," wrote one of the first explorers to try to duplicate Powell's 1869 journey, "when each dark new bend was a dark new mystery, must have been something to quite overpower the imagination, for then it was not known that, by good management, a boat could pass through this Valley of the Shadow of Death, and survive. Down, and down, and ever down, roaring and leaping and throwing its spiteful spray against the hampering rocks the terrible river ran, carrying our boats along with it like little wisps of straw in the midst of a Niagara."

On several occasions, moviemakers have gone to considerable trouble to re-create Powell's expedition. They have commissioned painstakingly crafted replicas of Powell's boats—built using nineteenth-century techniques—and hired professional river runners to play Sumner and Bradley and the others. But authentic as they may appear in their period costumes of grimy shirts and suspenders and torn pants, these expert boatmen can never truly put themselves in the shoes of Powell's pioneering crew. They know too much. They have put in years on the water, first of all, and they are the indirect beneficiaries of the accumulated wisdom of all their river-running predecessors besides.

A man who has grown up rich can give away his fortune, but he can never experience the world as it appears to someone who grew up poor. Nor can a veteran boatman today forget all he has learned and imagine what it was like to run a wild river in the days when almost no one had done such a thing. Like mountaineers who compete to achieve first ascents of unconquered peaks, river runners today do still roam the world in search of first descents, on rivers that plummet far more steeply and pose far greater dangers than the Green or the Colorado. But a modern-day first descent, because it is a venture undertaken only by experts who have logged thousands of demanding river miles, is *less* of a venture into the unknown than was Powell's. "Basically, Powell and his men were on their first river trip," says Brad Dimock, a highly regarded boatman who rowed a "Powell boat" for a *National Geographic* film. "We ran all the rapids—Powell portaged as many as he could—but we had 120 or 130 years of white-water skills on our side, plus we knew within an inch where every rock in the river was."

The value of knowing the river so well can hardly be exaggerated. It made for confidence and security, first of all, but it had more tangible importance as well. Even in the hands of modern-day experts, Powell's boats remained hard to maneuver. The key to making it successfully through a rapid was to cut the maneuvering to the minimum possible, which meant understanding where to enter the rapid in the first place and at what angle. A good start made a good run possible; a bad start made disaster almost certain.

An experienced boatman scanning a rapid quickly identifies what are, in effect, entrance ramps. But that recognition only comes with practice and experience. A novice might find the ramp but take it at the wrong angle and careen out of control, or never identify the proper entrance in the first place, barrel through a "Wrong Way" sign, and spin into a guardrail. Veteran boatmen almost never make such blunders. They are experts at reading rivers in general, and, in the case of the boatmen replicating Powell's expedition, were rereading a text they had long since committed to memory.

"We knew intimately what was around every bend," says Bruce Simballa, a retired boatman who participated in an IMAX movie about Powell's trip. By the time he rowed for the cameras, Simballa was a veteran of about twenty-five Grand Canyon trips. Kenton Grua, who rowed in the same boat, had made some seventy Grand Canyon trips.★ "When you do the river all the time, you have little markers," Simballa goes on. "You can be a mile above a rapid and look over at a rock along the shore, and by seeing the river's level on that rock you'll know *exactly* what you'll find around the next bend."

On June 10, 1869, Powell and his men had no time to think ahead to any challenges beyond the immediate ones. "The river in this cañon is not a succession of rappids as we have found before," Bradley lamented, "but a *continuous* rapid." The result was a seemingly endless stretch of brute labor. The supplies (and sometimes the boats as well) had to be carried across a field of enormous fallen boulders. At one point, when the men were lining the boats down a rapid, "poor *Kitty's Sister* got a hole stove in her side" by a rock. The damage could be repaired, but it meant more delay. After a full day's labor, the expedition had advanced perhaps a mile.

The next day was more of the same. "Rapids and portages all day," Sumner scribbled that evening before he collapsed into sleep. Bradley, as usual, was more inclined to linger lovingly over the awful details. He had been wet all day and would spend a wet night, for his extra clothing was wet, too. Beyond that, he had fallen while trying to keep his boat off a rock and had a bad cut over his left eye. "Have been working like galley-slaves all day," he wrote. "Have lowered the boats all the way with ropes and once unloaded and carried the goods around one very bad place. The rapid is still continuous and not improving. Where we are tonight it roars and foams like a wild beast."

Though Powell, Sumner, and Bradley each described the day's hard labor independently, their voices ring out in three-part harmony. Powell put a good face on a hard day, singing melody. "When night comes, . . . we are tired, bruised, and glad to sleep." Sumner contributed a brief, matter-of-fact rhythm

---

★Grua and two other boatmen hold the record for fastest trip through the Grand Canyon, set in 1983: 277 miles in just over 36½ hours. (Commercial rowing trips cover the same distance in about fourteen days.) The winter of 1983 had been marked by tremendous snowstorms in the Rockies. When the snow melted, the rampaging Colorado nearly destroyed Glen Canyon Dam. The river was officially closed to boaters, but Grua and his companions ran day and night (sometimes using handheld floodlights), swapping off at the oars. Despite flipping and nearly drowning in Crystal Rapid, they escaped unharmed but for a $500 fine.

accompaniment. "Camped . . . under an overhanging cliff." Bradley growled out a long, innovative bass line. Powell "as usual [had] chosen the worst camping-ground possible," a spot so miserable that even a dog would have sniffed in disdain and turned away.

The next day was the same thing yet again, Bradley noted, "only more of it." They advanced a paltry three miles, about the same as on the previous day. The lone bright spot was a brief river run, a rare chance to ride in the boats rather than line or carry them. But the pleasure was short-lived. The run, their longest in four days, lasted less than a mile. They made camp just above another monster rapid. It stretched a mile or so and looked impossible. "Still there is no retreat if we desired it. We must go on and shall—and shall no doubt be successful," Bradley wrote gamely.

They had pulled into camp early, at about three in the afternoon, to dry their clothes and food in the sun. Bradley immediately began fishing, which always cheered him up even when, as today, he failed to catch anything. Even so, only the day before he had been fretting about his injured eye and lamenting that he was fed up with always being wet. Now, with the sun beating down and drying out his clothes and his papers, life looked better. "My eye is very black today," he noted with wry pride, "and if it is not very *useful* it is very ornamental."

Powell, on the other hand, was uncharacteristically glum. He dealt with the river in a handful of words—"Still rocks, rapids, and portages"—and knocked off the whole day in three sentences. For Powell, for whom a bend in the river or a sunset was ordinarily worth at least half a dozen adjectives, this was remarkable. The last of the three sentences was one of the plainest that Powell ever wrote. "Everything is wet and spoiling."

At the time, it seemed a more or less innocuous observation, akin to a routine note—"the patient has a cough"—in a medical chart. Soon enough its true significance would emerge. This cough was tuberculosis.

Since Disaster Falls, the expedition had been beset with a kind of free-floating anxiety. A night or two after the loss of the *No Name*, the men had made an unsettling find while they were building their campfire. "We discover an iron bake oven, several tin plates, a part of a boat, and many other fragments," Powell wrote, "which denote that this is the place where Ashley's party was wrecked." Actually, Ashley and his men left the Green more or less intact. But *someone* had apparently come to a bad end, and the find was hardly encouraging.

Now a new fear was taking shape. To this point, none of the men had grasped the possibility that they could run short of food. The hunting had been poor all along, but that had seemed merely an annoyance, little more than a joke at the hunters' expense. No one anticipated that eventually the game (and the fish) might vanish altogether. The expedition's supplies still seemed bottomless. Early on, and without a second thought, Hall and Hawkins had thrown away hundreds of pounds of food to lighten their load. Two weeks later, the wreck of the *No Name* had cost one-third of the remaining supplies, but, with the endless portaging, the men continued to see their food stores as a burden rather than a blessing. "We have plenty of rations left," Bradley wrote the day after Disaster Falls, "much more than we care to carry around the rap-pids, especially when they are more than a mile long."

But Bradley, and everyone else, had missed the point. The men hated man-handling the boats and the tons of supplies because it was dangerous, exhaust-ing work. Their complaints were fully justified. But in emphasizing how hard the work was, the men neglected the other side of the coin—how *slow* it was. The expedition confronted two enormous problems. The rapids were danger-ous and getting worse, and the alternatives to running them—portaging or lining—were hard, slow work. For Powell and his men, the dilemma could hardly have been more stark. Move quickly or slowly. Drown or starve.

On June 13, the men stayed in camp to recover from their labors and to dry their clothes and gear. It was a Sunday, Bradley noted, the first one they had honored by resting. That was presumably coincidence, Bradley complained, "for [I] don't think anyone in the party except myself keeps any record of time or events." He was wrong about that—Sumner and Powell may have been writing in their own diaries at the same moment that Bradley set down his complaint—but he was right that for Powell, Sunday was just another day.

Powell had been raised in a devoutly religious household. Throughout his childhood, the family gathered twice a day to pray and read the Bible aloud. As a precocious young boy, Powell learned all four Gospels by heart. His father was a Methodist preacher who believed that the Bible was the unchallengeable word of God. The preacher and his equally pious wife named their son for John Wesley, the founder of Methodism, whose sermons put a heavy weight on the value of self-denial, sobriety, and hard work. Powell retained the discipline but abandoned the dogma, not by any overt act of rejection but simply by sloughing off the beliefs of his parents and moving on, as a snake unceremoniously abandons his outgrown skin. All religious doctrines, Powell came to believe, were "mythology."

Bradley was often surly, even without Powell's religious laxity to provoke him. On most topics, his grumbling was worth heeding. Now, without fanfare, the cantankerous ex-soldier set down a crucial observation. "Our rations are getting very sour from constant wetting and exposure to a hot sun," he noted in the same day's journal. "I imagine we shall be sorry before the trip is up that we took no better care of them."

Three men used the day off to go hunting. They returned, wrote Bradley, "as ever without game." The problem, as Bradley diagnosed it, was that the would-be hunters made so much noise they scared off every animal for miles around. "They seem more like school-boys on a holiday than like men accustomed to live by the chase," he groused, "but as I am no hunter myself I must not criticize others. Still as usual I have my opinion." Sumner had an opinion, too, but his was more forgiving. "There is nothing in this part of the country but a few mountain sheep, and they stay where a squirrel could hardly climb."

From the start of the trip, the men had complained about how maddeningly elusive these desert bighorn were. The men could see them from the river, standing in a line on a cliff ledge two or three hundred feet above the water, fat and tempting and as still as statues. But, as Walter Powell had noted a week or two before, whenever the boats pulled to shore the sheep "suddenly wheel[ed] around like a platoon of well-drilled soldiers," and leaped, in unison, to an even more remote perch.

The next day, June 14, began slowly. The sun beat down, and the men took refuge in the shade of some box elders and sewed up their torn clothes or made themselves moccasins or played cards or read or napped. Powell and Oramel Howland climbed a two-thousand-foot cliff and scanned the view; then, as they had the day before, they did their best to reconstruct the maps that had been lost with the *No Name*. Bradley worked on his journal and then caulked his boat's bulkhead. Sumner repaired a broken barometer. Walter Powell, proud of his deep, rich voice, amused himself, and perhaps the others, with his favorite songs.

They had camped just above still another roaring and seemingly impass-able rapid. "The prospects of success are not bright," Bradley had written two days before, when he contemplated it from above, but when they finally got under way, they managed to run it uneventfully. Their reward was a new rapid "five times as bad as the last."

Powell was too taken with the scenery to be much perturbed. "On the east side of the cañon, a vast amphitheater has been cut, with massive buttresses, and deep, dark alcoves, in which grow beautiful mosses and delicate ferns, while springs burst out from the further recesses, and wind, in silver threads, over floors of sand rock." In the meantime, the men had gear to haul. Struggling over the rocks with the cargo was backbreaking work, as always, but it went better than the lining. "My boat was sunk while being lowered over the rapid this morning," Bradley noted that evening, though it was not wrecked but only swamped. Bradley lost some books and photographs but managed to save his notes. He dried them and decided that from here on, for safety's sake, he would carry them in his hat.

The river grew steadily worse. "Here we have three falls in close succession," Powell wrote blandly, as if this were nothing out of the ordinary. "At the first, the water is compressed into a very narrow channel, against the right-hand cliff, and falls fifteen feet in ten yards; at the second, we have a broad sheet of water, tumbling down twenty feet over a group of rocks that thrust their dark heads through the foaming waters. The third is a broken fall, or short, abrupt rapid, where the water makes a descent of more than twenty feet among huge, fallen fragments of the cliff."

Bad as it looked, Triplet Falls, as Powell named it, was only a kind of appe-tizer for the diabolical main course to come. Today, professional boatmen regard both Triplet and Disaster Falls, upstream of it, as almost routine. That judgment is more a measure of how formidably skilled these men and women are than it is a statement about the rapids themselves. An amateur scouting the river from shore can hardly identify a path through Triplet, let alone imagine how he would coax a boat to follow it if he did find one. Pick up a stick and toss it into the river hoping for clues, and it bobs and twists until, suddenly and mysteriously, it dives toward the river bottom as if pulled by an unseen hand. The prospects for a boat seem just as bleak. "The main current is into the cliff," one modern-day writer observes, "and a boat pushed into this wall . . . flips in a second. The main push of the current bounces off the wall and shoots between it and a rock, forming a gateway that is just wide enough to wedge a boat but not let it through."

And that is only the first part of the first rapid, a mere warm-up. Powell never even thought of running Triplet Falls. "We make a portage around the first; past the second and third we let down with lines."

It would be hard to devise a sentence that described more work in fewer words. "Lining a boat is horrible, horrible work," says Michael Ghiglieri, a commercial boatman with decades of experience in the Grand Canyon and on rivers around the world. "You always think, 'Gee, that wouldn't be so bad, you'd just sort of push the boat out into the current and hang on to the ropes, and every once in a while if it didn't go where you wanted, you'd give it another push.' But it never works that nicely."

Ghiglieri pauses for a moment to recall exactly how things go wrong. "The boat's always getting wedged between boulders, or it does a kind of slingshot thing where it drifts out and then slams back against the rocks." His voice picks up speed, as if running a rapid itself. "Or it takes off downstream and there's somebody running along the shore holding on to a rope but they can't keep up and they're already holding the *end* of the rope so there's no slack to let out, and they trip and fall on their face on a boulder."

The thought of falling sets Ghiglieri off on a related tack. Having studied the Powell expedition and rowed the Grand Canyon more than a hundred times, he talks about Powell's boatmen as if he and they are old pals who have shared many a beer at the end of a long day on the river. "The thing to remember," he says, "is the sheer pain involved. People hear 'portaging' and 'lining' and they think of drudgery, but it's worse than drudgery. These guys were getting beat up and falling down and wrecking their feet—they were either barefoot or wearing really bad, worn-out shoes, with the soles flopping and the stitches coming out, because the leather couldn't take being wet all the time.

"So they're walking around boulders, it's muddy and slippery, they can't see the bottom, every step is difficult because who knows what's underneath that water, and they're yanking ropes and getting knocked over. All these guys got hurt—somebody *always* gets hurt. Maybe they didn't get crippled for life, but they got hurt. *And*, on top of all that, it's a tremendous amount of work."

One of the earliest accounts of a white-water trip, from an 1872 diary, describes river runners on the Colorado discovering precisely those hardships. "The boat at times will be wedged in between the rocks while we are tugging and pulling away; suddenly away she will go, dragging us after her, holding on for dear life. . . . 'Tis a wonder that some of us have not had a leg or two broken. All of us wear horrible scars from our knees downward to remind us of the days when we made portages."

Ghiglieri neglected mentioning still another hazard. One of the great dangers in white water is "foot entrapment." Someone falls out of a boat and, instinctively, tries to stand up and scramble to shore. If his foot gets wedged under a rock or caught between rocks, he can be trapped and, in moments, forced underwater by the current, held there helpless, and drowned. It can happen fast, and not only in fierce rapids or mighty rivers. On a canoe trip in 1972 on a fairly calm stretch of the Green, the future writer Anne Fadiman (then an eighteen-year-old) saw a companion fall from his boat into waist-deep water. The young man stood up, near a wave just downstream from a large rock, but one foot was caught between two rocks. Fadiman and her fellow novices tried to reach him but failed. "Thirty seconds passed, maybe a minute. Then we saw the standing wave bend Gary's body forward at the waist, push his face underwater, stretch his arms in front of him, and slip his orange life jacket off his shoulders. The life jacket lingered for a moment before it floated downstream, its long white straps twisting in the current. His shirtless torso was pale and undulating, and it changed shape as hills and valleys of water flowed over him, altering the curve of the liquid lens through which we watched him."

If they could not pull Gary to safety, his companions hoped, at least they could hold him upright in the river until help arrived. "But the Green River was flowing at nearly three thousand cubic feet—about ninety tons—per second. At that rate, water can wrap a canoe around a boulder like tinfoil. Water can uproot a tree. Water can squeeze the air out of a boy's lungs, undo knots, drag off a life jacket, lever a boot so tightly into the riverbed that even if we had had ropes—the ropes that were in the packs that were in the trucks—we could never have budged him." Gary drowned.

On the Colorado, a flow of a mere three thousand cubic feet per second would count as barely a trickle. A flow ten times that rate, thirty thousand cubic feet per second, is only middling. Before the dams, the spring floods routinely reached one hundred thousand cubic feet per second. The highest flow on record on the Colorado, in 1884, was three hundred thousand cubic feet per second. (The rising water submerged the official flood gauges, but a cat trapped in an apple tree served as a stand-in gauge.)

At most spots on the Green and Colorado, the river is so deep that a swimmer in midstream could not trap his foot if he wanted to. But for Powell's men, lining the boats often meant getting into the water along the shoreline and slipping and sliding over the rocks while wrestling with the boats and the current. Entrapment would have been a lurking hazard, although the men may never have known it, as if they were proceeding down an African river never having heard of crocodiles.

Powell's crew struggled through the three-part ordeal of Triplet Falls. Then came a brief break, a lull just long enough to permit a glimmer of hope that the worst was past. "We run down, three-quarters of a mile, on quiet water," Powell wrote, "and land at the head of another fall." This one dwarfed everything that had come before. "On examination, we find that there is an abrupt plunge of a few feet, and then the river tumbles, for half a mile, with a descent of a hundred feet, in a channel beset with great numbers of huge bowlders. This stretch of the river is named Hell's Half-Mile."

The boulder-strewn rapid is still considered one of the most difficult in America, in part because it is so long that plotting a route seems impossible. Powell and the others had occasionally talked about mile-long rapids, but they had been speaking loosely. Here was the real thing—not a series of rapids in close succession but a seemingly endless stretch of white foam. When river runners swap stories about particular rapids, they tend to focus on a handful of landmarks—stay out of the eddy at the bottom right of Granite Rapid, say. No such tip sheet will do at Hell's Half Mile. One acclaimed river runner compared the task of memorizing the half-mile rapid to committing the entire score and all the singing roles of *Parsifal* to memory.

The challenge for Powell and his men was less intellectually demanding. They began by carrying the cargo to the foot of the falls. That was the easy part. "Then we commence letting down the boats," Powell wrote. "We take two of them down in safety, but not without great difficulty; for, where such a vast body of water, rolling down an inclined plane, is broken into eddies and cross currents by rocks projecting from the cliffs and piles of boulders in the channel, it requires excessive labor and much care to prevent their being dashed against the rocks or breaking away."

Then came the third boat. "We are letting down the last boat," Powell wrote, "and, as she is set free, a wave turns her broadside down the stream . . . They haul on the line to bring the boat in, but the power of the current, striking obliquely against her, shoots her out into the middle of the river. The men have their hands burned with the friction of the passing line; the boat breaks away, and speeds, with great velocity, down the stream."

The *Maid of the Cañon* vanished out of sight.

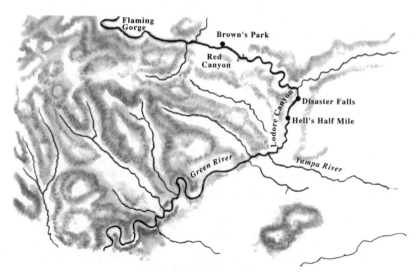

CHAPTER TEN

FIRE

"While letting the *Maid* down with ropes," Sumner wrote that evening, "she got crossways with the waves and broke loose from the five men holding the line, and was off like a frightened horse. In drifting down she struck a rock that knocked her stern part to pieces." The men, trying to coax the *Maid* around some rocks, had made the mistake of paying out too much line. The current had instantly snatched the boat for itself—a tug-of-war between five strong men and one angry river was not a fair fight—and flung it downstream.

The expedition had begun with four boats and had already lost one. If the loss of that first boat qualified as a disaster, it would be hard to find a word too strong for a second such loss. Sumner and Hawkins jumped into the *Emma Dean* and took off after the speeding runaway. The men were not hopeful, Powell least of all. "We gave up the *Maid of the Cañon* as lost," he wrote.

But this time the always unpredictable river seemed content merely to tease the upstarts who had presumed to challenge it, like a boxer who sees a chance for a knockout blow but passes it up in order to emphasize his dominance of his flummoxed rival. After half a mile's frantic chase, Sumner and

Hawkins found the *Maid of the Cañon* caught in an eddy, beaten up but still afloat and spinning placidly.

The unexpected rescue lifted everyone's spirits. "Rejoicing that we had not lost her entirely," as Howland put it, the men set to repairing the *Maid*. "She got one pretty severe thump in her stern, which stove a hole and made her leak badly," Howland wrote, "but at noon we repaired that, and she is now as good as ever."

They camped that night, June 16, at the mouth of a small brook and dared to hope that their fortunes had turned. "We hope and expect that the worst of this cañon is over," Bradley wrote, "for the softer rock is getting near the water and the softer the rock the better the river generally." They had been laboring through red sandstone but seemed to be coming to a section of white sandstone, which was more prone to crumble under the river's incessant assault and thus less likely to form formidable rapids.

Powell, always intrigued by geology, set off exploring. This was scenery on a grand scale, with "cliffs and crags and towers, a mile back, and two thousand feet overhead," and a dozen mini-waterfalls cascading over the rocks, and pines and firs and aspens. High above the river, the cliffs were buff and vermilion and bathed in sunlight; far below, they were red and brown and deep in shadow. But the view was as dismaying as it was inspiring. "The light above, made more brilliant by the bright-tinted rocks, and the shadows below more gloomy by the somber hues of the brown walls, increase the apparent depths of the cañons, and it seems a long way up to the world of sunshine and open sky, and a long way down to the bottom of the cañon glooms."

"Never before," Powell continued, "have I received such an impression of the vast heights of these cañon walls." But perched on an overlook halfway up to "the world of sunshine and open sky" and hopeful that the river would soon grow accommodating, Powell quickly brushed aside any dark thoughts. "We sit on some overhanging rocks, and enjoy the scene for a time, listening to the music of falling waters away up the cañons. We name this Rippling Brook."

Despite Bradley's forecast to the contrary, the river the next day was nearly as unruly as ever. Fortunately, at least some of the rapids were manageable, a welcome change after days of lining and portaging with scarcely any progress to show for it. "Ran many little rappids," Bradley wrote, and Sumner's slightly fuller account shows what the men were now prepared to consider "little." At the first rapid of the day, for example, "the freight boats went through in good style, but the *Emma*, in running too near the east shore, got into a bad place and had a close collision, filling half full, but finally got out, all safe." There followed a stretch of easy water, then a bad rapid that had to be lined, then another easy stretch, and finally a rapid with a drop of about 12 feet in a length of 150 feet. Two of the boats made it through cleanly, "but poor *Kitty's Sister* ran on a rock near the east side and loosened her head block and came down to the other boats leaking badly."

After pulling ashore for a slapdash recaulking of the *Kitty Clyde's Sister* (enough "to keep her afloat for a while"), they set out again. The *Emma Dean*, leading the way as usual, "got into a complete nest of whirlpools." The *Emma* managed to free herself after a hard struggle, but it seemed too risky for the bigger boats. At the end of a long day, then, there was no choice but to line both the heavy boats past the rapid. When the tired crew finally made camp— and this on a day they all hailed as a welcome break from the rigors of the previous week—they had made about five miles.

The campsite was a pretty one, a little spot about a quarter of a mile long and fifty yards wide and thick with willows, cedars, pine trees, sagebrush, and grass. Just downstream was another rapid they would deal with in the morning, after a night's sleep. Everyone had taken special note of the trees and grass, not only as a welcome contrast with the endless vistas of bare rock but because they planned to use pitch from the pines to caulk the leaks in their boats. They had covered only a handful of miles, but the rocks had taken their toll and all the boats needed attention.

Hawkins, the cook, started a fire "in a nice little cove in the brush and rock" and began to prepare lunch. Powell ventured off to explore. The men unloaded the boats to dry out their gear. Suddenly a gust of wind swept up the river and sent embers everywhere. Hawkins had set up shop a little *too* near the brush. In an instant, the entire camp was aflame. Hemmed in by towering stone cliffs, the men could not run inland, so they raced for the boats.

The next moments were chaos. The men tried, more or less simultaneously, to leap in the boats, tear off their burning clothes, smother the flames in their singed hair and beards, and grab the oars. Hawkins grabbed as much of the mess kit as he could carry (in case they could not make it back to camp) and ran for his boat. But the flames had gotten there first and had already burnt through the ropes he had tied up with. Leaping for the boat as it drifted away from the shore, Hawkins lost his balance and fell into ten feet of water. (As many a surprised swimmer has learned, in a big river a step off the beach can be like a step off a table.) He surfaced a moment later, empty-handed and sputtering, and swam after the boat. It was about thirty feet off, and Hall, at the oars, was doing his best to keep it in position upstream of the rapid. Hawkins dragged himself aboard, his clothes no longer aflame but the mess kit gone.

Hall and Hawkins and the others pulled back into shore a short distance downstream, but the wind had spread the fire there, too, and they had to pull away again. Now they found themselves caught in the current and racing headlong into a rapid that no one had scouted. At just this moment, Powell, climbing the cliffs above camp, looked down and saw the men in their boats. Unable to see the flames from where he stood, he watched in astonishment and perplexity as the boats ricocheted through the rapid, like drunks trying to navigate a revolving door.

All three boats made it through safely, the boats intact but the men in poor shape. "One of the crew came in hatless, another shirtless, a third without his pants, and a hole burned in the posterior portion of his drawers," Oramel Howland reported with malicious good humor, "another with nothing but drawers and shirt, and still another had to pull off his handkerchief from his neck, which was all ablaze. With the loss of his eyelashes and brows, and a favorite moustache, and scorching of his ears, no other harm was done."

Everyone clambered back over the rocks to camp to see what gear could be salvaged. Not much. Hawkins was the goat of the episode, but no one seemed much perturbed by the loss. "Our plates are gone; our spoons are gone; our knives and forks are gone," Powell wrote, and then, uncharacteristically, he ventured a small joke: "'Water catch 'em; h-e-a-p catch 'em.'" Howland drew up a more detailed inventory. They had already lost a portion of the mess kit at Disaster Falls. Now the total that remained, for ten men, was: "One gold pan, used for making bread. One bake-oven, with broken lid. One camp-kettle, for making tea or coffee. One frying-pan. One large spoon and two tea spoons. Three tin plates and five bailing cups." From here on, the bailing scoops would have to serve as coffee and tea mugs, one for every two men.

In addition, Howland noted, there was "one pick-ax and one shovel." The last two items were reserved for special occasions. "When a pot of beans, which by the way is a luxury, is boiled in place of tea or coffee," he explained, "our cook sometimes uses the latter article for a spoon, and the former to clean his teeth after our repast is over."

These were hardy men, and they viewed setting themselves afire mainly as a comic interlude. Even Bradley—and it was the long-suffering Bradley whose prize mustache had gone up in flames and whose ears were scorched—seemed more irritated by the loss of the mess kit than by his own injuries. Sumner saw it all as capital fun. "Had supper," he wrote, "and laughed for an hour over the ludicrous scene at the fire."

After the grinding labor of portaging, the fire seemed not so much a near catastrophe as a sign that the men were untouchable. For the moment, at least, they felt invincible. They had survived the Civil War, they had survived Disaster Falls, they had survived the *Maid's* dash for freedom, and now they had survived a fire. Let the Green do its worst!

The next day found the river as high-spirited as the men. "Had a splendid ride of six miles," Sumner noted, and Bradley celebrated "a run of almost railroad speed." This railroad comparison was a favorite of all the men and one they apparently intended literally. Earlier in the trip, Powell had made the identical comparison, and Walter Powell would echo the same phrase later on. Once, when Oramel Howland's boat sped by the *Emma Dean*, which had already pulled to shore, "those in her said we passed them as rapidly as a railway train at its highest speed—sixty miles an hour . . . [and] this was slow to some rapids we have run since."

This was off by more than a little. Even the fastest rivers are slow. In most places, the Green and the Colorado travel at only five or six miles an hour, a bit faster than a fast walker. In the narrowest stretches of the biggest rapids in the Grand Canyon, if the river is running at high volume the current might briefly touch thirty miles an hour. But, as a rule, the current in a world-class rapid is somewhere in the vicinity of twenty miles an hour. If there were a track alongside a rapid, a top sprinter could keep up with a boat for a short distance and a Sunday bicyclist could pass it without much trouble.

Then why do rapids feel so fast? Sumner and Bradley and the others never knew the pleasure of zooming down the highway with the top down, but living in a faster world hasn't taken the thrill from river running. Take a race-car driver out from behind the wheel and push him into white water, and his heart, too, will try to thump its way out of his rib cage.

The key factors are acceleration, drop, and proximity—the river is mere feet away. (Subliminal thoughts of drowning may add a soupçon of nervous excitement, too.) The acceleration hits like a punch to the gut. In the Grand Canyon, for example, the Colorado River is essentially a series of relatively flat, slow-flowing ponds that spill over into steep, violent rapids. The water in a rapid may travel ten times faster than the water in the pond just upstream. The nearness of the water magnifies the feeling of speed. Psychologists talk about "optic flow": Sled down a hill and the ground, only inches away, seems to fly by. Look out an airplane window at the ground thousands of feet below, and the plane seems to be crawling.

There is a moment of extraordinary tension in a boat poised at the top of a rapid. ("I believe I would have given everything I possessed to have been able to turn back," admitted one novice somehow dragooned into running Hell's Half Mile in 1926, "but there was no turning back then.") You hang there on the lip, irrevocably committed to the surging waters ahead but still inches beyond the rapid's grasp. The fascination is akin to what some people feel in looking over a balcony railing or a cliff edge and wondering what it would feel like to fall. An instant from now, you will know.

Until that moment, you can at least assure yourself that this is not really going to happen, that there is still a way out. You are leaning out over the balcony railing, it is true, but, after all, you are only *looking*.

And then the railing disappears.

When it comes to picturing a river's speed, our usual habits of thought mislead us. Because rapids throw problems at us by the second, not by the hour, talk of "miles per hour" gets us off on the wrong foot. Twenty miles per hour is a shade less than thirty feet per second, and this alternative way of putting it is far more relevant. At thirty feet a second, in a rapid marked by waves and whirlpools and glistening boulders, the pace seems breakneck.

And we tend to think of cars when we hear "miles per hour." But a car is a heavy, sturdy cage with brakes and seat belts, and a road is broad and smooth. To get the feel of white water, come out from inside the car. Toss someone the keys and take a seat on the hood. Head down a tilted, twisted road studded with sharp-edged rocks and pocked with eight-foot holes. For best results, find a road that folds itself into ten-foot waves.

•   •   •

When Powell ran the Colorado, there was no such thing as river-running technique. He and his men had to invent their own. In two crucial ways, their tactics
differed from those that all modern boatmen use. First, Powell's boats had two
men at the oars. Today's boatmen are soloists. Second, Powell's crew ran the rapids
"backward," facing upstream and backing into the dangers lying in wait for them.
Boatmen today do row "backward" in a river's flat, easy stretches, so that they can
make good time in slow water, but they turn to face downstream when they
come to a rapid. (Once in the rapid, there may be maneuvers that call for turning backward briefly, perhaps to spin away from a looming rock or a hole.)

The risk in running rapids backward is plain enough. Having two men at the
oars, on the other hand, seems innocuous. The trouble came in putting the two
together—Powell's men were rushing backward into trouble *and* there were two
pairs of hands at the controls. Emergencies, even ones faced head-on, are not a
time for discussion—imagine what would happen to a race car with two steering wheels and two drivers. In the case of a rowboat moving backward through
a rapid, how would the two men know, for example, that *now* was the instant
they both needed to pull as hard as they could on their right-hand oar?

But Powell had no choice in the matter. Heavy boats meant two-man
teams. Hard-to-maneuver boats built for speed cannot dance a water ballet;
they must blast their way through rapids as fast as they can go. The reason
is that if a boat in a river is to maneuver through white water with any
degree of success, there are only two choices: It must move either faster or
slower than the current. If the boat simply moves at the river's speed, like a
drifting stick, it will go where the river takes it. With boatmen rowing backward in hard-to-steer boats, Powell had to opt for speed, with all the risks that
came with it.

To race toward a threat that you could hear but not see was terrifying, like
hurrying blindfolded toward a bear snarling in his cave. "I pulled the bow
oars," one early river runner wrote, "and my back was toward the terrific roar
which, like the voice of some awful monster, grew louder as we approached.
It was difficult to refrain from turning round to see what it looked like now . . .
We kept in the middle of the stream, and as we neared the brink our speed
began to accelerate. Then of a sudden there was a dropping away of all support, a reeling sensation, and we flew down the declivity with the speed of a

locomotive. The gorge was chaos. The boat rolled and plunged. The wild waters rolled over us, filling the open spaces to the gunwale."

Then, finally, someone had a better idea. By most accounts, the greatest innovation in the history of river running was the brainstorm of a balding, bushy-mustached trapper and prospector from Vernal, Utah, named Nathaniel "Than" Galloway. Sometime toward the end of the nineteenth century, some twenty years after Powell, Galloway did something that no one had ever done before. As he approached a rapid in his rowboat, he turned around to face forward so that he could see where he was going.

Sometimes the best ideas are the simplest. Galloway sat facing the stern, with his back to the bow, like everyone else who rowed a boat. The difference was that he turned the boat around and ran the rapids stern-first. That way he could see the dangers ahead of him, and, more important, he could do something to avoid them. Where Powell's crew raced their way blindly downstream, Galloway moved slowly by rowing *against* the current, ferrying his way across the river and dancing past holes and boulders. If he did happen to hit a rock, he would at least be moving relatively slowly.

Now picture Powell's boats. The *Emma Dean*, Powell had noted proudly, was "every way built for fast rowing." So were the freight boats. Any collision would catch the men at the oars unprepared, since they were pulling blindly with all their might, and it would be a high-speed crash besides. Powell's strategy was to use boats that were heavy and sturdy enough to survive collisions. Galloway's innovation was to use lighter, maneuverable, flat-bottomed boats and avoid the rocks in the first place.

The flat bottom sounds like an afterthought, but it was crucial. "With that flat bottom, especially if it's raised up out of the water at either end," says Brad Dimock, a boatman and white-water historian, "you can be turned three-quarters to the current and it's not going to catch on anything and spin you." Powell's boats were at the opposite extreme; rather than a flat bottom, they had a keel that was essentially a single twenty-two-foot-long two-by-eight running the length of the boat. As soon as *that* boat turned even slightly sideways to the current, the river would grab the keel and whirl the boat around. "The genius of the Galloway approach," says Dimock, "isn't just that you're looking downstream. It's that even though you're *consistently* off a straight line to the current, so that you can ferry back and forth, the current won't spin you. If you tried that in a Poweli boat, you'd be smashed to bits."

So, on entering the obstacle course that is a rapid, Powell had no choice but to run blind and at full speed. The result was "calamitous," in Dimock's judgment, but it is hard to blame Powell. Galloway's seemingly obvious ideas had never occurred to any of the mountain men who wandered the West before Powell, and for two decades after Powell, his successors missed them as well. It was not until Galloway made a successful trip through the Grand Canyon in 1909, and two brothers named Ellsworth and Emory Kolb took up his technique for a Grand Canyon trip of their own in 1911, that the "facing your danger" technique finally grew common.

On June 18, 1869, with the river flowing smooth and fast, it was easy to dismiss thoughts of danger. The expedition had reached the junction of the Green River and the Yampa (sometimes called the Bear), which flowed into the Green from the east. This was a checkpoint, if not quite a milestone. In his trip West in 1868, Powell and his guides had explored some stretches of the canyon cut by the Yampa. To have reached almost-familiar territory seemed a good omen. The cliffs were lower here, too, rising only some four hundred feet instead of two thousand or more. As the walls fell, the men's spirits rose. Powell decided that the low walls signified the start of a new canyon and, more important, the end of the much-feared Lodore.

All the signs seemed good. After the rigors of the previous ten days, the men needed a break from their labors, and this seemed an ideal spot. They camped on a bit of land between the two rivers and settled in happily. "Opposite the mouth of Bear River there is the prettiest wall I have ever seen," Sumner wrote. "It is about three miles long and five hundred feet high, composed of white sandstone, perpendicular and smooth, as if built by man." After a series of highly satisfactory experiments involving whoops and shouts, the men named it Echo Rock. (It is known today as Steamboat Rock because of a vague resemblance to a steamship run aground in the desert.)

Most of the crew were busy fishing, Bradley chief among them. The fish were temptingly big, so large and lively that time and again they broke free just as Bradley brought them to the surface. ("Bradley was much provoked," Sumner noted happily.) When the count reached four lost hooks and three broken lines, Bradley settled down to do battle in earnest. By twisting four lines into one and fashioning a two-inch-long hook, he finally hauled in a ten-pounder and retired satisfied.

With the rigors of Lodore behind them, the men were moved to valedictory thoughts. Powell seemed to waver back and forth between grim memories of hardships endured and strained attempts to convince himself that the recent ordeal had actually been a splendid jaunt. "This has been a chapter of disasters and toils," he wrote, "notwithstanding which the cañon of Lodore was not devoid of scenic interest, even beyond the power of pen to tell. The roar of its waters was heard unceasingly from the hour we entered it until we landed here. No quiet in all that time. But its walls and cliffs, its peaks and crags, its amphitheaters and alcoves, tell a story of beauty and grandeur that I hear yet—and shall hear."

Oramel Howland betrayed no such mixed emotions. Safe on shore for the time being, this survivor of Disaster Falls professed a yearning for more rapids. "A calm, smooth stream, running only at the rate of five or six miles per hour, is a horror we all detest . . . ," he wrote. "Danger is our life, it seems now, almost. As soon as the surface of the river looks smooth all is listlessness or grumbling at the sluggish current . . . But just let a white foam show itself ahead and . . . jokes generate faster and thicker than mosquitoes from a bog, and everything is as merry as a marriage bell."

Bradley knew better than to provoke the river gods with blasphemous wishes for rough water. The Yampa looked to be about 120 yards wide and 10 feet deep, virtually as big as the Green, and Bradley declared hopefully that this indicated good times to come. "I predict that the river will improve from this point," he wrote, "for the more water there is the wider channel it will make for itself and the less liability will there be of its falling in and blocking up clear across."

In any case, whatever lay ahead was certain to be an improvement on Lodore. That canyon, Bradley observed, had been "the worst by far, and I predict *the worst we shall ever meet.*"

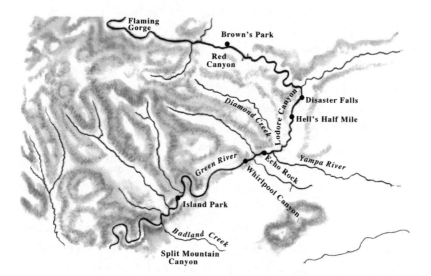

CHAPTER ELEVEN
THE FIRST MILESTONE

They celebrated their escape from Lodore with a happy, lazy interlude of a day or two in camp. Lazy, at least, in comparison with what they had just accomplished. The weather was good and spirits were high. Powell, Bradley, and Oramel Howland explored a short way up the Yampa. The cliffs were light gray sandstone, in some places forming walls about a thousand feet high and in others jagged, sloping terraces that extended for a mile or more. The three men fought their way upstream against the current, making only four or five miles' progress in several hours of hard work. ("When we have rowed until we are quite tired," Powell wrote, as if he had been hard at it, "we stop.") The return trip, with the current this time, took only twenty minutes.

Everyone else had been content to stay in camp, fishing or snoozing. The following day, a Sunday, was even quieter. The most taxing project the men took on was scrawling their names on Echo Rock. Bradley spent part of the day trying to salvage a photo album he had brought with him. He had taken great pains to keep it dry, but now he saw that water had spoiled most of the pictures. "Mother has but one eye while all that is left of Aunt Marsh is just *the top of her head,*" he lamented. "Eddie has his chin untouched while Henry loses nearly all his face . . . One of Lucie's lost a nose but luckily it was the poorest one and I have a good one left."

Powell, as always, had been more restless than anyone else. He spent the day climbing to the top of the canyon to see what he could see. On an exceptionally clear day, he could make out the Sweetwater and Wind River Mountains 100 miles to the north, the Wasatch and Uintas to the northwest, and the Rockies more than 150 miles to the east. Far below him he could see the river gleaming.

The next day was business as usual. The men launched the boats at seven in the morning and found themselves back in the hard, red sandstone they feared. The river cut through "a narrow, dangerous canyon full of whirlpools," Sumner noted, "through which it is very hard to keep a boat from being driven on the rock." They had already weathered countless hazards, but this new canyon would be the worst place yet for a smashup. "If a boat should be wrecked in it," Sumner went on, "her crew would have a rather slim chance to get out, as the walls are perpendicular on both sides and from 50 to 500 feet high."

Powell, who usually focused his descriptive energies on rapturous accounts of the scenery, echoed Sumner's fears. "The walls are high and vertical; the canyon is narrow; and the river fills the whole space below," he wrote, "so that there is no landing-place at the foot of the cliff." This was new, and it wasn't good. From the start, Powell's plan had been to run those rapids that seemed doable and to portage or line those that seemed too dangerous to run. But now the river had revealed a new trick. Portaging and lining were only possible if there was some kind of beach or rocky shore along the river's edge. With a river enclosed between sheer walls, plan B was suddenly irrelevant.

Since their first few days on the river, Powell and his men had spent most of their time hemmed in by towering cliffs. But even though in most places it would have been difficult to escape over the walls, there was at least a bit of leeway *within* the canyons—there had always been the possibility of heading to shore to detour around a rapid.

Until now. All along, they had been in the predicament of mice trapped in an endless hallway. Even worse, it was a hallway with a channel of water rushing down the middle. But it had always been possible to sneak to safety by moving away from the water. Now, at least temporarily, that option had been snatched away. Now the hallway was flooded wall to wall.

They were at the river's mercy, and the river was not feeling merciful. The Yampa had added its flow to the Green, making for far more water than the expedition had yet seen. Bradley's guess that this would prove good news was not panning out. "All this volume of water," Powell wrote worriedly, "confined, as it is, in a narrow channel, and rushing with great velocity, is set eddying and spinning in whirlpools by projecting rocks and short curves."

The Balkans, someone once observed, are an example of what happens when too much history is squeezed into too small a space. Here the Green taught an analogous lesson. "The cañon is much narrower than any we have seen," Powell wrote. "With difficulty we manage our boats. They spin about from side to side, and we know not where we are going, and find it impossible to keep them headed down the stream."

That made for "great alarm," Powell conceded, but soon enough the mood lifted. The boats were as out of control as twigs in a stream, but they were still afloat and intact. Perhaps the river was merely boisterous and rowdy rather than bad-tempered. "It is the merry mood of the river to dance through this deep, dark gorge," Powell wrote, "and right gaily do we join in the sport." But then, almost without warning, the emotional weather changed again. "Soon our revel is interrupted by a cataract; its roaring command is heeded by all our power at the oars, and we pull against the whirling current."

The *Emma Dean* pulled to the cliff that walled in the river on the right, and Sumner and Dunn rowed hard against the current to try to hold their position along the rock wall. Fifty feet downstream, a rapid roared. Powell signaled the freight boats to pull over where they could. The *Kitty Clyde's Sister* ducked into an alcove along the right-hand wall and sat in an eddy, a bit upstream of Powell. Caught by herself, the *Maid of the Cañon* drew near the cliff on the opposite side of the river from the other two boats and fought to hold her position. Like soldiers trying to hide from rooftop snipers, the men clung grimly to the walls.

But they could not hide forever, and from where they stood, they had no means of scouting ahead. Powell spotted a horizontal crevice in the rock face, about ten feet above the water and a dozen or so feet downstream from the

Emma Dean. They drew their way cautiously toward the crack in the rock. One of the men scrambled up to it, the others flung him a rope, and he tied up the boat. Powell climbed to the crevice, too, and found that there was room to crawl upstream. That was the wrong direction, away from the rapid, but in a short distance the crevice opened up, and Powell found he could pick his way higher up the cliff face. In fifty feet, he came to a rock shelf, a kind of thin, natural catwalk. Then it was a matter of following the shelf back downstream to a position even with the rapid. From there, Powell clambered down a pile of rocks to river level.

The *Maid* still had to cross the river, to join the *Emma Dean* along the right-hand cliff. She made it. The *Sister*, in the meantime, was on the correct side of the river but still hidden in her alcove, out of sight and out of hearing, the farthest away from the rapid of the three boats. Powell, now on foot at the rapid, spotted yet another crevice. This one stretched in the direction of the *Sister*. Powell crawled along the crevice and, when he had come as close to the *Sister* as he could, yelled with all his might. Eventually someone made out his voice above the water's noise. Powell shouted orders to the *Sister's* crew to move downstream to join the other two boats. Afraid to commit to the river, the *Sister* crept along the cliff face, the men grabbing desperately at every crevice and knob they spied. Finally, the three boats were together a few yards above the rapid. "Now, by passing a line up on the shelf, the boats can be let down to the broken rocks below," Powell wrote. "This we do, and, making a short portage, our troubles here are over."

The "troubles" had taken up three or four hours, but no one made much of them. In their journals that evening, Bradley and Sumner each devoted more space to the afternoon's fishing. By this time, a bad time in the boats hardly counted as news; a good meal was a front-page story. Oramel Howland was the hero of the day. He had set aside his mapmaking duties "and soon had a score of large trout," Sumner noted happily, "the first we have been able to catch so far." When Sumner recorded the great event in his journal, he paid tribute to his friend by referring to him not simply by name but as "Mr. Howland."

The good fortune continued into the next day, June 22. It began in fine fashion with a breakfast of fried trout followed by what Sumner, in the *Emma Dean,* called "a splendid run of six miles through a continuous rapid." For the freight boats, the run was not quite as splendid as all that. "One of the boats in trying to make a landing could not be held when she touched," Oramel Howland recalled, and had instead spun down the river through the next rapid, out of control and dragging 120 feet of line from the bow. Even so, everyone muddled through.

Soon after, lured by the sight of sheep and deer tracks on a sandy beach at the foot of a rapid, the men pulled to shore. The hunters set out optimistically, but, here at least, life was as frustrating as always. "They *hunted* with their usual success," Bradley grumbled. (In camp later that day, Bradley would set out in search of easier prey. He returned proudly bearing four quarts of currants.)

By one o'clock, the men had finished lunch and returned to the river. If the hunting had left anyone out of sorts, the bad moods vanished quickly. Powell was positively exuberant. "Into the middle of the stream we row, and down the rapid river we glide, only making strokes enough with the oars to guide the boat. What a headlong ride it is! Shooting past rocks and islands! I am soon filled with exhilaration only experienced before in riding a fleet horse over the outstretched prairie. One, two, three, four miles we go, rearing and plunging with the waves."

Powell was, admittedly, a man who could be thrown into ecstasies by a fossil or a fern. The previous day, describing a section of river that had threatened to drown him, he had written delightedly that "the waters waltz their way through the cañon, making their own rippling, rushing, roaring music." On this glorious day, however, everyone shared his enthusiasm. Sumner, ordinarily a man of few (and sardonic) words, sounded fully as romantic as his leader. The expedition was, he wrote excitedly but not quite accurately, "dancing over . . . waves that had never before been disturbed by any keel."

They ran, Sumner wrote, in "splendid style." Occasionally their style proved not quite a match for the river's power, but the men continued undaunted. Toward the end of one long rapid, for example, Sumner described "a place about a hundred yards long that had a dozen waves in it fully ten feet high." There was no place to land, and so Sumner and Powell and Dunn rode through, bucking and leaping and trying their best to hit the waves straight on, the boat filling nearly full of water but the men finally emerging safe and gleeful, though "looking like drowned rats."

As the boats sped along, the canyon walls grew gradually lower until, at about four o'clock, the men suddenly emerged into what Sumner described as "a splendid park." This was the third time that the dependably caustic

Sumner had used the word "splendid" in a single journal entry, and even Bradley was only slightly more subdued. "We came out into an exceedingly beautiful valley full of islands covered with grass and cottonwood," he wrote. "After passing so many days in the dark cañons where there is little but bare rocks we feel very much pleased."

Since passing through the Gates of Lodore, Powell and his men had been traveling through territory that, though it remains nearly empty to this day, is now familiar to many tourists. This early part of the route cut across the vast area that stretches across northwestern Colorado and northeastern Utah and is now known as Dinosaur National Monument. Slightly farther south, Powell would follow the Green through what is today Canyonlands National Park. With their cliffs and buttes and endless erosion-carved vistas, these are the sort of landscapes that inspire modern Americans to say they love the desert. And, indeed, it is easy to love, especially from inside an air-conditioned car or from a lounge chair on a patio with a margarita in hand, salt glistening on the rim, and an endless water supply available a few steps away at the turning of a faucet.

But Powell and his crew, whose notions of landscape had taken form in the rolling fields of Wisconsin and Illinois, the dappled forests of Vermont and Massachusetts, or the snowy peaks of the Rockies, found little to admire. Surfeited with rock, they reveled in the simple pleasures of flopping down on soft grass and stretching out to rest in the shade of a tree. What desert connoisseurs like Edward Abbey would later see as starkly beautiful struck these men as barren and lifeless. Land, it went without saying, should be useful—it should be fertile, or rich in timber or minerals, or, at the very least, suitable for grazing. To heap superlatives on these raw stones would be to prefer a skeleton to a lush nude.

Nature herself, Powell implied, seemed to have little use for the desert. The clifftop plateaus high above the river were home to majestic elk and noble eagles, but the desert below was a kind of nightmare zoo where "rattlesnakes crawl, lizards glide over the rocks, tarantulas stagger about, and red ants build their play house mountains." Occasionally a scrawny rabbit might flit by, chased by a mangy wolf, "but the desert has no bird of sweet song, and no beast of noble mien."

"The whole country is utterly worthless to anybody for any purpose whatever," Sumner concluded flatly, "unless it should be the artist in search of wildly grand scenery, or the geologist, as there is a great open book for him all the way."

Powell, a bit of an artist and more than a bit of a geologist, found himself as intrigued as Sumner's words implied. He tried, with only sporadic success, to convey his fascination to his crew. On June 23, their first day out of the most recent canyon, most of the men stayed in camp to patch up the boats, which were leaking again. Powell and Bradley hiked off to examine a puzzling fold in the rocks and hunt for fossils.

The others preferred faster-moving quarry. "Wonderful to relate," Bradley noted sarcastically, the hunters finally succeeded. Hawkins brought down a deer, a fine, fat buck that had been standing two thousand feet above the river. "He stopped to take a look at me," Hawkins recalled, "and I shot just as he stopped and broke his neck." With Goodman carrying one of the forequarters, the proud hunter managed to bring both the deer's hindquarters to camp. They left the other forequarter hanging in a tree, thinking they might fetch it later. To celebrate his hunter's triumph, Powell named the spot where he had bagged his deer Mount Hawkins. The others, in hopes of game and glory of their own, all clamored for a crack at the wildlife. "The men are on tiptoe, and each swears by everything he can name that some *little innocent* deer must die by his hand tomorrow," Bradley teased. "We shall see."

The excitement over Hawkins's kill showed how heartily sick everyone was of beans and bacon and rice, the drab fare that Powell archly called "our *cuisine*." But as welcome as it was, fresh meat from a single deer was only a diversion. Divided among ten hungry men, it would vanish in a day or two. "Have spread the rations to dry and find one sack of rice spoiled," Bradley had observed only the day before, but then he had waved the problem of short supplies aside. "We are glad to get rid of it, for our boat is too much loaded to ride the waves nicely but is all the time growing lighter as we eat the provisions."

The day after Hawkins' successful outing was spent near camp. As usual, Powell set out exploring. Bradley, convinced that the hunters had succeeded only by a fluke, headed up the cliff to bring in the deer's forequarter. He found it, untouched, at a spot that the barometer showed was 2,800 feet above the river rather than the 2,000 feet they had guessed at. "I am so used to climbing . . . now that I hardly notice it," Bradley observed, "yet it came very hard at first."

After climbing the cliffs, Bradley had taken the opportunity to see what the river had to offer. The view was encouraging if you looked far downstream but worrying if you focused closer in. "The river about four miles below here cuts this . . . mountain chain in two and comes out on the other side," Bradley noted. That was not how rivers were supposed to behave—rivers did not slice mountains apart—but once the Green reemerged on the far side, it seemed to settle down and behave. "We can see for 50 or 60 miles that it is all valley and island covered with cottonwood groves and the cañon cannot be very long," Bradley wrote hopefully.

Oramel Howland compared the colors of the cliffs to "a muddy looking rainbow." Powell, who saw the world through rosier lenses, was more effusive. "The park is below us," he wrote from a cliff-top perch, "with its island groves reflected by the deep, quiet waters. Rich meadows stretch out on either hand, to the verge of a sleeping plain, that comes down from the distant mountains. These plains are of almost naked rock, in strange contrast to the meadows; blue and lilac colored rocks, buff and pink, vermilion and brown, and all these colors clear and bright."

From his vantage point high above the river, Powell could see the canyon they had just floundered through. They had named it Whirlpool Cañon. For Powell it was something out of a Gothic novel, "a gloomy chasm, where mad waves roar." But their new camp, which they had named Island Park, was a haven that dispelled any melancholy thoughts. From high above, with the sun beating down and the river seeming "but a rippling brook," it was easy to hope that the worst danger lay behind them.

They set out eagerly the next day, knowing that it was not far to their first true milestone, the junction of the Uinta River and the Green, where they planned to stop for several days. They would be in the Uinta Valley, territory they all knew well. It was, moreover, as much of a crossroads as any place in this lonely corner of the world could be, for the valley marked one of the few places where the Green could be crossed. Nearly a century before, in September 1776, Spanish missionaries, guided by Indians, had made the first documented crossing of the Green near this river junction. They had been searching (futilely, it would turn out) for a new route from Santa Fe to California. In the nineteenth century, wagon trains heading West had crossed the Green at the same spot.

Powell and his men would be sticking with the Green, not fording it. For them, the beckoning prize was the Uinta Indian Agency, a small headquarters for a newly established reservation. The area "reserved" for the Indians stretched across two million bleak and barren acres for which the federal government could imagine no use. The immense tract of land, Brigham Young had assured his followers, was "one vast contiguity of waste."

The agency consisted of a couple of wooden huts and a handful of employees. But for Powell and his crew, who had not seen another human being since their departure from Green River Station a month before, it offered the promise of a link with the outside world. They could replenish their supplies, but that was the least of it. Like campers or soldiers or prisoners, the men longed for a chance to read letters from home and send heartfelt replies.

Before they could take that welcome break, Powell and his men would have to negotiate the curious geological beast that Powell would later name Split Mountain Cañon. Here the Green ran headlong into a mountain ridge, "splitting [it] for a distance of six miles nearly to its foot," in Powell's words.

Since early in the trip, the Green's course had perplexed them all. The river seemed to flow without any design whatever, violating laws of nature and common sense willy-nilly. Oramel Howland decided that the Green acted not out of ignorance but out of malevolence. "The river seems to go for the highest points within the range of vision," he wrote, "disemboweling first one and striking for the next and serving it the same, and so on, indefinitely." The Green tore into a mountain at one point, made one sharp turn after another, "whirling, splashing and foaming as if in fury to think so tiny an obstacle should tower 3,000 feet above it to check its progress."

For Howland, bright and curious though he was, this was a puzzle more than a preoccupation. Powell's interest, in contrast, was anything but casual. The prospect of investigating just such geological riddles was one of the lures that had spurred him to risk his life in canyon country in the first place. For the time being, though, he would have to put such pleasures to one side.

On June 25, Powell and the men set out at seven in the morning. "We enter Split Mountain Cañon," Powell wrote, "sailing in through a broad, flaring, brilliant gateway." They ran two or three rapids and portaged two more but then decided to stop for the day because Sumner had fallen ill. (Sumner was no more inclined to coddle himself than anyone else. "One of the men sick," he noted curtly in his journal.) By evening, he felt better. Powell had spent the previous couple of days not quite well himself.

The next day began with a portage, which was as miserable as ever, and then continued with a portage in the rain, which was worse. While the men worked, Powell climbed the hills looking for fossils. "Spent two hours to find one," Sumner complained, "and came back to find a peck that the men had picked up on the bank of the river." By mid-afternoon, the portages were complete. Four fast river miles brought everyone out of the canyon, and then, wrote Sumner, "all at once the Great Uinta Valley spread out before us as far as the eye could reach. It was a welcome sight to us after two weeks of the hardest kind of work, in a canyon where we could not see half a mile, very often, in any direction except straight up."

Spurred by the sight of their goal, "all hands pulled with a will," Sumner wrote, "except the Professor and Mr. Howland. The Professor being a one-armed man, he was set to watching the geese"—in the hope of roast goose for dinner—while Oramel Howland was exempt from rowing because he had to map the country they were passing through. Sumner's brief mention of Powell's injury was rare. In all the first-person accounts of the expedition, including Powell's own, this passing remark was one of perhaps half a dozen such comments.

The slyly mocking tone of Sumner's references to "the Professor" was telltale, a not-so-subtle reminder that it was Sumner, not Powell, who truly knew the West. In the next sentence, Sumner changed the allusion but not the message. "Our sentinel," he wrote, "soon signaled a flock of geese ahead."

For once the hunters, who had come in for so much teasing, aimed true. When the crew made camp shortly afterward, they carried ten fresh-killed geese. Bradley managed a perfunctory grumble (the geese were "very poor at this season"), but everyone was in high spirits. The men calculated that the day's run had brought them thirty miles, and Sumner was nearly well again. Bradley even seemed to have taken on a bit of Powell's fondness for geology. "Found a fine lot of foccils in the last cañon and have added three new varieties to our number and found them in great abundance," he noted contentedly.

The next day, June 27, lived up to the promise of its predecessor. The boats sped "down a river that cannot be surpassed for wild beauty of scenery," Sumner wrote, "sweeping in great curves through magnificent groves of cottonwood. It has an average width of two hundred yards and depth enough to float a New Orleans packet. Our easy stroke of eight miles an hour conveys us just fast enough to enjoy the scenery, as the view changes with kaleidoscopic rapidity."

Only two weeks before, the men had spent a day of killing labor and limped into camp in the evening with only three miles' progress to show for it. Now they reckoned they had covered sixty-three miles in a single day and, as a bonus, had bagged eight more geese. Though hardly the fat, domesticated birds of a Dickensian Christmas feast—these geese were scrawny, wild, half-grown things with puny pinfeathers that left them unable even to fly to safety—the birds provided diversion, if not quite sport, and they had the great virtue of not being bacon.

The "easy stroke" downriver that Sumner celebrated was not quite as easy as all that. For weeks, the expedition had been bedeviled by rapids; now the men were finding that rowing in flat water was perhaps as bad, if not as dangerous. As the valley broadened, the river grew wide and sluggish. "We have had a hard day's work," Bradley wrote, "which comes harder to us than running rapids." And though Bradley conceded that the valley they were traveling through was indeed "beautiful" and filled with green islands, he pointed out that the bluffs were still "dry and barren" and the camp was swarming with mosquitoes.

It would take more than mosquitoes to mar the big picture. The expedition had set out a month and three days before. They estimated that they had already come roughly 350 miles, perhaps one-fourth or one-third of the distance to their final destination. (In fact, they had covered 258 miles. The next major milestone, the point where the Green and the Grand joined to form the Colorado, was another 245 miles downstream.) A progress report on the journey so far, an outsider might have guessed, could have gone either way. On the one hand, the expedition had made it safely for hundreds of miles, through countless rapids, with everyone not only alive but cheerful. On the other hand, they had already lost one boat, three men had nearly drowned, game was hard to find, the food had begun to spoil, and there was every reason to fear that the worst part of the journey had yet to begin.

But as Powell and his crew approached this first milestone, with its promise of news from the outside, no one seemed ambivalent. Not even the ever-wary Bradley could hide his optimism. "We must be very near the Uinta River which everybody said we could never reach," he crowed, "but everybody will be mistaken for we are nearing it so fast and so easily that we are certain of success."

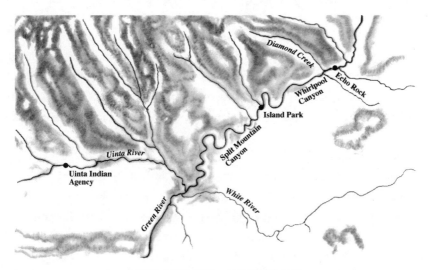

CHAPTER TWELVE

HOAX

At about three o'clock on the afternoon of June 28, Powell and his men pulled ashore near the mouth of the Uinta River. They made camp under a large cottonwood tree on the Green's west bank. This was open country and congenial scenery. Across the river lay "a splendid meadow," Bradley wrote, ". . . without exception the finest mowing land I ever saw, as smooth and level as a floor and no rocks."

As pleasant as the Uinta Valley was, especially in contrast with the barren canyons they had been traveling through, the men had little interest in looking around. Wagon trains en route from Denver to Salt Lake City had once crossed the Green here, in low water, but Sumner noted that the wagon trail was "not much of a road" and lately seemed to have been neglected by all travelers except "wolves, antelope, and perhaps a straggling Indian." Tired but restless, the men were induced to linger only by the prospect of letters from home. "Hope to receive a good lot and think I shall," Bradley wrote.

All the men started in on letters of their own. Andy Hall wrote to his brother. Young Hall sounded as breathless as a teenager who had just zoomed upside down and backward on a six-story roller coaster. "We had the greatest ride that ever was got up in the countenent," he exclaimed. "the wals of the canone where the river runs through was 15 hundred feet in som places."

Powell, Oramel Howland, and Sumner wrote personal letters and also composed accounts of the trip to send to newspapers. The newspaper stories served a twofold purpose. They delivered the welcome news that the men were still alive, and they shone the light of publicity on the expedition. For Powell, who had high hopes that this venture would launch his career, the public relations value of the story was not to be slighted. Powell wrote up the trip himself, and Howland (a printer and editor) happily set to work on a letter to his colleagues at the *Rocky Mountain News*. Sumner was a harder sell. In response to Powell's insistent arm-twisting, he grudgingly worked up his notes into a story for the *Missouri Democrat*. "I have written this with many misgivings," he complained, "being more used to the rifle, lariat and trap, than the pen."

He signed his account "Jack Sumner, Free Trapper," the title a proud declaration that Sumner was not a hired drone on a company payroll but a freelancer who survived by his own wits and skill. Powell sent a brief note with Sumner's diary to smooth its way into print. His letter gives a hint of the patronizing attitude that the men often complained about. "I send manuscript journal of one of the trappers connected with the Colorado River Exploring Expedition," Powell wrote. "I think you will find them somewhat lively, and may be able to use them. Of course they will need 'fixing' a little, may be toning somewhat. Jack Sumner, the writer, has seen much wildlife and read extensively. He has prepared the manuscript at my request. Should you conclude to publish he will send more."

That is the entire letter. It contains nothing overtly objectionable—the tone of apology, which Sumner would have resented, was at least in part a business letter convention—but it does seem to betray a telltale lack. What is missing is any hint that Powell saw his men as more than hired hands. Sumner was, after all, crucial to the expedition's success. The lead man in Powell's own boat, he was responsible (with Dunn) for bringing Powell and the *Emma Dean* through the rapids right side up and in one piece. After Powell and Walter, Sumner was the highest-ranking man in the expedition. He and the Major spent hours together every day drenched by the same waves and threatened by the same rocks. Daily he risked his life (as did all the others) for the expedition as a whole. At Disaster Falls, he had set out alone to rescue the Howland brothers and Goodman from the island where they were marooned. Perhaps he could be excused for chafing a bit at "the Professor" for taking him for granted.

Powell was the head of a team of proud and touchy men. By now, they were not only proud and touchy but tired and wet and hungry. A prudent leader would have gone out of his way to make sure that they did not feel unappreciated as well.

That was not Powell's strength. Strong-willed and self-confident, he took for granted that others felt as he did. On June 29, he wrote a letter to a colleague at Illinois State Normal University. "The party has reached this point in safety and having run 4 cañons of about 25 miles in length each . . . ," he began. Then came a quick mention of some of the hardships they had overcome, and then an utterly characteristic coda. "Personally, I have enjoyed myself much, the scenery being wild and beautiful beyond description. All in good health—all in good spirits, and all with high hopes of success." As telltale as the upbeat tone was Powell's automatic assumption that what he felt "personally," "all" the others felt as well.

While Powell and his men sat writing cheery letters home, the nation awoke to startling news. "It is reported that the Powell exploring expedition was lost in the rapids of the Colorado river, with the exception of one man, who has come to Green River City," the St. Louis Democrat announced on June 28. "He stated that he had not embarked with the party in the rapids, but followed along the banks and saw the party perish."

As with many rumors, the story of Powell's drowned expedition had its roots in a true event. Just after Powell and his men launched their boats from Green River Station, a man named Theodore Hook (in some accounts his name is given as H. M. Hook) set out down the Green with an exploring party of his own. What a one-armed tenderfoot from back East could do, he could do better.

Hook was mistaken. Barely under way, in the chasm that Powell had named Red Canyon, he perished in a rapid. The expedition collapsed, and the rest of the party scrambled for home. Powell and his crew had taken only a week to reach Red Canyon. Word of the drowning spread at once, and somehow a true story about a prospecting party led by a man named Hook became a false story about an exploring expedition led by Powell.

Two years later, another expedition found grim evidence of the tragedy in Red Canyon. "I do not know how far they expected to go but this was as far as they got," wrote one member of the 1871 group. "Their abandoned boats . . . still lay half-buried in sand on the left-hand bank, and not far off on a sandy knoll was the grave of the unfortunate leader marked by a pine board set up, with his name painted on it. Old sacks, ropes, oars, etc., emphasized the completeness of the disaster."

As the rumor that Powell and his men had been lost flashed along the telegraph lines, it grew and mutated. On July 2, the *Omaha Republican* reported that nine of the ten members of the Powell expedition had drowned two weeks before. For "the particulars of their sad fate," the *Republican* relied on the report of a trapper and Indian fighter named William Riley, who had been told the story by the expedition's lone survivor. Riley had met this bereaved man, who gave his name as John Sumner, at Fort Bridger in Wyoming.

The crisis supposedly came at a spot "Sumner" called Hell's Gate, apparently somewhere below Brown's Hole. "The water is precipitated down this gorge at a velocity of forty miles per hour," the *Republican* reported, "hence not the shadow of a chance for escape presents itself." This was not a challenge to an intrepid explorer but an invitation to suicide. "No boat ever built could possibly pass over these falls without being shattered to atoms."

The *Republican* hurried through the rest of the story, as if saddened to linger over such testament to human folly. Powell, the *Republican* noted regretfully, had been bent on glory no matter the cost. "Major Powell trusted too much to his long experience and superior intelligence . . . Warned time and again by those familiar with the nature of his route, he turned a deaf ear to every voice and pressed on. For the judgment of his followers we can only say that they are deserving a monument for their fidelity and the steadfast manner in which they confided and clung to him, as if, indeed, he was the great oracle to guide them on."

What was Powell's story but the sadly predictable tale of a leader with a tragic flaw? "Our account is soon told. Ambition had a strong hold upon reason. Judgment was laid aside, and the Napoleonic Major, with his brave band of faithful companions, saving one who was ordered on shore . . . made every preparation, and then entered death's portals—the awful, treacherous portals of Hell's Gate."

Even in an era that preferred its rhetorical flourishes laid on with a trowel, this was a bit much. But the *Republican* was not quite done. No Fourth of July orator celebrating the patriots who had sacrificed their lives for their country could have outdone the fervor of the *Republican's* tribute to the misguided explorer and his loyal men. "We can only say, they must have died as they had lived—heroes all; yielding up their spirits with the same quiet indifference and pure faith manifested during the horrible descent of the rapids in Brown's Hole."

Tragedy draws crowds. The supposed loss of the expedition drew far more attention than its launching ever had. Emma Powell, whose faith in her husband was limitless, seems not to have given the story a moment's credence. From her Detroit home, she set out at once to shoot the rumors down. Yes, John Sumner was with her husband's expedition, Emma told the newspapers, but this supposed "Sumner" was highly unconvincing. "The John Sumner of the expedition in such a case would, I believe, have made his way or reported to Chicago or Detroit immediately. Sumner is a most reliable man and brother-in-law to William Byers, editor of the *Denver News*." The last remark was a pointed hint that readers in search of truth rather than sensation might wait to see what Byers's newspaper, properly called the *Rocky Mountain News*, had to say. (This was also the paper that employed Oramel Howland.)

The *Rocky Mountain News* did quickly contact Fort Bridger and learned that no one from the Powell expedition had ever come to the fort. So much for "Sumner."

But a lie, Mark Twain remarked, can travel halfway around the world while truth is putting on its shoes. The first disaster story was soon superseded by one even more outlandish, based on a *second* sole survivor. The new man, who may have heard "Sumner's" story, was called John A. Risdon. He spun a tale that left sobbing audiences pressing money on the devastated hero to pay his train fare home to Illinois.

Risdon's story sped across the nation's telegraph wires and into its newspapers. The *New York Times* carried long accounts, reprinted from Western papers, under such headlines as "The Loss of the Powell Expedition," and big-city dailies across the country did the same. Small papers plucked unattributed snippets from the bigger papers. The tiny *Appleton* (Wisconsin) *Post*, for example, reported that "Gen. Powell and 15 others, on an exploring expedition, crossing the Colorado River, was drawn into a whirlpool, and sunk forever. A man who was left to watch the mules, saw the sad catastrophe and he alone of all the party is saved. He was 150 miles from the nearest fort, but gathered up the baggage, and hastened to make known the sorrowful event."

On July 3, the *Chicago Tribune* told the story in breathless detail. A front-page headline read "Fearful Disaster," and sub-headlines spelled out more of the grim tale. "Reported Loss of the Powell Exploring Expedition Confirmed," the first one read, and then, "Twenty-One Men Engulfed in a Moment," and "Arrival of the Only Survival at Springfield" and "His Statement of the Manner of the Accident."

The long story, overflowing with specifics, was based on an account that Risdon had provided Illinois's Governor John Palmer. (Illinois had a special interest in Powell, a resident of the state.) The lone survivor's integrity shone forth. "Mr. Risdon is an honest, plain, candid man, and told his story in a straightforward manner." For all those onlookers who had heard Risdon choke out his anguished tale, the *Tribune* noted sadly, "the fate of Major Powell's expedition is left without a doubt, and another name is added to the long roll of martyrs to science."

With the governor hanging on his every word, Risdon conjured up imaginary rivers and mountains and waterfalls, as delighted with his own creativity as a six-year-old with a new box of crayons. His trademark was an endless store of detail, which, all his listeners agreed, endowed his tale with the unmistakable ring of truth. Not only had Risdon been a member of the Powell expedition, he had served three years under Powell's command in the Civil War, in Company B, First Illinois Artillery. Not only had he been with the expedition from the start, but he could rattle off the names of everyone in the party, in many cases down to the middle initial: there was Powell himself, of course, and also a pair of brothers, William C. Durley and Charles Durley, and Andrew Knoxton and T. W. Smith and William S. Dalton, and so on. The party also included "a half-breed named Chick-a-wa-nee," who served as a guide, and two teamsters, Fred Myers and Thomas Walch, who drove the supply wagon.

As Risdon told it, Powell and his crew had reached the Colorado River on May 7 or May 8, at a point near an Indian settlement the whites called Williamsburg. On May 16 or perhaps May 18—Risdon artfully "forgot" a name or date here and there, the better to enhance his credibility—Powell and his men set out to explore two tributaries of the Colorado, the Big Black and the Deleban. Powell's boats had been left behind in favor of a large birch-bark canoe the Indians called a yawl. The entire party, except Risdon, climbed into one yawl. The river was exceedingly rough, with a fall of 160 feet in a mile and a quarter. (Niagara Falls, by way of comparison, is 170 feet.) Risdon, who had been assigned to scout the Deleban from shore, tried to convince Powell not to challenge the river. "But Major Powell said laughingly in reply: 'We have crossed worse rapids than these, boys. You must be getting cowardly. If seven or eight men cannot paddle us across there, we will have to go under.'"

On they went, in a scene straight from a boys' adventure magazine. Twenty-five men in a single canoe "pushed out into the river with three hearty cheers, using seven paddles, the Major standing in the stern steering."

Risdon waved his hat at his departing friends. "You must be back in time for dinner," he called heartily, "for I will have a good lunch for you when you return."

The men knew better. "Goodbye, Jack," they called back. "You will never see us again."

A moment later Risdon saw the boat "commence whirling around, and like a living thing dive down into the depths of the river with its living freight, Major Powell standing at his post." Alone on shore, the only survivor gaped in horror. Even afterward, in safety, it was painful to conjure up that awful time. "For two hours," Risdon recalled, "I lay on the bank of the river, crying like a baby."

Risdon added still more detail. He told how he had searched the river for any sign of the drowned men, and how he had spotted a bag and dived into the river to rescue it, and how it had miraculously turned out to contain Powell's notes. Risdon had searched for another four long days. Sick with despair, he had reluctantly abandoned his quest. For eight days, he walked alone through empty country before reaching a military post where he finally delivered his dreadful news.

It was a fine yarn, except that not a word of it was true. (It was a mark of how little known canyon country still was that newspapers across the country described Risdon's imaginary geography without a second thought.) There was no guide named Chic-a-wa-nee, no Big Black or Deleban River, no canoe called a yawl, and no hapless crew of twenty-one or twenty-five squeezed into a single boat.

Skeptics surfaced almost at once. The specifics that had done so much for Risdon's credibility now conspired to trip him up. Powell's mother, in Wheaton, Illinois, told the *Chicago Evening Journal* that she did not believe a word of Risdon's account. Risdon had put Powell on the Colorado on a date when he was in fact just boarding the train in Chicago. Emma Powell was even more emphatic. "To one at all acquainted with the plans, aims and minutiae of the expedition," she wrote, "the whole story is glaringly false, and betrays entire ignorance of the matter."

A few days later, Emma presented her case in full. Risdon said he had been with Powell since July 10, 1865, but Powell had not made his first trip West until 1867. Emma knew several of Powell's men, but she had never heard of James A. Risdon or any of the other purported crew members. Risdon said the men had drowned on May 16 or 18, but Emma had a letter from her husband, sent from Green River Station, dated May 22. Four days *after* his supposed drowning, Powell had not even set out down the river.

Newspapers that only a day or two before had hailed Risdon as a hero now competed to condemn him. On July 6, Powell's hometown paper, the *Bloomington* (Illinois) *Pantagraph*, ran a story under the headline "The Pretended Loss of the Powell Expedition." Risdon's tale was "wholly false," the *Pantagraph* thundered, "the invention of a liar or a crazy man." The *Detroit Tribune* declared itself "justified in saying that the report of this man Risdon, beyond all reasonable doubt, is a tissue of fabrications from beginning to end." The *Rocky Mountain News* went even further. "Risdon ought to be hung."

In another two weeks, any lingering doubts were resolved. The letter Powell had written to an Illinois State colleague and mailed at the Indian Agency had finally arrived. It was immediately turned over to the newspapers. "Camp at Mouth of Uintah, June 29, 1869," the drowned man began cheerily. "Mr. Edwards: My dear sir: The party has reached this point in safety having run four cañons of about 25 miles each . . . All in good health—all in good spirits, and all with high hopes of success."

Meanwhile, his sad duty done, Risdon had bid the governor farewell and headed home. He stopped just long enough on the way to steal a horse from one man and some blankets from another. As inept a thief as a con man—he had already served one jail term for horse theft—he was quickly caught and almost as quickly recognized. Less than a month after it had begun, the career of the "last survivor" had ended, and John A. Risdon, alias Miller, alias Clark, stepped out of history's pages and into a Logan County, Illinois, jail cell.

Powell had hoped that his newspaper letters would find readers. Now his story had been trumpeted across the land. A nation hungry for heroes since the Civil War had found a new one. A public focused on the Western frontier, where the golden spike had been hammered home only the month before, had a reminder that much of the West was still wilderness, unknown and perilous. Alone in an empty desert and cut off from the outside world, Powell and his men had no idea that the country was buzzing about their fate.

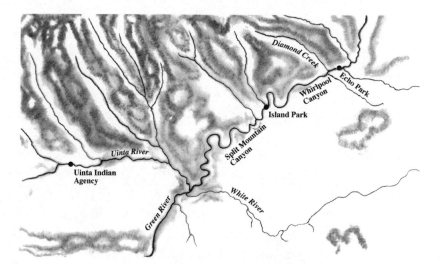

CHAPTER THIRTEEN
LAST TASTE OF CIVILIZATION

Mark Twain would not write *The Adventures of Tom Sawyer* for another seven years, but the overheated news coverage of the last moments of the Powell expedition anticipates the famous scene where Tom and Huck Finn, supposedly drowned, reappeared at their own funeral. If they managed to emerge alive from the Grand Canyon, Powell and his men might eventually have a chance to enact a similar coup. But first they had to survive.

The men whose agonized "deaths" had captured the nation's attention had in fact been suffering no torment except boredom. The expedition had temporarily split in two; five men had set off to the Indian Agency (near present-day Whiterocks, Utah) to exchange letters and buy supplies, while the other five stayed in camp, surly and moping. Walter Powell and Andy Hall had been the first to leave for the agency. It was a forty-mile trek, but forty miles on foot across the desert beat sitting in camp with nothing to do. (They had tried rowing up the Uinta River but the current was too swift to fight.)

Powell spent two days in camp pinning down his latitude and longitude, and then headed for the agency, too, along with Hawkins and Goodman. "It is a toilsome walk," Powell wrote, "twenty miles of the distance being across a sand desert." It was indeed a long slog and the men found themselves having to wade across the river several times, but it was not quite as arduous as Powell implied. "At noon met Hall coming for me with pony," Powell noted in his river diary, although he left Hall and the pony out of his published account of the expedition.

The remark about the pony is not especially significant in itself. But it marks a new stage of our tale, for it comes from the first entry in Powell's river diary, dated July 2. (If Powell kept a diary for the first month of the trip, it has never been found.) From this point on, as in the case of the disappearing pony, we will have the chance to compare the river diary that Powell wrote by the campfire each night with the more elaborate version of his story that he published years later.

The five men who had gone to the agency at least had a chance to see new faces and new scenery. ("It is rather pleasant to see a house once more, and some evidences of civilization, even if it is on an Indian reservation," Powell noted.) The forlorn men stuck in camp—Bradley, Sumner, the Howland brothers, and Dunn—had little to do but kill time and battle mosquitoes. "We feel lonesome," Bradley confessed, and the mosquitoes screaming around made life miserable. Bradley kept a fire going in front of his tent. The smoke helped keep the bugs away, but he complained that he had not seen mosquitoes as bold and as numerous as this even in Florida. These were muggers, not mosquitoes. "One of the men says that while out on the shore of the lake a mosquito asked him for his pipe, knife and tobacco and told him to hunt his old clothes for a match while he loaded the pipe."

Bradley tried berry picking to pass the time. Resolved to outlast the mosquitoes, he fashioned an elaborate costume. He draped a swath of netting over his hat and head and cinched it around his waist, donned gloves to protect his hands from bites, and completed his ensemble with a pair of knee-high boots to foil rattlesnakes. He emerged triumphant, with three quarts of currants.

The days in camp dragged on. Powell's expedition to the Rockies the year before had hit a similar dead spot. "I did nothing today," one of the crew had written then, "and the other boys helped me." By July 3, their sixth day stuck in the same spot, Powell's men felt just as antsy. Bradley, who had whiled away as much time as he could cleaning rifles and pistols, turned his thoughts to finding other entertainment.

With Hawkins, the cook, away at the agency, Bradley decided to fix a pot of beans. "Put on what I thought we could eat and set them to boiling," he wrote. "They boiled and swelled and I kept putting in hot water until I had a large bread-pan that will hold ten or twelve quarts solid full of beans." With the pan full to the brim, there was no space for water, so the beans scorched "a little." The rest of the crew, in the meantime, had set off duck hunting and fishing and currant gathering, partly to have something to do and partly to obtain the makings of at least a rudimentary Independence Day dinner. "We shall have a poor 4 of July anyway," Bradley fretted, "and we must make the best of it."

The Fourth bore out Bradley's glum forecast. Powell and the others had yet to return from the agency, which made for some anxious thoughts that something had gone wrong. Worry soon gave way to self-pity. Bradley, a man sentimental and tough in equal measure, tormented himself with thoughts of home and hearth. "Three successive 4ths I have been in the wilderness . . . ," he lamented, recalling the woes of his army days. "Where shall I be next 4th of July? Took a long walk tonight alone . . . and thought of home, contrasted its comforts and privileges with the privations we suffer here and asked myself why am I here? . . . With moistened eyes I seek again my tent where engaged with my own thoughts, I pass hours with my friends at home, sometimes laughing, sometimes weeping until sleep comes and dreams bring me into the apparent presence of those I love."

Bradley wasn't the only one taking stock. On July 5, at the agency, Frank Goodman told Powell that he had decided to leave the expedition. Goodman had nearly drowned at Disaster Falls and had lost all his gear and all his clothing except the underwear he was wearing. He had set out in search of adventure. By now, he told Powell, he had "seen danger enough." The men liked Goodman, but they were matter-of-fact about his departure. Bradley noted simply that he was "one of those that lost everything in the wreck," and Oramel Howland described him only as "one of the wrecked party [who] leaves us here." (Howland had been a member of the wrecked party himself.)

Sumner was perhaps a shade more sympathetic. "Goodman, having had all the experience his health called for, stopped at the post," he recalled later. "He had had several close calls, and possibly ran out of nerve. He was a fine singer of sea songs, and we missed him around the evening camp." Powell focused on the practical. "As our boats are rather heavily loaded," he observed, "I am content that he should leave, although he has been a faithful man."

Now they were nine.

Despite Goodman's departure and Bradley's melancholy, morale was not at all bad. Hall was the cheeriest of all. An excursion down a wild and untamed river was a rollicking good time, the kind of adventure a person didn't find every day. And unlike the surly mountain men, Hall had no quarrel with Powell's leadership. "The major is from near Blumington, Ill.," he wrote his brother. "I suppose you never herd off him and he is a Bully fellow you bett."

Powell himself felt pleased with how things were going. In the same letter where he had told his friend and colleague Edwards that he was alive and well (and thereby put an end to Risdon's hoax), he had delivered still more good news. "The boats seem to be a success; although filled with water by the waves many times, they never sink, the light cabins at each end acting well as buoys." Not sinking was indeed one criterion of success, although hardly a demanding one. By this time, Powell knew that the boats were not only hard to sink but hard to maneuver as well, and he suspected that the river would grow worse in the weeks ahead. Even so, Powell's optimism was genuine and unforced. He had the natural leader's knack for automatically framing every situation in such a way that some bit of welcome news dominated the picture. Gloomier possibilities, if they could be glimpsed at all, were relegated to the background.

But there was no denying how hard the trip had been. Even the rambunctious Hall hoped that the river had no surprises left. "i think that we ar now through the worst off the water now." Sumner held off from guessing about what lay ahead. "So far," he noted simply, "we have accomplished what we set out for."

On July 5, the same day that Goodman resigned, everyone else returned from the agency. The visit had been a disappointment. The reservation was nearly empty—most of the Indians had gone to see the newly completed transcontinental railroad, and Pardyn Dodds, the agency's top man, was off to Salt Lake City. What was worse, Dodds had gone to buy supplies because the agency had run low. Powell brushed aside the discouraging news (though even if there had been ample supplies on hand, he had little money to pay for them), but there was no denying that this was a blow. Nor, despite their hopes, did anyone receive mail from home. Bradley had been even more eager than the others for "letters from home and friends," and his diary entries for July 5 and 6 were brusque and grumpy.

Mail would have been a bonus but food was essential, and the agency represented the expedition's last certain chance to stock up. Powell bought what he could afford and managed to barter some of the expedition's coffee and tobacco for a bit more. In the end, Bradley noted, "[Powell] has got 300 lbs. flour which will make our rations last a little longer." Sumner was more caustic. "Major Powell was gone five days, and brought back a shirt-tail full of supplies," he complained later. "I thought at the time it was a damned stingy, foolish scheme, as there was plenty of supplies to be had, to bring back such a meagre mess for nine to make a thousand-mile voyage through an unknown canyon, but as I wasn't boss I supposed I ought to keep still about it."

Sumner's remark about "plenty of supplies" was unfair, but everyone shared his concern about whether they had enough food. Even Powell acknowledged that they might be headed for trouble. "Our rations . . . have been wet so many times that they are almost in a spoiling condition," he noted. "In fact we have lost nearly half now by one mishap or another."

At this juncture, as the men return to their boats, we should take a moment to consider one of the most vexing questions of the entire saga. It sounds arcane, but it is deadly practical. The question is this: Did Powell's boats have a steering oar at the stern, so that one man (facing forward) could steer while his two backward-facing companions rowed?

The great majority of accounts of Powell's 1869 expedition, even the best ones, simply take for granted that Powell's boats had steering oars. So do all the films, and even such quasi-official sources as a U.S. stamp issued in 1969 to commemorate the Powell centennial. That stamp, labeled "John Wesley Powell 1869 Expedition," depicts two men rowing and Powell himself at the steering oar.

A steering oar (also called a sweep oar) would have been an immense help. The man at the sweep would have been facing forward, so that he could have seen the hazards lying in wait downstream. More to the point, he would have had a chance of avoiding them. Picture, in contrast, the plight of two men at the oars of a heavy, hard-to-maneuver boat without a sweep, pulling with all their might into a rapid they could only glance at.

With so much in favor of steering oars, it seems that Powell's boats *must* have had them. Brad Dimock, a white-water historian, has rowed modern replicas of Powell's boats. Such replicas have been built more than once, and all the replicas have had sweep oars. "They're much more difficult to run without a sweep," Dimock says. "I mean really *much, much* tougher, to the point that it's almost completely impossible. And that's an interesting point, because Powell started without them."

To crew a sweep-oared boat, you need three men, one at the sweep and two at the oars. "But when Powell started out," Dimock goes on, "he had ten men and four boats, and you do the arithmetic on that and you don't have sweep oars." At the start of the Powell expedition, only the *Emma Dean* and the *No Name* carried three men.

Sumner and Dunn were at the oars in the *Emma Dean*. Powell was the third man, and in theory he might have been able to handle a sweep, but if it happened, neither he nor anyone else ever saw fit to mention it. For a man of Powell's bravura, that seems unlikely. (Powell's problem would have been keeping his balance while being buffeted by ten-foot waves. To keep from being washed overboard, a boatman with two arms would hold the sweep with one hand and a strap of some sort with the other. With his good hand on the sweep, Powell would have needed to rig some kind of loop that could have supported him by the stump of his amputated arm.)

The *No Name*, too, carried three men, Goodman and the Howland brothers. Oramel Howland, though, had his maps to prepare. Howland rode downstream, Sumner wrote, "perched on a sack of flour in the middle of one of the large boats, mapping the river as we rowed along." If the *No Name* did have a sweep oar, then, as in the case of the *Emma Dean*, no one ever said so. That would be a curious oversight, especially considering all the talk about the *No Name* when the boat was lost at Disaster Falls.

But Dimock's remark about Powell starting out without sweep oars raises a broader question—did Powell's boats *ever* have sweep oars? After the *No Name* sank, after all, the arithmetic worked in Powell's favor. Then there were ten men (or, after Goodman left, nine) and only three boats. These were resourceful, quick-thinking men. Seeing the predicament they were in, did they cut a twelve- or thirteen-foot-long sweep oar from a promising piece of driftwood and fashion an oarlock for it?

Regan Dale, a highly regarded river runner with many years' experience, feels certain that Powell had sweep oars, mainly because he cannot imagine surviving the rapids without them. Dale rowed Powell replicas for IMAX in 1983 and for German public television in 1999. He recalls the experience with a kind of wary respect, the way you might remember basic training. "They're definitely not the kind of boats that you'd want to do very much white water in," he remarks. "They'd barely pivot. They went in a straight line really well, but they didn't pivot at all, so it was really hard to hit waves straight on." The boats hardly pivoted, Dale emphasized, even *with* a sweep oar. "Without the sweep," he says flatly, "it'd be impossible to turn them."

Dimock and Dale are stars of the river-running fraternity. When boatmen of their skill and experience say they cannot conceive of navigating rapids in Powell's boats without sweep oars, it becomes hard to imagine that nineteenth-century novices could do it.

And so we are confronted with two possibilities. One is to reason, with Dimock and Dale and many other boatmen, that Powell must have hit on the steering-oar idea sooner or later. (But then why did Sumner describe Howland riding downriver perched in splendor on his flour sack, weeks beyond Disaster Falls?) The second alternative is simpler. Perhaps the river really was next to unmanageable in long, heavy boats without sweep oars. Perhaps that is why Powell portaged and lined whenever he could and why, in the many rapids that he found himself forced to run, he floundered out of control so often.

We cannot resolve the uncertainty by turning to drawings or photographs of the 1869 boats because, as noted earlier, no such depictions exist. (One major source of confusion is that the earliest photographs of boats on the Green and Colorado, from 1871, *do* clearly show steering oars.) And the journals, which one might hope would clear up the mystery, leave matters ambiguous. Neither Bradley nor Sumner ever mentioned sweep oars. Nor did Powell in his river diary, nor any of the others in the letters or newspaper articles they wrote at the time. (On the other hand, we know from accounts written many years later that Powell wore a life jacket, and no one ever mentioned that at the time either.)

In the river-running community, a handful of historically minded boatmen and a few white-water historians have mulled over the sweep-oar question for years. There is no consensus, but many on both sides are big names in their fields; the pro-sweep and anti-sweep camps could each muster an all-star team. (See the notes for a closer examination of the controversy.)

But nearly all outsiders have written as if the existence of sweep oars is settled fact. That is to prejudge a question that remains defiantly open. And that, in turn, is to risk underestimating the challenges and the dangers that Powell and his men confronted. We do not know whether Powell's boats had sweep oars. We do know that even for today's best boatmen, as one of them observes, "It would have been a *tremendous* handicap not to have had a sweep."

On July 6, Powell and his men were finally ready to set out again. They pulled into the river at ten in the morning. "Was exceedingly glad to get away for I want to keep going," Bradley wrote. "Don't like long stopping." Sumner liked it even less. "After 7 days of weary, useless waiting," he moaned, "we are at last ready to cut loose from the last sign of civilization for many hundred miles."

Only two miles later, they spotted one more sign of civilization, a garden growing on a small mid-river island, with no gardener in sight. Fed up with beans, flour, and salted meat, the men raced ashore to grab what they could. Andy Hall suggested that potato tops made good greens. Powell took pains, in his account of the episode, to explain that they were no raiding party. The year before, he wrote, he and his guides had met a hunter named Johnson who had told Powell of his "intention to plant some corn, potatoes, and other vegetables on this island in the spring, and, knowing that we would pass it, invited us to stop and help ourselves, even if he should not be there."

Bradley told a simpler tale. "We had read that 'stolen fruit is sweet' and thought we would try it," he wrote. Sumner seconded Bradley. "The Professor, Dunn and Hall stole their arms full of young beets, turnips, and potatoes." The men returned to the boats and pulled to shore a mile downstream to enjoy their plunder in the shade of a cottonwood tree. Oramel Howland passed up the feast altogether, and Bradley took one taste and threw his meal away, but everyone else dug in.

Back on the river and only another mile downstream, they began to repent of their crime. "Such a gang of sick men I never saw before or since," Sumner wrote. "Whew! It seems I can feel it yet. . . . [Hall] ripped out an oath or two, and swore he had coughed up a potato vine a foot long, with a potato on it as big as a goose egg." Sumner noted, deadpan, that "Hall was somewhat given to exaggeration, and he might have stretched the matter a bit." Powell was in as rough shape as the others. The men pulled to shore, he wrote, "and we tumble around under the trees, groaning with pain, and I feel a little alarmed, lest our poisoning be severe."*

By evening they had begun to feel better. Sumner remarked that, all things considered, he "didn't think potato tops made a good greens for the sixth day of July." Bradley, as pleased with himself as a hale man on an ocean liner full of seasick passengers, noted smugly that he expected "we shan't eat any more potato-tops this season."

They set out the next morning at seven o'clock. Almost without noticing it, they had entered another canyon. The water was quiet, "with no more current than a canal," and the river cut great sweeping curves as it swung back and forth between stone walls. The open valley and the "splendid meadow" that Bradley had delighted in only days before now seemed a remote memory. The cliffs grew steadily taller. By ten o'clock, when the men stopped to take measurements, the clifftop towered 1,050 feet above the river.

Side canyons were rare here and the walls almost continuous. The expedition was truly cut off now, as if traveling through a meandering stone maze. The cliffs were perpendicular in some places, eroded into great sloping terraces in others. Powell dubbed one especially striking formation Sumner's Amphitheater.

At one point, Bradley went off on his own while the others took measurements. "I . . . put my name on a flat stone with name of expedition and date and fastened it up very strong," he wrote. "Think it will stand many years. It is the first time I have left my name in this country for we have been in a part where white men may have been before but we are now below their line of travel."

The nine remaining men of the Colorado River Exploring Expedition were well and truly on their own.

---

*Powell was right to be alarmed. Potato greens, according to the writer and river runner Ellen Meloy, are relatives of deadly nightshade and, like it, contain the hallucinogen solanine.

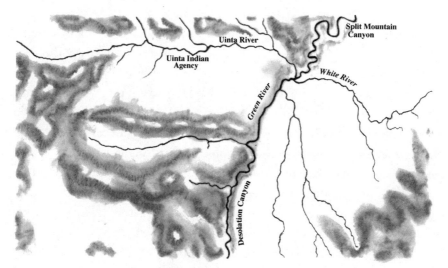

Split Mountain
Canyon

Uinta River

Uinta Indian
Agency

White River

Green River

Desolation Canyon

CHAPTER FOURTEEN

TRAPPED

The next day, July 8, brought scene after scene of discouraging bleakness. The river grew rougher, and the canyon was a study in grays and browns. Erosion had done its work, yielding great piles of broken rocks and strangely carved crags and towers. "The walls are almost without vegetation," Powell noted with dismay. "A few dwarf bushes are seen here and there, clinging to the rocks, and cedars grow from the crevices—not like the cedars of a land refreshed with rains, great cones bedecked with spray, but ugly clumps, like war clubs, beset with spines. We are minded to call this the Cañon of Desolation."

Powell and Bradley set out, as usual without ropes or other safety gear, to climb the cliffs to measure their height. The routine outing, on the morning of July 8, turned into one of the strangest episodes of the entire journey. "We start up a gulch," Powell wrote, "then pass to the left, on a bench, along the wall; then up again, over broken rocks; then we reach more benches, along which we walk, until we find more broken rocks and crevices, by which we climb, still up, until we have ascended six or eight hundred feet; then we are met by a sheer precipice."

Stymied for a moment, the two men soon found a route they could try. They made their arduous way upward, Powell in the lead. Making matters more complicated, they were carrying a barometer to measure the elevation when they finally got to the top. On the hardest parts of the climb, this made for a laborious kind of baton passing—while Bradley held the barometer, Powell inched his way upward a foot or two and then waited as Bradley passed the barometer up. Once Bradley had climbed over him, Powell would return the barometer to Bradley and scramble on a step or two farther, and so on.

With excruciating slowness, they made it nearly to the top. "Here, by making a spring," Powell went on, "I gain a foothold in a little crevice, and grasp an angle of the rock overhead. I find I can get up no farther, and cannot step back, for I dare not let go with my hand, and cannot reach foot-hold below without." This easy-to-overlook reference to "my hand" (an ordinary climber would have said "my left hand") was one of the few times Powell reminded his readers that he was not only taking on a mighty river and thousand-foot cliffs but doing it one-handed.

Picture his predicament—Powell was clinging to a rock face hundreds of feet above a river, keeping his balance only by grasping a protrusion in the rock. He could not see a nearby ledge that he could climb to, and, unlike any other trapped climber, he could not even probe the rock face for a new hold with one hand while holding on with the other.

Powell called to Bradley for help. After a moment, Bradley found a route that let him scramble to a ledge above Powell, but he could not reach him. Bradley looked for a stick that he could extend to Powell or a tree he could break a branch from. Nothing. He considered lowering the barometer case, but Powell didn't think he could hold on to it.

They needed a plan, and fast. "Standing on my toes," Powell wrote, "my muscles begin to tremble." Rock climbers call this involuntary spasming "sewing machine leg." It throws off a climber's balance and saps his strength, and it can spread to other overburdened muscles. "It is sixty or eighty feet to the foot of the precipice," Powell went on. "If I lose my hold I shall fall to the bottom, and then perhaps roll over the bench, and tumble still farther down the cliff."

Bradley devised a desperate scheme. He was dressed in only a shirt and long underwear. Stripping off his drawers, he lowered them toward Powell. They hung straight down from the ledge Bradley stood on, dangling in the air above Powell's head.

And, because Bradley's ledge overhung Powell's, the makeshift rescue line was behind the Major. Now came the key moment. To grab the drawers, Powell would have to release his handhold, lean back into space, and find the tattered underwear before he fell, empty-handed, to his death.

He let go, groped for the lifeline, grabbed it, and then hung on one-handed. Bradley struggled to haul Powell upward, like a fisherman who had hooked a monster (except that this fish prayed mightily for the line to hold). Each man concentrated all his efforts on his own grim test of strength: Powell clutched the underwear in his left hand, desperate not to lose his grip; Bradley, at 150 pounds not much bigger than Powell, at 120, strained every muscle to reel him in. Powell maintained his hold, and Bradley pulled him to safety.

The story is a true cliffhanger, and it sounds more like a scene from a dime novel or a *Perils of Pauline* film than like real life. Indeed, Powell's first detailed account of his river trip—published in *Scribner's Monthly*, one of the most popular magazines of the day—ran with a series of dramatic illustrations. "The Rescue" showed Bradley lifting Powell to safety. (Though Powell barely mentioned his amputated arm, the artist had no such qualms. He also took the liberty of conjuring up a pair of pants for Bradley to cover his nakedness.)

Powell described the adventure not only in *Scribner's* but again in his *Exploration of the Colorado*. Unaccountably, he set the story at Steamboat Rock, a hundred miles upstream and not in Desolation Canyon at all. But we know it happened, for Bradley recorded it in his diary on July 8, in the all-in-a-day's-work tone that was de rigueur whenever any of the men found himself describing his own accomplishments. "Climbed the mountain this morning," Bradley wrote, "found it a very hard one to ascend but we succeeded at last. In one place Major having but one arm couldn't get up so I took off my drawers and they made an excellent substitute for rope and with that assistance he got up safe."

With Powell safe at last, he and Bradley resumed their climb to the top as if everything had proceeded according to plan. The view from the summit can hardly have helped settle their shaky nerves. From a thousand feet above the river, the outlook was "wild and desolate," with sharp, jagged peaks in all directions. Bradley judged they were about halfway through the canyon "but not the worst half," for the rapids had grown more threatening throughout the day. Bradley ranked one of them, yet again, as "the worst we ever run."

Even so, Bradley was eager to carry on. It would be too much to say that he had come to welcome rapids, but there was no denying that running a rapid woke a man up. "It is a wild exciting game," Bradley declared, "and aside from the danger of losing our provisions and having to walk out to civilization I should like to run them all for the danger to life is only trifling."

Bradley's exhilarated tone was more significant than the words themselves. Powell's men knew by now that nearly everything was against them—they had no life jackets and the wrong boats and not enough food and too little experience. *And* they were trying to do what no one had ever done. The one thing they had going for them was courage. These were men who genuinely saw the prospect of being flung overboard into a wild river as a splendid "game." The question was how far their courage could take them.

No one could find lining or portaging rapids exhilarating. But it was up to Powell, not Bradley, to decide whether they would run a rapid or struggle around it instead, and Bradley's taste for "wild excitement" played no role in Powell's decision-making. Late that afternoon, Powell made his usual call, and the men camped at the head of another "unrunnable" rapid. Before knocking off for the day, the crew managed to line the *Emma Dean* and one of the freight boats past the white water and carried about half the supplies on a path along the shore. They left the rest for the following morning. "I should run it if left to myself . . . ," Bradley wrote. "Major's way is safe but I as a lazy man look more to the ease of the thing." Bradley had hated the army, but, for now at least, he obeyed orders like a good soldier.

It would be hard to find men less lazy than Bradley and the others. Their eagerness to challenge the river reflected impatience and growing confidence, not sloth. The men were still raw, but by now they had seen scores of rapids and had begun to break the code. They had figured out early on to look for a V-shaped tongue of smooth water and follow it downstream, and now they had detected other patterns. Solitary bits of commotion in the river—an isolated patch of churning, splashing water, say—usually meant trouble. Regular features—a line of five waves, for example—were marginally less risky.

They had learned something, too, of the ways of water and rock. Pourovers,

for example, form where the water flows over a barely submerged rock. From
upstream, the boatman sees a smooth line of water. An untutored eye might
see an invitation, a bit of order amid the whitecapped anarchy. In fact,
pourovers can form "holes," and holes can be deadly. Water passing over a rock
or a boulder is suddenly confined to a smaller "channel," so it speeds up. As it
plummets off the boulder's far edge, gravity speeds it further. Smacking into
the river, this mini-waterfall then dives downward, creating a "hole" where it
hits, and often, on the hole's downstream side, a breaking wave that can reach
fifteen or twenty feet. The true trouble begins as the river rushes to fill the
hole. The result is a kind of perpetual motion nightmare, as a steady stream of
falling water continues to re-dig the hole and the river labors just as insistently
to fill it back up. A boat or a swimmer trapped in a hole can be recirculated
endlessly, held in a remorseless, watery fist.

First, the trapped swimmer is driven beneath the river's surface by the
water crashing onto him as it cascades over the boulder. Then his natural buoy-
ancy and his life jacket spit him up to the surface, where he will try desperately to take in a gulp of air while he has a chance. A lucky swimmer may
break free, but if the hole is deep and powerful, he will likely be caught in the
backwash of water moving *upstream* to fill in the hole. Then, like a prisoner
strapped to an underwater Ferris wheel, he will find himself carried back to
the starting point and driven under the surface a second time by the water
pounding down from the rock above. The cycle repeats endlessly, while the
swimmer grows weaker and more frantic.

For the same reason, fishermen standing near the base of dams sometimes
drown. A man-made dam is perfectly smooth and regular, so the water flowing over it can form a hole with no "weak points" where a swimmer can pop
free. It is a particularly cruel death because the instinctive and desperate urge
to stay on the surface and breathe is precisely what allows the hole to retain
its grip. A swimmer trapped in a bad hole may be unable to break away while
wearing his buoyant life jacket. If he could somehow make the jacket vanish
for a split second, he might be able to sink deep beneath the surface and escape
from the hole, but the last thing he would want is to have to swim the rest of
the rapid without a jacket to keep him afloat.

A boulder standing high out of the water creates trouble of a different sort.
Here the water that hits the sides of the rock flows around it while the water
that smacks head-on into the rock's midsection "pillows" up for a moment

before finding its path around the sides. The problem is that the "pillow" extends considerably farther upstream than a beginner might guess, and the pillow behaves like an extension of the rock. A boat that nestles up against *this* pillow can find itself broadside to the current, the boat's downstream side up out of the water and its upstream side held under the river's surface by the onrushing current. The river will happily leap aboard that upstream edge, like a malicious giant jumping on a child's seesaw, and a boat can flip in an instant.

Despite the men's increasing familiarity with the river's ways, Powell continued to insist on lining and portaging. What Powell took to be prudent, his men saw as maddeningly overcautious. What they saw as bold, he took to be reckless. This was trouble. Every day brought a profusion of new rapids. If the muttering about how to deal with them grew into outright rebellion, the expedition might well fall apart.

The disagreement between Powell and the crew did not reflect differences in temperament so much as differences in responsibility. Powell was as bold as any of his crew. But as leader and organizer of the trip, he had different responsibilities and priorities than his men. That made for built-in conflict. "If you're the leader and you have the welfare of the whole expedition in mind," explains Michael Ghiglieri, a boatman who has led trips on rivers around the world, "then you know that prudence is the wiser course, however much pain is involved. But if you're one of the crew, and you've got big, nasty bruises on both shins and your knees are bleeding and your toes are cracking from fungal infections and your hands are raw from ropes burning through them, and then the leader says, 'Let's line this one,' that's hard to take."

Lining and portaging are painful, and in addition they are painfully slow. A rapid that might take thirty terrifying seconds in a boat can eat up agonizing hours on foot. "Lining is galling work," says Ghiglieri. "It's also fatiguing, it's also hurting, and it delays everything. It's all bad. The only thing that's good about it is that when you finish you probably still have all your stuff. So for the men, there's a tremendous temptation to just get in the boats and go. But for Powell, the pull was just as strong the other way. He learned the hard way right off the bat, at Disaster Falls, what happens when things screw up. And if he lost one more boat, he was out of business, it was over. It was two strikes and you're out, not three."

For the expedition as a whole, a fragile coalition to begin with, here was still another potent source of dissension. As the men grew ever more tired and hungry and beat-up, it seemed certain to grow uglier.

A gale of hot, dry wind blasted through camp on the evening of July 9, making the prospect of the next morning's portaging all the more burdensome. The air was so thick with blowing sand that the men covered their heads to breathe. The rapids' unending roar ratcheted the tension up another notch. "We need only a few flashes of lightning to meet Milton's most vivid conceptions of Hell," Bradley lamented.

As if insulted that Bradley had taken it lightly, the river the next day was in as foul a mood as the men. The morning of July 9 was "the wildest day's run of the trip thus far," a nerve-racking journey through "a succession of rapids or rather a continuous rapid with a succession of cataracts for 20 miles." The men ran some of the rapids and lined two. Twenty-four hours earlier, Bradley had waved aside the dangers of hiking out of the canyon and had clamored for a chance to run rapids. Today he had the hurry knocked out of him. "We are quite careful now of our provisions as the hot blasts that sweep through these rocky gorges admonish us that a walk out to civilization is almost certain death," he wrote, "so better go a little slow and safe."

The men spent July 10 in camp making observations and repairing the boats. Sumner was practicing with the sextant, learning how to determine latitude and longitude. Powell and Oramel Howland climbed the cliff with a barometer, measuring the thicknesses of the various strata as they climbed. This called for an odd combination of athleticism and scientific precision. A barometer's reading changes with altitude but also with the weather; in order to make sure that the difference between the readings on two barometers is a true indication of the differences in altitude between them (and has not been skewed by a change in the weather), the measurements must be simultaneous. A man in camp took a barometer reading at half-hour intervals. Powell and Howland, in the meantime, raced ahead with *their* barometer, struggling to reach the base of one rock level or the top of another precisely on the half hour so that they could take a reading of their own. Later they would compare the two sets of figures. Thin strata found Powell and Howland waiting cockily for the half hour to arrive. On thick strata, they ran and stumbled to meet their deadline.

Early in the afternoon, they reached the plateau, four thousand feet above the river. (Hikers who have struggled up the manicured Bright Angel Trail in the Grand Canyon know that this was no casual jaunt.) Howland set off in pursuit of some deer he had seen in the distance, and Powell went off to climb a mountain. By the time they met again (Howland was empty-handed), it had grown late, and the sun was sinking fast. Running and jumping their way downhill, the two men made it as far as they could. But darkness caught them still high on the cliff side, and they had to feel their way the rest of the trek home, guided only by the light of the campfire in the distance. "A long, slow, anxious descent we make," Powell wrote.

Only two days before, Powell's exasperating caution had driven the crew to the edge of their patience. Now, as if defying anyone to pigeonhole him, the mercurial leader had provided a matching display of recklessness. It was not merely that racing down a cliff in fading light was asking for trouble. Powell was a fine climber, but he knew he was adding *extra* risk to what was risky enough in any case. A companion on a different expedition has left us a record of climbing with Powell: "On the way back the Major's cut-off arm was on the rock side of a gulch we had followed up, and I found it necessary, two or three times, to place myself where he could step on my knee, as his stump had a tendency to throw him off his balance. Had he fallen at these points the drop would have been four hundred or five hundred feet. I mention this to show how he never permitted his one-armed condition to interfere with his doing things."

Bradley, who had been in camp throughout Powell and Howland's cliff-side adventure, might have preferred to be caught in the dark himself. Instead, he had spent the day working on his journal and doing his best to endure Andy Hall's exuberance. Playful as a puppy, the boisterous Hall had been singing. Hall had "a voice like a crosscut saw," in Bradley's judgment, but he was not bashful about displaying it. Over and over again came the booming chorus: "When he put his arm around her she *bustified* like a forty pounder, look away, look away, look away in Dixie's land."

The next day was a Sunday. To Bradley's dismay, Powell again showed no inclination to honor the Sabbath. Instead, the men set out early and, in about three-quarters of a mile, came to the day's first rapid. Bradley had scouted it the day before—it struck him as "bad" but runnable despite "very heavy" waves—and Powell evidently shared his judgment. Everyone made it through, but Powell's boat, the *Emma Dean*, broke one oar and lost another. That left her with only two oars rather than four, so that only one man could row. They had no spare oars and could not find any pieces of driftwood suitable for sawing, but they figured they could limp on. Downstream, when they reached a point that permitted an easy climb to the plateau, they could hike to the top and cut new oars from a tree.

Soon they approached another rapid. Standing up to study it, Powell decided that it could be run. Once in the rapid, when it was too late for second thoughts, he saw that the channel veered sharply to the left and the river hit hard against the cliff face that blocked its way. Powell gave the order to land upstream from the cliff, but it was too late, especially with only one man at the oars. The boat swept downstream, overpowered and out of control, headed toward a boulder in midstream. Somehow they made it by, but a wave rebounded off the boulder and filled the *Emma Dean* to the gunwales. Then another wave hit her. Bradley, watching from his own boat, saw the *Emma Dean* "rowled over and over" and Powell, Sumner, and Dunn flung into the river. "I soon find that swimming is very easy, and I cannot sink," Powell wrote later. Powell had a life jacket, but swimming one-armed through the swirling chop can hardly have been easy.

By the time Powell located the *Emma Dean*, she was twenty or thirty feet away, still afloat but tumbling, right side up one minute and upside down the next. Powell swam toward her, trying to keep his head above water but occasionally finding himself pummeled by giant waves and gasping for breath. Finally he reached the boat and found Sumner and Dunn there already, doing their best to hold on. Dunn lost his grip and disappeared underwater, but Sumner quickly hauled him back to the surface. They were drifting faster now, approaching another rapid. Holding the boat and kicking with all their strength, the three men struggled toward shore, the rapid ever closer. Finally, still clinging to the boat and just upstream of the rapid, they managed a crash landing into a pile of driftwood.

Once ashore, the men made camp and built an enormous fire to dry their clothes. It had been a hellacious day. They had advanced less than a mile, and Sumner and Dunn might well have drowned. "We broke many oars and most of the Ten Commandments," Sumner observed. "Major Powell said he lost three hundred dollars in bills. I lost my temper and at least a year's growth—

didn't have anything else to lose." The only good news was that there would be no need to climb a cliff to look for wood to make oars. The driftwood provided plenty of raw material.

Powell put the day's losses at two guns, one barometer, and some blankets. The gear had been thrown from the open compartment of the *Emma Dean* when the boat rolled over. "The guns and barometer are lost," Powell wrote, "but I succeeded in catching one of the rolls of blankets, as it drifted by, when we were swimming to shore." (The last claim sounds unlikely, even for as fine an athlete as Powell. Could a one-armed man swimming in high waves and strong current while helping to pull a boat to shore *also* nab a floating bedroll? Powell may have indulged in an understandable bit of exaggeration in describing how he helped rescue the *Emma Dean*. Bradley's account sounds more plausible: "Major had to leave the boat and swim to land, as he has but one arm and her constant turning over made it impossible for him to hold onto her with one hand, but the other two [Jack and Dunn] brought the boat in below safe." While he swam, Powell may well have spotted his bedroll and grabbed it.)

In the privacy of his diary that night, Bradley gloated just a bit. Perhaps Powell would be less inclined from now on to treat Sunday as just another day. "Sunday again," he wrote, "and Major has got his match."

Bradley's own match came soon enough. At noon the next day, just when everyone was beginning to relax after successfully running several rapids, "we came suddenly upon an old roarer." Boulders strewn along the channel's left side pushed the water right, beneath a rock ledge that jutted out from the cliff face. They stopped to scout, and, curiously, Powell decided that this was one they could run. The plan was to stay as far left as possible, skirting the rocks but staying well clear of the overhanging ledge on the river's right side.

The *Emma Dean*, running first, swallowed a wave but made it safely. So did the *Kitty Clyde's Sister*. Last in line came the *Maid of the Cañon*, with Bradley, Walter Powell, and Seneca Howland. The *Maid* got too far to the right and found herself swept down a chute into a cauldron of waves rebounding off the right-hand wall. A giant wave knocked Bradley overboard, but somehow one foot got snagged. The boat rushed along willy-nilly, Walter Powell flailing helplessly at the oars and Bradley dragging behind the boat like a tin can tied to a car's bumper. Unable to pull his foot free and unable to lift himself back into the boat, he could do no more than struggle to get an arm to the surface so that he could grab the gunwale and lift his head above the water for a moment.

The boat seemed only seconds from smashing into the jutting rock ledge on the river's right side. "To us who are below," Powell wrote, "it seems impossible to keep the boat from going under the overhanging cliff." At the last instant, Walter Powell, a man of bull-like strength, managed to pull away from the cliff. Then, with the boat finally in flat water, he grabbed Bradley and dragged him safely aboard.

Bradley brushed it all aside. No harm done, he wrote that night, "except a glorious ducking and a slight cut on one of my legs."

To everyone's immense relief, this fearsome rapid seemed to mark the end of what Powell had named the Cañon of Desolation. Their emergence into more open country, they dared to hope, would also bring a change of fortune.

CHAPTER FIFTEEN

"HURRA! HURRA! HURRA!"

It was not to be. On the afternoon of the same day that Bradley had washed overboard, they found themselves entering a new canyon, this one cut in gray sandstone. They named it Coal Cañon. (Today it is known as Gray Canyon.) The trouble began almost at once.

"About three o'clock in the afternoon," Powell wrote, "we meet with a new difficulty." The river filled the entire channel, hemmed in on both sides by vertical cliffs. Downstream a rapid waited. Portaging was out of the question, for there was no possibility of climbing past the rapid (to say nothing of climbing while carrying hundreds of pounds of supplies or a boat). Lining the boats along the edge of the rapid seemed impossible, as well, because there was no shoreline where the men holding the ropes could stand. And the rapid seemed too formidable to run.

Eventually they came up with a plan for a new kind of lining. It was complicated, as schemes to maneuver boats and ropes often are, but the difficulty was nothing in comparison with the danger. Every boatman knows that long ropes, heavy boats, and strong currents are a bad combination. Ropes can snap, boats can lurch, men can slip. A man caught on the downstream side of a rope tied to a careening boat can be pulled underwater and drowned before he has time to cry out.

The strategy here involved a kind of high-stakes game of leapfrog. The men landed on a boulder in the river and, holding a rope tied to the *Emma Dean*, paid out the rope to its full length. The *Emma Dean* strained at her line like a German shepherd watching a squirrel just out of reach. But here was the clever bit—the *Emma Dean* had brought with her another long line, this one tied to the second boat (which was still waiting back upstream). On a signal, the second boat pushed out into the current, yanking against the line held by the men in the *Emma Dean*. In a moment, this line, too, was taut. Now *two* boats sat in mid-rapid, fighting against the lines constraining them. The *Emma Dean* was the farther upstream of the two, still held by the men standing on the boulder; the second boat sat farther downstream, held by the *Emma Dean*. A replay of the same trick brought the third boat into the game. It ended up the farthest downstream of all, held by the second boat.

It is hard to imagine a more precarious setup than three boats lashed one to the next in a wild rapid. Now came the hard part. The men in the third boat, who were the nearest to the flat water below the rapid, managed to pull into a cove along the left-hand cliff wall. Safe at last, they tied their boat up securely. Then they turned their attention to reeling in the other two boats as they swung by. The last task would be to rescue poor Bill Dunn, who had been left by himself, upstream on the boulder where the whole adventure had begun.

Dunn was a strong swimmer whose skill had earned him this dubious honor. He jumped into the river, swimming with one hand and holding on to the line that trailed behind the *Emma Dean* with the other, a tail to the *Emma Dean's* kite. Having the line to hold was a mixed blessing. It meant, on the one hand, that if the men could drag in the *Emma Dean*, they would bring Dunn in as well. But it also meant that Dunn was perpetually in danger of snarling himself in the rope and being dragged underwater. He tried to aim somewhere in the direction of the cove where the third boat had tied up, in the hope that he could be pulled to safety. And so he was.

What remained, Powell noted nonchalantly, was only "a short portage."

·   ·   ·

It had been a grueling day, and night brought no solace. "We camp on a sand beach," Powell wrote. "The wind blows a hurricane; the drifting sand almost blinds us; and nowhere can we find shelter." Even for guests on the most luxurious of commercial river trips today, the sand can become oppressive if the wind is blowing hard. Sleeping bags become scratchy and gritty, and so do bathing suits. Eyes itch and burn. Salad, steak, even pudding all crunch disconcertingly. (To prepare for a river trip, one outfitter suggests, it is advisable to place pans full of sand around your home and then turn on a dozen or so industrial-strength fans.) Throw in a thunderstorm or a few too many dousings toward the end of a long day on the water and everyone's nerves start to fray. Honeymooners who have been cooing at each other begin to snarl instead. Doting parents snap at their children, and boatmen scowl at any passengers foolish enough to venture a question.

For Powell and his men, who were immeasurably more rugged than modern travelers, the sand was merely an irritation. But, having suffered irritation aplenty, they were in a mood to take it personally. "The wind continues to blow all night," Powell wrote. "The sand sifts through our blankets, and piles over us, until we are covered as in a snow-drift. We are glad when morning comes." This last, mild-sounding comment bears noting. These were men who referred to a near-drowning as "a glorious ducking," and who hardly saw a reason to refer to a missing arm at all. We can presume that if they admitted they were "glad" to see morning come, then in truth the night must have been close to unbearable.

Morale was fast becoming a problem. After forty-nine days on the river, the men's reserves of optimism and enthusiasm were wearing thin. They were exhausted, bruised, sick unto death of beans and bacon, and endlessly worried about what lay ahead. By now, hunger and anxiety had become chronic, a kind of maddening background drone like a mosquito's buzzing. The last few days had heaped new woes on top of the old ones. The *Emma Dean* had catapulted Powell, Sumner, and Dunn into the river only two days before. Bradley had washed overboard the next day. That same day had brought the trickiest lining experience of the entire trip. Even the weather, with its baking sun and gusting winds and skin-flailing sandstorms, seemed newly malevolent.

And then, just when they most needed the river to relent, it did. "Pulled out early," Sumner wrote on July 13, "and had an exciting ride of 18 miles, passing 19 rapids." The river was swift and easy, the rapids big enough to be exhilarating but not so ominous as to present any great danger. For several days, the men's diary entries had talked of swamping and foundering and trudging and slipping and falling. Now, suddenly, it was as if a curtain had risen on a new play. "We glide along, mile after mile," Powell noted with delight.

Did this represent a genuine change of heart on the river's part—by now it was almost impossible to avoid thinking in such terms—or was something more malevolent at work? Perhaps the river, like a cat with a particularly amusing mouse, had merely put a halt to today's torment in order to enjoy tomorrow's all the more. No matter. The mice were too relieved to ask questions.

On the dismal day when the *Emma Dean* had flipped, they had advanced less than a mile. Now they were on pace to gain forty miles in a single day. By noon on July 13 they had emerged from Coal Cañon. Instead of cliffs of sandstone and shale and vistas of brown and gray, the expedition found itself in what Bradley called "a barren parched dessert" swept by hot, swirling winds.

As the sun beat down on the bare sand, the haze of heat imparted a kind of shivering instability to the blue and gray and brown buttes and to snow-capped Uncompahgre Peak far off to the east. "Plains, and hills, and cliffs, and distant mountains seem vaguely to be floating about in a trembling, wave rocked sea," Powell wrote, "and patches of landscape will seem to float away, and be lost, and then reappear." Sumner, as usual, resisted the desert's spell. It was, he noted curtly, "as desolate a country as anyone need wish to see."

But the fast pace had revived the men's spirits, at least temporarily. "We have run so many rapids we pay but little attention to them," Bradley remarked, "and the weather and water boath being warm we rather enjoy getting wet and laugh at the one who gets wettest." It helped, too, that they were nearing one of the crucial milestones of the trip. Somewhere, and soon, their rudimentary maps told them, they should reach the spot where the Green and the Grand joined to form the Colorado.

On the afternoon of July 13, the men came to one of the few spots where the Green could be forded. (It is today the site of the town of Green River, Utah.) Though the Green itself was largely uncharted, this crossing had long been known and formed part of the Old Spanish Trail from Santa Fe to California. When Powell and his men passed by, they found fresh footprints and a number of log rafts, newly made by Indians, tied up at the river's edge. "This is the place," Powell wrote, "where the lamented Gunnison crossed, in the year 1853, when making an exploration for a railroad route to the Pacific coast." The unfortunate Gunnison had been head of a surveying team looking for a railroad route across the Green. Pahvant Indians had killed him and seven others of his party and butchered the bodies. Powell decided not to linger.

The river seemed almost still now—so much for gliding along—and the men had to row hard in the broiling heat. Mid-July in the desert is the worst of the worst, but the crew toiled like slaves and made thirty-three miles. "As we get farther south," Bradley noted, "the weather gets hotter and the sun shining on the sand-stone heats the whole cañon like an oven." It hardly cooled at night, and the men felt as if they might suffocate. Still, Bradley insisted, neither the work nor the heat had sapped anyone's enthusiasm. If a stranger somehow appeared in camp, he would note at once "the cool deliberate determination to persevere" that possessed every man in the expedition.

Perhaps so, or perhaps Bradley felt momentarily in need of a self-administered pep talk. For in the same day's diary, he struck a considerably more somber note. "The whole country is inconceivably desolate," he wrote, "as we float along on a muddy stream walled in by huge sand-stone bluffs that echo back the slightest sound. Hardly a bird save the ill-omened raven or an occasional eagle screaming over us; one feels a sense of loneliness as he looks on the little party, only three boats and nine men, hundreds of miles from civilization, bound on an errand the issue of which everybody declares must be disastrous."

July 15 was just as hot and just as hard as its predecessor. At midday, the men pulled out a thermometer—one hundred degrees. They passed the mouth of the San Rafael River, which joined the Green from the west, and stopped for an hour or two to explore. Powell's description makes the brief stop sound pleasant as a picnic—"Arrowheads are scattered about, many of them very beautiful. Flint chips are seen strewn over the ground in great profusion, and the trails are well worn"—but Sumner left a more troubling, though cryptic, account. "Another chapter of the Powell-Howland squabble was commenced as we left the camp near the San Rafael," he wrote. "A sad, bitter business. I wish I had put a stop to it long before I did." Sumner gave no details, but Oramel Howland was in charge of mapmaking, and it seems likely that Powell had criticized him for doing poor work.

The river that once had swept the boats along at what seemed a locomotive's pace now seemed not to be moving at all. Rowing downstream was like rowing across an endless lake. Or, maybe, worse, for the river had revealed a new trick. As if to compound the misery caused by the merciless sun and a barely detectable current, the Green found a new way to toy with its victims.

"We go around a great bend to the right, five miles in length," Powell wrote, "and come back to a point within a quarter of a mile of where we started. Then we sweep around another great bend to the left, making a circuit of nine miles, and come back to a point within six hundred yards of the beginning of the bend." Powell seemed fascinated. "There is an exquisite charm in our ride to-day down this beautiful cañon," he wrote. "It gradually grows deeper with every mile of travel; the walls are symmetrically curved, and grandly arched; of a beautiful color, and reflected in the quiet waters in many places." The crew was less enchanted and more succinct. "River very crooked," Sumner snapped.

For the men, laboring to the edge of their strength and nourished on only beans and rice (even the scrawny geese had vanished), this was exquisite torture. "The sun was so hot we could scarcely endure it," Bradley wrote, "and much of the time the cañon was so closely walled in that the breeze could not reach us." It had not been a terrifying or dramatic day, like some others, but it had been miserable in a grinding, spirit-sapping way. "We have worked as hard as we could to get only 25 miles . . . ," Bradley went on. "Most of us have been unwell today from eating sour beans for supper last night."

Powell struck a different note. "We are all in fine spirits, feel very gay, and the badinage of the men is echoed from wall to wall," he wrote. Never much of a judge of other people's moods, he may simply have missed the men's sullenness. But he was writing years after the fact, with a sharp eye on history. More likely, he made a pragmatic decision that a portrait of indomitable pioneers would have more appeal than a tableau of weary, grouchy men clutching their bellies and moaning.

In honor of this latest canyon's graceful curves and twists, Powell named it Labyrinth Canyon.

Powell himself was indeed in fine spirits. As always, his fascination with the natural wonders around him lifted his mood. A geologist, and therefore a connoisseur of rock, Powell was nearly overwhelmed by the panorama before him. The sandstorms that had swept through camp only days before seemed to have vanished, from the canyons and from Powell's mind. "The landscape everywhere, away from the river, is of rock," he marveled. "Cliffs of rock; tables of rock; plateaus of rock; terraces of rock; crags of rock—ten thousand strangely carved forms. Rocks everywhere, and no vegetation; no soil; no sand."

Still new to such scenes, Powell found this "strange, weird, grand" region of the world hard to take in. More than a century later, a river runner and writer named Ellen Meloy spoke to Powell's puzzlement. The rivers and canyons of the Utah deserts, so new to Powell, were a kind of second home to Meloy. "What strange anomaly is a *desert river*?" she asked, as if speaking directly to the Major. "Bound by deep sandstone gorges for much of their course through Utah, the Green and Colorado water little along their banks. In many stretches there are no banks, merely a precipitous jumble of wall and cliff, a boulder or sandbar to stand on but no place to park towns or pasture or croplands. Think of the rivers not as a Tigris-Euphrates floodplain but as stone-encased, sky-vaulted blood vessels, their fluid pulled swiftly seaward by gradient and gravity."

Powell was just learning to think in these unfamiliar terms, but for a geologist the world could hold no better classroom. He was not heedless of his men's concerns so much as preoccupied with his own far different ones. Like a person absorbed in a concert on the radio, he focused so intently that much of the everyday life around him faded into a kind of background murmur.

On July 16, the expedition reached the end of Labyrinth Canyon. But almost at once the river cut into still another canyon, this one with walls of red sandstone. Powell called it Stillwater Canyon, an apt name but a sign of frustration, too. The men were working without letup, and their labor never seemed to bring them any closer to their goal. Where was the Grand River?

And then, suddenly, at five-thirty in the afternoon on July 16, after days of frying in the sun, there it was. "Hurra! Hurra! Hurra!" Bradley exulted. "Grand River came upon us or rather we came upon that very suddenly and to me unexpectedly." It was not at all what the men had pictured. At least for the half mile or so they could see downstream, the Colorado—the name of the merged Green and Grand—was "calm and wide" and stately. They had been told to expect an "impossible unpassable succession of foaming and raging waterfalls and cataracts," Bradley noted, but they found merely a broad and sedate river. The Green ran muddy and red-colored, the Grand ran clear, and the two rivers flowed side by side, like a young couple wary of committing to marriage.

"It is possible we are allured into a dangerous and disastrous cañon of death by the placid waters of this cañon," Bradley reminded himself, but he was too glad to have made it this far to continue thinking along those lines. "Here we float . . . [at a spot] never before beheld by white men," he wrote proudly.

This was almost true, but not quite. Denis Julien, for example, was a fur trader who roamed the West in the 1830s. Almost nothing is known about him even today (and Bradley had never heard of him), but later travelers found his name scratched on rocks up and down the Green and Colorado, as if Julien was a kind of canyon-country Kilroy. On a sandstone rock in Hell Roaring Canyon, off Labyrinth Canyon, a scratched inscription reads, "D. Julien, 3 Mai, 1836." He cut his name into another rock twenty miles up the Green on May 16, 1836, and on other trips he left signatures far upstream, where the White River joins the Green, and far *downstream* as well, at the foot of Cataract Canyon. How he made his way, and what became of him, no one knows.

Sumner tried hard to seem unimpressed at finally meeting the Grand—"not timber enough within 10 miles to supply one family 6 months," he griped— but even he was struck by the scene in front of him. Here was "an apparently endless cañon in 3 directions—up the Grand, up the Green, and down the Colorado; the walls 1250 ft. high."

Powell was the most excited of all. Even the geological passages in his journal, purportedly nothing more than somber explanations, conveyed unmistakably the notion that everything led up to *this*. Powell had begun his account of the expedition not with a description of himself or his men but with several pages on the natural forces that had shaped the Western landscape.

He began with rivers. Through the course of the winter, Powell wrote, snow accumulated high in the mountains. Then, "when the summer-sun comes, this snow melts, and tumbles down the mountain-sides in millions of cascades. Ten million cascade brooks unite to form ten thousand torrent creeks; ten thousand torrent creeks unite to form a hundred rivers beset with cataracts; a hundred roaring rivers unite to form the Colorado, which rolls, a mad, turbid stream into the Gulf of California."

That "mad, turbid stream" had been beckoning them from the moment they had left Green River Station nearly two months before. Now they were poised to go where no one, as far as they knew, had ever gone.

This was what they had come for.

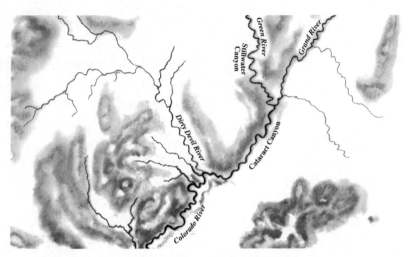

CHAPTER SIXTEEN

OUTMATCHED

The euphoria at finally reaching the Grand quickly burned off under the piti-less desert sun. There was no shade in camp, but the exhausted men could not push on until they had rested a bit and patched their boats. No one knew what games the river would get up to next, but it seemed prudent to assume the worst. The men would be riding on the merged back of two mighty rivers, where before there had been only one, and the newcomer, the Grand, was the bigger of the two. Just downstream this merged monster hurled itself into a narrow, stone chasm. Trouble seemed inevitable.

But trouble turned up ahead of schedule, when the men took advantage of their break to look closely at the food stores. "Our rations, . . . we find, are badly injured," Powell wrote. "The flour has been wet and dried so many times that it is all musty, and full of hard lumps." Sumner had been the one to make the unhappy discovery. The only food left for the rest of the trip was "about 500 pounds of flour and a little bacon and dried apples," he wrote, and the flour was "a miserable mess of green fermentation."

Sumner and Oramel Howland sifted the flour through an improvised sieve made from mosquito netting. Almost half the total, two hundred pounds, was beyond salvaging and had to be thrown away. They put the rest on a sheet to dry in the sun. By Powell's reckoning, that left food for two months. The arithmetic was discouraging. At the start, seven weeks before, they had packed supplies intended to last ten months. In less than two months, then, they had used or lost or ruined eight months' worth of food. From here on, they would be down to near-starvation rations—Powell's estimate that five hundred pounds of food would last two months meant that each man would get less than a pound per day. They had no idea how much farther they had to go.

The pressing question was not Where had it all gone? but How much longer would it last? If ten months' supplies had provided food for two months, how long would two months' supplies last? With the looming prospect of worse rapids just ahead, this was bad news at a bad time, and everyone seemed edgy. It had stormed the night before, and the lightning and thunder, Bradley wrote, "seemed as if commissioned to make doubly desolate this regeon set apart for desolation."

Powell and Oramel Howland set to squabbling once again. The trouble between the two had begun weeks before, when each had blamed the other for the loss of the *No Name* at Disaster Falls. "From that date there was more or less rag-chewing on the part of Major Powell, nearly always directed at Howland," wrote Sumner. Most often the disputes centered on Howland's mapmaking. Only a few days before, at the mouth of the San Rafael, Powell had rebuked Howland for not keeping his charts up-to-date. Now, with the worrisome discovery that they were low on food, Howland took the opportunity to snipe about Powell's foolishness for not having bought more supplies back at the Indian agency.

Sumner sided with his old friend Howland against their boss. "Howland would sometimes get a little behind in platting up his survey work, as he had his other work to do the same as the rest of the men," Sumner wrote. "That always brought censure from Major Powell." Sometimes, though, Sumner admitted, Howland "got things somewhat confused." Sumner tried to smooth things over, but Powell insisted that "discipline" must be maintained. Sumner didn't buy it. "Military martinets and civilians very often disagree," he growled.

Food became a preoccupation. The easy confidence that the hunters could augment the supplies by bringing in fresh meat had long since vanished. Often the canyons and cliffs revealed no signs whatever of game (and, to the men's frustration, the river rarely gave up its fish). When the men did spot an occasional deer or sheep, it almost always fled before they could get near enough to bring it down. The hunters' few triumphs would, in ordinary circumstances, hardly have been worth noting. On the first day in camp after they had reached the Grand, for example, the men managed to kill two beavers. "They are quite decent eating and the tails make excellent soup . . . ," Bradley wrote. This was not so much a tribute to the tastiness of beaver flesh as a reminder that, in comparison with beans and smoked, salted, half-rancid bacon, almost anything qualified as banquet fare. "We find any kind of fresh meat palatable," Bradley went on, "even the poor . . . geese, so poor they can't fly and just off the nest, go good to us when long fed on bacon."

In the past, when the men had fretted about what lay ahead, their fears had nearly always centered on how bad the next rapids might be. Now a new fear squeezed its way next to the old one. From here on, the twin specters of drowning and starving would never be far off, like a pair of malevolent spirits haunting the canyon cliffs. Powell and the crew examined the vexing matter of supplies from every angle, though they tended to tiptoe around the real issue. They focused their complaints on the miserable quality of the rations, not on their fear of running out of food altogether. On July 19, for example, after a long, hot, miserable day in camp (for several days the temperature had hovered around 100 degrees), the men sat down to another dismal meal. Like castaways on a desert island or polar explorers on the ice, they found their thoughts flying to favorite dishes they vowed they would never again take for granted. "While we are eating supper," Powell wrote, "we very naturally speak of better fare, as musty bread and spoiled bacon are not pleasant."

For Powell, a man who truly saw the skies as not cloudy all day, this was a remarkable concession. After the sorry meal, Hawkins wandered off to take measurements with a sextant. He had never shown any interest in scientific chores and Powell, befuddled, asked what he was up to. Hawkins explained that he was trying to find the latitude and longitude of the nearest pie.

The joking and grousing about the food masked darker concerns, but those were the stuff of private journal entries and not a fit subject for campfire conversation. There had been no choice about stopping to repair the boats, but an emergency stop was one thing and dawdling was quite another. They *had* to make better time, Bradley wrote, "for we cannot think of being caught in a bad cañon short of provisions."

The men had no illusions about how dangerous the river was, but since they had to confront those dangers sooner or later, sooner was better. Only Powell seemed immune from the urge to hurry. He had wanted to camp at the junction of the Grand and Green until August 7, when a solar eclipse was forecast. Armed with tables that pinpointed the time of the eclipse, Powell hoped to make a precise calculation of his longitude. But the eclipse was three weeks off, and with the food supply so low, even Powell conceded that three weeks in one spot was out of the question. Still, the junction of the Southwest's two most important rivers was a crucial spot, and no mapmaker knew its proper position to within 100 miles. Everyone agreed to stay at the junction until they had mapped its location carefully.

In the meantime, the men sulked and sweated in the sun and worked on the boats. The Indian name for this area, Toom-pin Tu-weap, meant "Land of Standing Rock," and the view from the river was of countless buttes perched on the canyon rim like colossal sentries from an alien world. To Bradley, with his religious cast of mind, the stone formations looked like church steeples. "Sunday again," he wrote on July 18, "and though a thousand spires point Heavenward all around us yet not one sends forth the welcome peal of bells to . . . remind us of happier if not grander scenes."

While the men worked on the boats, Powell spent most of his time exploring. Dazzled by the views from the canyon rims, he felt little of his men's frenzy to get under way. Instead, like Marco Polo gazing in awe at wonders that his homebound audiences had never dreamt of, Powell found himself trying to come to terms with the strange new landscape before him. He began by emphasizing what the canyon lands were *not*. Nothing in the teeming cities of the East or the rolling farmlands of the midwest prepared the eye for this desert landscape. Both the scenery itself, and that scenery's epic scale, were utterly unfamiliar. "We must not conceive of piles of boulders, or heaps of fragments, but a whole land of naked rock," Powell cautioned his readers, "with giant forms carved on it: cathedral shaped buttes, towering hundreds or thousands of feet; cliffs that cannot be scaled, and cañon walls that shrink the river into insignificance, with vast, hollow domes, and tall pinnacles, and shafts set on the verge overhead, and all highly colored—buff, gray, red, brown, and chocolate; never lichened; never moss-covered; but bare, and often polished."

On July 20, Powell and Bradley climbed the cliffs to reconnoiter. ("Climbed 'Cave Cliff' with Bradey," Powell wrote in his river diary. It is perhaps a sign of Powell's high-handed manner that, although Bradley had long since proved himself an invaluable member of the expedition and had even saved Powell's life, Powell never spelled his name right.) The temperature at camp was ninety-five degrees in the shade. The climb could hardly have been more difficult. Bradley and Powell set off up a gulch and climbed over and around boulders for an hour until they found their way blocked. They thought they saw a route up and to the left and followed it for half an hour, but found that it made a dead end. "Then we try the rocks around to the right, and discover a narrow shelf, nearly half a mile long. In some places, this is so wide that we pass along with ease; in others, it is so narrow and sloping that we are compelled to lie down and crawl." They continued along, stopping occasionally to peek down over the shelf's rim to the rolling river eight hundred feet below.

The two men continued climbing and found themselves in an area honeycombed with vertical caves or crevices, each one a kind of twisting tunnel with a natural skylight at the top. There seemed to be no paths leading up, but "we determine to attempt a passage by a crevice, and select one which we think is wide enough to admit of the passage of our bodies, and yet narrow enough to climb out by pressing our hands and feet against the walls. So we climb as men would out of a well. Bradley climbs first; I hand him the barometer, then climb over his head, and he hands me the barometer. So we pass each other alternately"—Powell doing all this, it went without saying, with one hand—"until we emerge from the fissure, out on the summit of a rock. And what a world of grandeur is spread before us!"

Powell was ecstatic. He could see the Green in its narrow, winding gorge, and the Grand in what seemed a bottomless canyon, and occasional glimpses of the Colorado as it swept for mile after mile through the next canyon to come. The rocks were a match for the rivers. "Away to the west are lines of cliffs and ledges of rocks—not such ledges as you may have seen where the quarryman splits his blocks, but ledges from which the gods might quarry mountains . . . and not such cliffs as you may have seen where the swallow builds its nest, but cliffs where the soaring eagle is lost to view ere he reaches the summit."

Powell's delight and astonishment at the sheer scale of the West's natural wonders echo that of his contemporary Melville, celebrating the grandeur of whales. Each man felt that the glories of his subject could scarcely be compressed to human scale. "Give me a condor's quill!" Melville cried. "Give me Vesuvius' crater for an inkstand!" and Powell was no less fervent.

Bradley, characteristically, was harder to enthrall. "The scenery from the top," he wrote, "is the same old picture of wild desolation we have seen for the last hundred miles."

The difference in tone reflected a genuine difference in style between Powell, a romantic who saw the world as if it were a Thomas Moran landscape, and Bradley, the prototype of the down-to-earth, pomposity-piercing common man. The difference in temperament was magnified by the contrasting ways in which the two men composed their journals. Bradley, who had next to no impulse to rhapsodize in the first place, jotted down his notes at the end of long, draining days. Powell, inclined by nature to operatic excess, wrote at his leisure years later, with all the time in the world to buff and hone his memories.

As practical as he was in many ways, Powell had more than a little of a Don Quixote in him. Embarked on the quest of a lifetime, he found himself endlessly dazzled by sights of almost supernatural splendor. Bradley took the Sancho Panza role, seeing scrawny chickens and angry publicans where Powell marveled at giants and castles and princesses.

Finally, on July 21, the repairs and the measurements were complete, and the expedition set off downstream once more. Powell now seemed as eager as the men to be heading into the unknown. "We start this morning on the Colorado," he wrote excitedly, but almost at once the new river showed that they had been right to fear it. "Rapids commenced about two miles from the junction and have now become continuous," Bradley wrote. "We can't run them or rather we don't run many of them, on account of our rations. We are afraid they will spoil and if they do we are in a bad fix."

Afraid to run the rapids, the men had to line or portage them. It was even crueler work than usual. "Two very hard portages are made during the forenoon," Powell wrote. As lavish as he was with the likes of "grand" and "superb," Powell was just as stingy with "hard" and "difficult." But to unload the boats and carry hundreds of pounds of supplies along a rocky trail in 100-degree heat, and then to stagger along under the weight of three waterlogged wooden rowboats besides, was enough to sap the spirit of the strongest man. To know that the reward for doing it once was to repeat the entire process around the next bend was nearly unbearable.

In a crushing day, the expedition advanced a total of eight and a half miles. The men lined several rapids and portaged four. In one of the few rapids they *had* tried to run, the *Emma Dean* had been swamped yet again. "We are thrown into the river, we cling to her, and in the first quiet water below she is righted and bailed out," Powell wrote, "but three oars are lost in this mishap." The two other boats, having witnessed the *Emma Dean*'s struggle, pulled ashore upstream of the rapid, but it took them all afternoon to portage it. "Have made two portages within 100 yds. above," an exhausted Bradley wrote at day's end, "and there is another waiting not a hundred yds. below."

Even when the men had dragged themselves ashore at the end of the day, their labors were not quite done. The camp they had found was so boulder-strewn there was scarcely room to lie down. The remedy was to dig out the sand that had collected near the biggest boulders and carry it to a spot where it could be patted into a pallet of sorts. Bradley managed a wry summary of this first, discouraging day. "So I conclude the Colorado is not a very easy stream to navigate," he observed.

Back on the river the next morning, the men quickly came to a huge pile of driftwood along the riverbank. They set to sawing new oars for the *Emma Dean* from some cottonwood logs. The boats already needed repairs again, too, for they had taken a pounding in the previous day's giant waves. They had advanced a paltry mile and a half downstream. Powell and Walter set out to measure the height of the cliffs and, while they were at it, to gather some sap for recaulking the boats from the pine trees growing on the rim.

On the third day on the Colorado, July 23, the Powell expedition found itself outmatched again. "We come at once to difficult rapids and falls, that, in many places, are more abrupt than in any of the cañons through which we have passed," Powell wrote. Bradley agreed. They had never seen rapids like these, either in number or in severity. "All the way rapid," Bradley wrote, in a kind of private pidgin. "Much of the last three miles we have let down with ropes." The canyon seemed to be closing in ominously. "Rapids get worse as we advance and the walls get higher and nearly perpendicular."

The rapids continued without letup. In a single half-mile stretch, Sumner reckoned, they ran five bad rapids and portaged four. The hard day's labor won them only five and a half miles. "We camp tonight above a succession of furious cataracts," Bradley wrote. "There are at least five in the next mile around which we shall have to make portages." Powell's descriptions were, as usual, more lyrical and less focused on the chores to come. "Our way . . . is through a gorge, grand beyond description," he wrote, with nearly vertical walls rising some sixteen hundred feet above the river. In a section where the river ran swiftly but relatively peaceably, "we seem to be in the depths of the earth, and yet can look down into waters that reflect a bottomless abyss."

They *were* in the depths of the earth, or near enough. What in ordinary circumstances would be merely a mishap could be deadly here. Twenty years later, one of the first men to try retracing Powell's route made camp in this same difficult canyon. Climbing near camp, he lost his balance and wedged his leg into a tight spot between two boulders. "It was only a few seconds, but it seemed an hour while I was waiting to hear those bones snap," he wrote later. "In that short time, I realized the whole situation; where we were, the height of the limestone cliffs, the distance to outside assistance, the heat of the summer, and that when those bones gave way—the end. Not a sudden blotting out of existence, that was not what I feared, but with a mangled leg, a sure, but lingering, death in that hot desolate canyon."

The towering cliffs had caught Bradley's eye, too. He tried to convince himself that they did not mean what they seemed to. "We have as yet found no place in the Colorado where we could not land on either side of the river," he wrote, "for though the walls come quite close to the water yet there has always been a strip of fallen rocks or a sand bank." If there was no shore to pull to, there would be no choice but to run the rapid, no matter how dangerous.

Only three days before, the men had clamored to return to the river. Now, camped above a long line of intimidating rapids, sweltering in temperatures that topped 100 degrees during the day and barely fell at night, exhausted by the endless lining and portaging, they were beginning to unravel. The horrific new canyon—they named it Cataract Canyon, to acknowledge its fierce rapids—had spooked everyone.

As it should have. Cataract has long been known as the "graveyard of the Colorado," the melodramatic name more than justified by dozens of drownings over the years. Approach a professional boatman and ask point-blank if Cataract scares him. (It is a breach of etiquette for an outsider to pose such a question, as it would be for a civilian to pipe up at a VFW lodge, but let us ask regardless.) "Hell, yes!" he will answer at once, and indignantly. "I may be addicted to adrenaline," the tone implies, "but I'm not *crazy*."

Two of the earliest and boldest river runners, Ellsworth and Emery Kolb, set out down the Green in September 1911. Like Powell, they left from Green River Station, Wyoming, their destination the Grand Canyon. The Kolb brothers were photographers who would do anything for a picture. They lugged a movie camera along on their river trip and filmed their flips and crashes and their good runs, too. This first-ever white-water documentary made their career; until 1976, when he died at age ninety-five, Emery showed the flickering, black-and-white film four times a day in his studio at the South Rim of the Grand Canyon.

The brothers nearly added their names to the roster of Cataract Canyon's victims. "We always thought we needed a certain amount of thrills to make life sufficiently interesting for us," Ellsworth wrote. "In a few hours' time in the central portion of Cataract Canyon, we experienced nearly enough thrills to last us a lifetime. In one or two of the upper canyons we thought we were running rapids. Now we were learning what rapids really were."

Modern boats (and life jackets) have failed to tame Cataract. A fatal accident in 1997 was more or less typical. The victim was a forty-year-old boater named Melvin Fisher. "Fisher's raft flipped in Little Niagara, . . . dropping seven people into the water," according to a police blotter–style accident report. "This hole has been the site of several other fatalities. . . . Fisher washed two miles through four more rapids before his party could catch up. He was found blue and pulseless."

Faced with Cataract Canyon's roiling waters, Powell and his men spent half the afternoon and evening of July 23 in worried conversation about their prospects for surviving the expedition. They carefully examined the barometric readings to see how much their elevation had dropped since the start of the trip and how much farther it still had to fall before they would finally emerge from the canyons. "The conclusion to which the men arrive seems to be about this," Powell wrote, "—that there are great descents yet to be made, but, if they are distributed in rapids and short falls, as they have been heretofore, we will be able to overcome them." On the other hand, they might come to "a fall in these cañons which we cannot pass, where the walls rise from the water's edge, so that we cannot land, and where the water is so swift that we cannot return." What then?

For men whose fate was not in their own hands, such speculation was irresistible but useless. From the start, everyone had understood that the crucial question was whether the river dropped toward sea level in a series of many short rapids or a few Niagaras. Hundreds of miles downstream, in the midst of the worst rapids so far, they were no nearer an answer. All they could do was echo Powell's plaintive query, "How will it be in the future!"

Even when it came to matters of life and death—perhaps *especially* when it came to matters of life and death—the men rejected anything that smacked of earnestness. "They speculate over the serious probabilities in jesting mood," noted Powell, whose own tendency was to discuss serious matters in a voice best accompanied by cracks of lightning and ominous minor chords. Bradley professed to welcome any challenges the Colorado could muster. "Let it come," he wrote. "We know that we have got about 2500 ft. to fall yet . . . and if it comes all in the first hundred miles we shan't be dreading rapids afterwards for if it should continue at this rate much more than a hundred miles we should have to go the rest of the way *up hill* which is *not often the case with rivers*."

Bradley seemed pleased with his joke—the underlined words (here in italics) in his journal served as a kind of elbow to the ribs—but it was only a joke, a convenient way to keep the unthinkable at arm's length. Like the others, Bradley knew there was no choice but to make the best of what had begun to look like a very bad predicament. All they could do was hope that the river would show mercy.

They set out early the next day, fighting the most dangerous rapids they had seen. The rapids were produced by huge, sharp-angled blocks of rock that had broken off the cliffs and fallen into the channel, forcing too much water through too small a space. "Among these rocks," Powell wrote, "in chutes, whirlpools, and great waves, with rushing breakers and foams, the water finds its way, still tumbling down." Four times in less than three-quarters of a mile the men found themselves forced to portage. "Had to take everything around by hand and around the second we had to carry our boats over the huge bowlders which is very hard work as two of them are very heavy, being made of oak," Bradley wrote.

There was plenty of drama to complement the hard work. "*Kitty's Sister* had another narrow escape today," Sumner wrote. "While crossing between rapids Howland broke an oar in a very bad place and came very near being drawn into a rapid that would smash any boat to pieces."

In all, they advanced less than a mile and camped above yet another rapid. "They tell me [it] is not so bad as the others but I haven't been to look at it yet," Bradley wrote. "They don't interest me much unless we can run them. That I like, but portage don't agree with my constitution." By contrast, the unwaveringly cheerful Andy Hall found even these endless days agreeable. Bradley, as naturally gruff as Hall was sunny, watched his young colleague with admiration and puzzlement. "Andy has been throwing stones across [the river] for amusement tonight," Bradley remarked. At least, he might have added, Andy was not singing.

Sumner, exhausted though he was, managed to spare a moment to admire the scenery. The canyon walls, he wrote, were "¾ blue marble, the remainder grey sandstone, lightly touched with red by a thin bed of red shale on the top." (The colors conjured up, for one modern-day river runner and writer, thoughts of "God gone mad with the Play-Doh.") A closer look dispelled any dreamy speculation about the beauty of their surroundings. "Driftwood 30 ft. high on the rocks," Sumner noted. "God help the poor wretch that is caught in the cañon during high water."

He certainly would have little chance of helping himself. When the Colorado is in flood through Cataract, rapids that ordinarily are distinct merge into white-water marathons. One stretch called Mile Long Rapid ends in a rapid called Capsize that features a giant, boat-flipping hole smack in the middle of the river. Waves can tower twenty feet and can flip even the enormous motor-powered rafts that ride through ordinary rapids as imperturbably as city buses plow through puddles in the street. For oar-powered boats, capsizing is closer to the rule than the exception. In one trip in high water a few years ago, the last boatman in a line of ten watched as eight of his nine predecessors flipped. In some high-water years, the National Park Service stations a motorboat and a paramedic just below the biggest rapids to deal with the inevitable near-drownings.

On the evening of July 24, 1869, when another man might simply have flung himself on the ground in nervous exhaustion, Powell climbed a boulder near the river's edge and spent an hour gazing at the water in fascination. Watching the waves in the gathering darkness, he was still studying, still categorizing, still enthusiastic, above all, still curious. "The waves are rolling, with crests of foam so white they seem almost to give a light of their own. Near by, a chute of water strikes the foot of a great block of limestone, fifty feet high, and the waters pile up against it, and roll back. Where there are sunken rocks, the water heaps up in mounds, or even in cones. At a point where rocks come very near the surface, the water forms a chute above, strikes, and is shot up ten or fifteen feet, and piles back in gentle curves, as in a fountain."

Decades later, when boatmen began to run rapids for sport, they would name the watery formations that Powell had identified and add others to the menagerie. Powell's crests of foam, for instance, capped tall, pyramid-shaped breaking waves that are now called haystacks. The water he saw piling up against the face of an enormous limestock boulder was a dramatic example of a pillow. The mounds or cones were mini-horizon lines where the river poured over a barely submerged rock and dropped abruptly into a hole. The ten- or fifteen-foot-tall fountain marked a spot where a stretch of particularly fast-moving water collided with a rock, forming what today's river runners call a rooster tail.

Powell had not found every entry that would someday have a place in a white-water glossary, but he had made a considerable start. A complete inventory could come later; the key accomplishment was sensing that there was system underlying what appeared to be chaos. A white-water Mendeleyev, Powell shared his Russian contemporary's yearning to uncover the natural world's hidden order and pattern. The difference was that Powell conducted his research not at a chemist's bench but at the bottom of a thousand-foot canyon.

Green River Station, Wyoming
Territory, where the trip began. This
photo, taken in the winter of 1868,
shows the transcontinental railroad, still
under construction. Powell's boats
arrived by train from Chicago on May
11, 1869, one day after the golden
spike was hammered home. *(Courtesy of
the Oakland Museum of California)*

In an era when most whites disdained
Indians as savages, Powell had a fascination with Native American cultures,
languages, and myths. Anthropology
(and geology) became lifelong passions.
He is shown with Taugu, a Paiute
chief. *(National Anthropological Archives,
Smithsonian Institution)*

Powell's service record. The entry for April 1862 reads, "Wounded in right arm at Pittsburg Landing (Shiloh Tenn). Arm has been amputated." *(Courtesy National Archives)*

Powell, at right, with his younger brother, Bramwell, in 1865. Powell turned his left side to the camera. He had lost his right arm at Shiloh. *(Used by permission, Utah State Historical Society, all rights reserved)*

Union forces at the Hornets' Nest at Shiloh, a scene of desperate and chaotic fighting. The Confederates launched at least seven assaults against the Hornets' Nest. At four o'clock, when Powell raised his right hand to give the order to fire, a minié ball plowed into his forearm. *(The Seventh Regiment Fund, Inc.)*

Desperate to join the fighting, Powell had rushed to the Hornets' Nest on his own initiative. He was unaccounted for and later, despite his wound, had to prove that he had not run off. In this affidavit, Lt. Joseph Mitchell swears that he fought with Powell and saw him "taken on a Hospital Boat to Savannah Tenn. where his arm was amputated." *(Courtesy National Archives)*

Powell carried this locket with photos of himself and his wife, Emma, on his 1869 expedition. Powell was immensely proud of his lovely, adventurous wife, who nursed him at Shiloh and accompanied him on his first excursions into the West. Powell named his lead boat the *Emma Dean* in her honor. *(Bruce Dale/NGS Image Collection, Courtesy the Darrah Collection)*

Powell's crew. Jack Sumner **(upper left)** was proud, combative, and capable, the first man Powell recruited. He saw Powell as a tenderfoot ignorant of the West. Billy Hawkins **(upper right)** was camp cook, at $1.50 a day. Andy Hall **(lower left)**, the youngest member of the crew, maintained his good cheer even in the expedition's direst moments. Walter Powell **(lower right)**, was Powell's melancholy, half-mad brother. According to Sumner, Walter was "about as worthless a piece of furniture as could be found in a day's journey."

J. W. Powell the party of the first part, to furnish boats, supplies, ammunition etc, sufficient for the use of the expedition. This agreement to go into effect the first day of June, eighteen hundred and sixty-nine, and not to continue over one year. Should it be necessary to proceed on the journey without delay on account of Disaster to boats or loss of rations then the time specified for hunting may not be required by either party, nor shall it be deemed a failure of contract to furnish supplies should such supplies be lost in transit—

J. C. Sumner
William H. Dunn
O. G. Howland

J. W. Powell
In Charge of Col River Ex. Ex.

Deer	$1.25 each	Otter	$3.5 each	Bear (grown grizzly)	$10.00
Sheep	1.25 "	Beaver	1.00 "	" cub	1.00
Antelope	1.00 "	Wildcat	.50 "	" grown Common	5.00
Elk	2.00 "	Porcupine	.50 "	" cub	1.00
Wolf (gray)	1.00 "	Squirrel	.35 "	" grown black	3.00
" coyote	.50 "	Rabbit	.35 "	" cub	1.00
fox (cross)	1.50 "	Woodchuck	.35 "		
" red	.75 "	Badger	.50 "		
Mink	1.50 "	Weasel	.35 "		
Martin	1.50 "				

And all other skins at proportionate rates

Three of the crew were hired to hunt, along with their other duties. The careful list of fees ("Deer $1.25 each, Sheep $1.25") proved almost irrelevant. Game was so scarce that the danger of starving grew almost as great as the danger of drowning. *(Special Collections, University of Arizona Library, Dellenbaugh Collection)*

Frank Goodman, an Englishman wandering the West in search of adventure, joined Powell's expedition at the last minute. Two weeks later, his boat cracked apart in a rapid and he nearly drowned. At the first opportunity, Goodman left the expedition. He settled in Utah, married, and raised a large family. *(Courtesy of the Huntington Library, San Marino, Calif., from print owned by Mrs. E. G. Evans)*

This modern-day photo shows two dories just above Badger Rapid in the Grand Canyon. Until the last second, a passenger poised to enter a rapid can convince himself that he is as safe as if he were standing on a balcony and looking out over the railing. Then the railing disappears. *(From* The Hidden Canyon: A River Journey © *John Blaustein, 1999)*

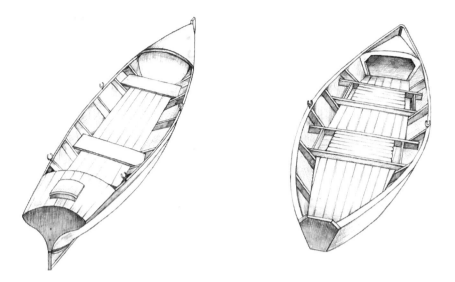

Powell's boats (left) were Whitehalls, long, heavy rowboats built for speed. With round bottoms and prominent keels, the boats were fast and hard to knock off course. In lakes and harbors, where Whitehalls rose to fame, the design was ideal. In whitewater, the lack of maneuverability was disastrous. Modern white-water dories (right) are shorter and flat-bottomed, so the boatman can pirouette his way through a rapid. Powell's boats, reinforced for strength, were so heavy that it took four men to carry an empty boat. Modern dories are lightweight, built to dodge rocks rather than survive collisions. *(Illustrations by Jordan Cutler)*

A modern dory about to enter Lava Falls, one of the most formidable rapids in the Grand Canyon. One person, facing forward to see where he is headed, does the rowing. Powell's boats needed two oarsmen, and they rowed backward, blind, into danger. *(From* The Hidden Canyon: A River Journey © *John Blaustein, 1999)*

This drawing of Disaster Falls, from *Scribner's Monthly*, shows the first great crisis of the trip. In an instant, the expedition lost one boat (of four) and one-third of the supplies. Three men barely escaped drowning. They had been underway two weeks. *(Courtesy of Library of Congress, Manuscript Division)*

George Bradley kept the most detailed diary of anyone on the trip, and somehow he wrote without anyone seeing him. A disgruntled career soldier, Bradley said he would "explore the river Styx" if it would get him out of the army. *(Courtesy of Library of Congress, Manuscript Division)*

Sockdolager Rapid, in the Grand Canyon. The name was nineteenth century slang for a knockout punch. Powell and his men made a terrifying run through Sockdolager. In this journal entry, Bradley's description of "emphatically the wildest day of the trip so far" referred to Sockdolager. This early photograph shows the Kolb brothers in Sockdolager, in 1911. (The arrow indicates their boat.) *(La Rue, E. C. U. S. Geological Survey)*

On July 3, 1869, the *Chicago Tribune* reported a "Fearful Disaster." All but one member of the Powell expedition had drowned. The sole survivor told the tale. The "survivor" turned out to be a con man, but word of the "deaths" drew more notice than the launching of the expedition ever had. *(Copyrighted 1869 Chicago Tribune Company. All rights reserved. Used with permission.)*

Missouri Historical Society, St. Louis

Kansas State Historical Society

Early river explorers. In 1822, a fur trader named William Ashley took out a newspaper ad **(upper left)**. Among those who responded, several later became famous "mountain men," including Jedediah Smith and Jim Beckwourth. The photo shows Beckwourth **(upper right)**, "a tough hombre, a dare-devil, a thug, and a liar." Ashley himself later led a small party down the Green. They nearly drowned in what Beckwourth called the "Green River Suck," a waterfall supposedly 250 feet high.

Courtesy of the Huntington Library, San Marino, Calif., from print owned by Emma White Smith

Reprinted from Pioneer Nevada, *Courtesy Harold's Club*

James White **(left)**, a prospector, was found naked and incoherent on September 8, 1867, clinging to a raft about one hundred miles below the Grand Canyon. Fleeing Indians, White had made a crude raft and launched himself into the Colorado. His rescuers claimed, almost certainly mistakenly, that White had floated all the way through the Grand Canyon.

The first man after Powell to lead an expedition through the Grand Canyon was Frank Brown (**left**), a cheery businessman who had visions of building a railroad down the Colorado and through the Grand Canyon. The inscription (**below**), cut by a crewman named Peter Hansbrough, records Brown's fate. Hansbrough himself drowned five days later.

Emery and Ellsworth Kolb (Emery is dangling from the rope) were daredevil photographers, based at the Grand Canyon, who would do anything for a picture. The brothers were among the first to duplicate Powell's trip with no deaths, in 1911. *(Kolb Collection, Cline Library – Northern Arizona University)*

Robert Stanton, Frank Brown's second in command, took over after Brown drowned. In 1889, after three drownings, Stanton and his near-mutinous crew hiked out of the Grand Canyon. As they climbed, they saw Brown's lifeless body float downstream. *(Copyright New York Public Library, image courtesy of the Huntington Library, San Marino, Calif.)*

Carrara, P., U.S. Geological Survey

Steamboat Rock, in Echo Park on the Green River. Powell set out to climb it, he wrote, but found himself trapped. Bradley took off his drawers, lowered them to Powell, and lifted his one-armed leader to safety. The rescue was real, but the misadventure occurred a hundred miles from Steamboat Rock. Powell made a mistake or decided that so dramatic a tale deserved a fitting setting.

THE RESCUE.

Courtesy of Library of Congress, Manuscript Division

FIRE IN CAMP.

On June 17, the men were pleased to find a camp where trees and bushes provided some welcome shade. Hawkins, the cook, built a fire that spread out of control, and the men raced to the boats, their hair and clothing ablaze. Adrift and alit, they careened through a rapid. *(Courtesy of Library of Congress, Manuscript Division)*

One of the earliest photos of the Grand Canyon, by Jack Hillers, in 1872. The Colorado River, which carved the mile-deep canyon, looks tiny and insignificant. *(Hillers, J.K., U.S. Geological Survey)*

This Gothic rendering of the Grand Canyon appeared in Powell's account of his expedition, in *Scribner's Monthly*. *(Courtesy of Library of Congress, Manuscript Division)*

The last entry from Powell's river diary, written on August 28 at Separation Rapid. Writing awkwardly with his left hand, Powell scrawled "Boys left us ran rapids Bradey boat broke camp on left bank camp 44." Bradley (not "Bradey") might have drowned when his "boat broke." That adventure, he wrote, "stands A No. 1 of the trip." *(National Anthropological Archives, Smithsonian Institution)*

HERE · ON · AUGUST · 28
1869
SENECA HOWLAND, O.G.HOWLAND,
AND
WILLIAM H. DUNN
SEPARATED · FROM · THE · ORIGINAL
POWELL · PARTY, CLIMBED · TO
THE · NORTH · RIM · AND · WERE
KILLED · BY · THE · INDIANS.

FOR · FURTHER · AUTHENTIC
INFORMATION · SEE "COLORADO
RIVER · CONTROVERSIES"
OBTAINABLE · FROM · UNIVERSITY
LIBRARIES.

THIS · CENOTAPH · WAS · PLACED
AND · DEDICATED · IN · 1939 · BY · LATER
COLORADO · RIVER · VOYAGERS.

The plaque commemorates the three members of the expedition who left the river at Separation Canyon and took their chances on hiking to safety. *(Grand Canyon National Park)*

An early photo of Separation Rapid, taken in 1923. It was the worst-looking rapid they had seen, "appalling" to behold. Powell stayed awake all night, pacing back and forth on the small sand beach (top of picture), trying to decide whether to brave the rapid or to hike out of the canyon. *(LaRue, E.C., U.S. Geological Survey)*

After the expedition's end, good-natured Andy Hall resumed his career as a mule driver. In 1882, he was killed by thieves who stole a payroll he was carrying. The thieves were quickly caught and dispatched "in an orderly lynching." *(courtesy Ellen Watters)*

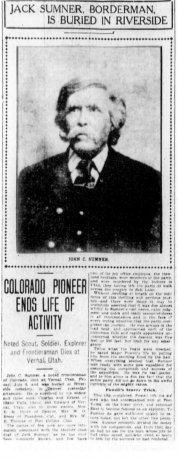

Jack Sumner turned to prospecting and died in 1907, a poor, embittered man convinced that Powell had robbed him of the pay and the glory that were due him. His obituary described Sumner as the expedition's "real leader" and denounced "Powell's Scurvy Trick" in stealing credit from him.

Image reproduced from the Brand Book, Courtesy the Stanton Collection, New York Public Library.

Courtesy Colorado Historical Society

This 1873 print of "American Progress" (sometimes called "Manifest Destiny") depicts the unstoppable advance of farmers, pioneers, railroads, and telegraph lines moving West. Indians (far left) flee in dismay. Powell argued endlessly against this vision of progress and made countless enemies in the process. *(Reproduced from the Collections of the Library of Congress)*

Powell at his desk in Washington, D. C., in 1894. For years one of the most powerful bureaucrats in Washington, Powell insisted that the West was fundamentally unlike the East, because it was so dry. Furious with this heresy, Western boosters eventually drove Powell out of town. Decades after his death, Powell's views found new favor. *(National Anthropological Archives, Smithsonian Institution)*

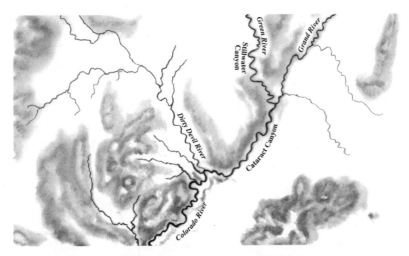

CHAPTER SEVENTEEN

FLASH FLOOD

In years to come, many river runners would find themselves staring at the same waves that had commanded Powell's rapt attention. ("As soon as we had decided on a channel we would lose no time in getting back to our boats and running it," wrote Emery Kolb, "for we could feel our courage oozing from our finger tips with each second's delay.") The rapids in Cataract Canyon are numbered, and Powell had arrived at numbers 21, 22, and 23, the most feared of all. In recognition of their special status, they are also known as Big Drop 1, Big Drop 2, and Big Drop 3. The worst is Big Drop 3; its most harrowing feature, though it is the only path to salvation, is a smooth and skinny raceway that slithers between two enormous holes. Since the 1950s, it has been known as Satan's Gut.

On July 25, Powell and his men tried to run Big Drop 1. This was an uncharacteristic decision, and, it immediately became clear, an unwise one. "The *Emma Dean* is caught in a whirlpool, and set spinning about," Powell wrote, "and it is with great difficulty we are able to get out of it, with the loss of an oar." Chastened by the near-crisis, Powell insisted on returning to lining and portaging. For the rest of that long day, the men inched their way down-stream. In Powell's account, there are none of his customary lyrical descriptions or upbeat pronouncements. "We camp on the right bank," he wrote, "hungry and tired."

Forty years would pass before someone managed to run the Big Drops. The run is still intimidating today, especially in high water. When Cataract is in flood, boatmen can lose track of where they are. Unable to identify which rapid they are in, they career along until they reach the Big Drops, which are unmistakable. No longer lost, they are in deep trouble nonetheless.

At flows higher than about forty thousand cubic feet a second, stopping between the Big Drops to scout and make plans becomes impossible, and the three rapids merge into one giant, interminable horror called simply Big Drop.

The first, and easiest, stretch features a series of giant standing waves that, if all goes well, should do no more than fill your boat. Then comes a series of unrunnable holes that must be dodged (in a boat quite likely hobbled by a load of water), and then, if the boat is still right side up, *another* set of towering waves. The only "choice" is to ride through these waves, but hidden among them is a joker—an "exploding" wave that breaks and crashes at irregular, unpredictable intervals. When this fickle beast seizes on a boat with the bad fortune to venture near at feeding time, the boat capsizes, regardless of its size. The boatman, in the meantime, has yet another task. At the same time as he is fighting through these waves, he must maneuver his way to the right, to avoid an enormous "pourover" hole that lies just beyond them.

The writer and boatman Pam Houston wrote a prize-winning short story that deals with running Cataract's rapids in flood, among other things. The white-water passages, Houston says, are not fiction at all but an exact depiction of a near-catastrophic trip of her own. Houston and her passenger made it through the towering waves of Big Drop 1 uneventfully and almost dry (unlike Powell). The trouble began at Big Drop 2, which has as its central feature a gigantic boulder that forms a pourover called Little

Niagara.* "I was worried about a funny little wave at the top of 2 on the right-hand side," Houston wrote, "a little curler that wouldn't be big enough to flood my boat but might turn it sideways, and I needed to hit every wave that came after it head on."

She decided to miss the wave by scooting a bit to the right, wondered if she had overcompensated, and realized in one stomach-turning moment that

> we *were* too far right, way too far right, and we were about to go straight down over the seven-story rock. We would fall through the air off the face of that rock, land at the bottom of a seven-story waterfall, where there would be nothing but rocks and tree limbs and sixty-some thousand feet per second of pounding white water which would shake us and crush us and hold us under until we drowned.
>
> I don't know what I said to Thea [her passenger] in that moment, as I made one last desperate effort, one hard long pull to the left. I don't know if it was *Oh shit* or *Did you see that* or just my usual *Hang on* or if there was, in that moment between us, only a silent stony awe.
>
> And as we went over the edge of the seven-story boulder down, down, into the snarling white hole, not only wide and deep and boat-stopping but corkscrew-shaped besides, time slowed down to another version of itself, started moving like rough-cut slow motion, one frame at a time in measured stops and starts. And of all the stops and starts I remember, all the frozen frames I will see in my head for as long as I live, as the boat fell through space, as it hit the corkscrew wave, as its nose began to rise again, the one I remember most clearly is this:
>
> My hands are still on the oars and the water that has been so brown for days is suddenly as white as lightning. It is white, and it is alive and it is moving toward me from both sides, coming at me like two jagged white walls with only me in between them, and Thea is airborne, is sailing over my head, like a prayer.

---

*Houston estimates the boulder's size as "seven stories," which is almost certainly a considerable overestimate. "You could never get a group of boatmen to agree on the actual size," one river runner told me, but the experts I surveyed offered estimates that ranged from fifteen or twenty feet to fifty-plus. The rest of Houston's account struck the experts as terrifyingly accurate. Little Niagara is "one hell of a scary place," "a place you want to avoid at all costs," "possibly the biggest white water in North America," "horrifying."

> Then everything went dark, and there was nothing around me
> but water and I was breathing it in, helpless to fight it as it wrapped
> itself around me and tossed me so hard I thought I would break
> before I drowned. Every third moment my foot or arm would catch
> a piece of Thea below me, or was it above me, somewhere beside
> me doing her own watery dance.

In 1969, Gaylord Staveley, a professional boatman with more than a dozen years' experience on the Colorado, decided to celebrate the centennial of Powell's 1869 expedition by retracing his journey (using modern river-running techniques and lighter, more maneuverable boats). Cataract Canyon provided the expedition's first real danger. Staveley's party made it safely by Big Drop 1, where Powell had run into a whirlpool, and by Big Drop 2, where Houston had plunged over a colossal boulder. Staveley scouted Big Drop 3 from shore while the others in his group ate breakfast. He returned without an appetite.

The rapid was formed by a kind of natural dam made "not of smooth concrete but of cruel, mighty rocks," Staveley wrote. "The river above was nearly brought to a standstill by their close-set, steeply coursed arrangement from bank to bank. Then, when it finally pitched over the edge, the waiting rocks instantly tore it to shreds." Ellsworth Kolb, the turn-of-the-century photographer and river runner, had blanched at the same sight. Worst of all, Kolb wrote, were the "jagged rocks, like the bared fangs of some dream-monster."

Satan's Gut, gleaming and glossy as a scar, marked a boat-wide course between two "bottomless, thrashing abysses." Staveley and his fellow boatmen stood above the rapid, trying to commit its features to memory. One man methodically threw pieces of driftwood into the maelstrom, one after the other and at various distances from shore, to see what became of them. Most swept over the edge and disappeared from sight. Only those pieces that went straight down Satan's Gut reappeared downstream. The visual cue to watch for, the boatmen decided, was a particular wave, about three feet long and six inches high, capped with white foam. The left side of that wave marked the entrance to Satan's Gut. Staveley took the first boat.

> In the instant I could look down on the rapid before dropping in, I
> realized the water had carried my fourteen-hundred-pound boat a
> little differently than it had the arm-sized chunks of wood. With inches

as important there as whole boatwidths would normally be, I was in the wrong place—we should be taking the middle of the wave instead of the left one-third! These realizations were projected against my stare down—down, down, down—into a monstrously thrashing void just three feet off my left stern quarter. There was, I know, one stroke, one deep, desperate holding stroke from my feet against the bulkhead up through pushing legs and taut stomach muscles and shoulders and elbows and then a prolongation of it by hauling the oars back just as far as I could lean. I may have gotten part of a second one like it. They would have been upstream strokes, weakening the right oarpull toward the end, to get both holding and right-quartering action. Satan's Gut was on my right, a half boatwidth away!

It will always be vivid, I think, the memory of that foam-filled abyss, and how we hung above it, and the stroke slowed us a little, and then a wondrous vagary of last-chance current skirting the brim to go down the Gut took us with it. We were in and then out of danger in perhaps five seconds.

For the Powell expedition, danger had become a traveling companion who showed no signs of leaving. Powell had portaged Big Drops 2 and 3, but the portages were dangerous and exhausting, the boats were leaking badly again, and the river was "still one foaming torrent" as far downstream as anyone could see.

Nor did trouble confine itself to the river. The cliffs and side canyons proved as life-threatening as the rapids. On July 26, after struggling by the Big Drops, Powell, Bradley, Seneca Howland, Hall, and Walter Powell set out up a side canyon, hoping to climb to the top and collect tree sap to caulk the boats. Eventually they found themselves in an enormous amphitheater topped with unclimbable walls. The men split up, each trying to find his own route to the top. Powell, characteristically, decided to spend the time hunting for fossils instead. Then, spotting a rock slide that seemed to offer a path up, he changed his mind and began climbing again. He emerged onto a narrow shelf, followed it a short distance, and found a narrow, vertical fissure leading up to another shelf perhaps forty feet higher on the cliff face.

"I have a barometer on my back, which rather impedes my climbing," Powell wrote. "The walls of the fissure are of smooth limestone, offering neither foot nor hand hold. So I support myself by pressing my back against one wall and my knees against the other, and, in this way, lift my body, in a shuf-fling manner, a few inches at a time." Powell crept upward about twenty-five feet, and then the fissure widened ever so slightly.

Unable to climb any higher (because he could not press his knees hard enough against the far wall to gain any purchase), Powell tried instead to retreat the way he had come. He found he could not move lower without falling. Unable to move higher or lower, he found he could at least move *horizontally*, into a kind of alcove. "So I struggle along sidewise, farther into the crevice, where it narrows. But by this time my muscles are exhausted, and I cannot climb longer; so I move still a little farther into the crevice, where it is so narrow and wedging that I can lie in it, and there I rest."

After five or ten minutes gathering his strength, Powell crabbed his way sideways back to the main chimney. This time, he managed to make it upward past the wide spot in the fissure and finally emerged onto the upper shelf. After another hour's climbing, Powell struggled out onto the summit. On top at last, he could finally turn his attention to the point of this whole excursion, gath-ering resin from the piñon pines to seal up the leaks in the boats.

"But I have with me no means of carrying it down," Powell realized. Considering that he had just risked his life precisely so that he could retrieve tree sap, this was a remarkable oversight. Powell tried, fruitlessly, to improvise. "The day is very hot, and my coat was left in camp, so I have no linings to tear out." Then, finally, he had a brainstorm. "It occurs to me to cut off the sleeve of my shirt"—which was, as always, dangling uselessly—"tie it up at one end, and in this little sack I collect about a gallon of pitch." A missing arm, to hear Powell tell it, seemed a stroke of fortune.

All that remained was to return to camp with the pitch. But few excursions that involved Powell were routine. Powell was a magnetic man, and, somehow, he was as adept at drawing trouble as he was at drawing people. Things hap-pened when he was around.

For an hour or so, Powell had gathered pitch, measured the altitude with his barometer, and wandered about the cliff top sightseeing. "Suddenly I notice that a storm is coming from the south. I seek a shelter in the rocks; but when the storm bursts, it comes down as a flood from the heavens, not with gentle drops at first, slowly increasing in quantity, but as if suddenly poured out."

Powell was drenched, almost washed away. Half an hour later the clouds had passed, the sun shone again, and Powell started his downhill climb.

Desert thunderstorms are unlike anything familiar in more evenly watered regions of the globe. "People die in *chubascos* [the proper term for what are commonly called monsoons]," writes the natural historian Craig Childs, "when twenty minutes earlier they didn't even think there would be weather. Most of a year's precipitation can easily be unloaded in six minutes, while one mile away the ground might not even be dampened." The storms truly come out of the blue, often in groups "like packs of feral dogs," and then they vanish as suddenly as they appeared.

The rains bring floods. "The waters that fall, during a rain, on these steep rocks, are gathered at once into the river," Powell noted, awestruck. "They could scarcely be poured in more suddenly, if some vast spout ran from the clouds to the stream itself." Tiny streams that an eight-year-old could hop over become angry rivers. A pickup truck parked in a dusty, eroded gulch can suddenly be swept away like a twig blasted by a fire hose.

"I have stood in the middle of a broad sandy wash with not a trickle of moisture to be seen anywhere, sunlight pouring down on me and on the flies and ants and lizards, the sky above perfectly clear," the writer Edward Abbey recalled, "listening to a queer vibration in the air and in the ground under my feet—like a freight train coming down the grade very fast—and looked up to see a wall of water tumble around a bend and surge toward me."

In canyons, the flood system plays out at its most dramatic. "Canyons are basically nets that catch water," Childs explains. "Branches and fingers and tributaries scour the land above, sending everything down, so that when a storm passes, all of its rainwater is driven toward a single point. Water can run from tens of miles down hundreds of feeder canyons, spilling into deeper and deeper, fewer and fewer canyons until the volume of the flood has jumped exponentially into one final chasm where everything converges."

Flash floods are common in the West. In 1997 and again in 1999, for example, flash floods hit Bright Angel Trail, the most popular hiking trail in the Grand Canyon. The nine-mile trail runs from the South Rim at the top to Phantom Ranch at the bottom. Often it is as crowded (and as cosmopolitan) as a Manhattan sidewalk, bustling with tourists chatting in French and German and Japanese as well as English. But the floods—and the refrigerator-sized boulders they toss about—make plain that the Grand Canyon is not a stage set. Two people were killed in 1997; four were injured in 1999, and another twenty-seven were evacuated from the canyon by helicopter.

The floods move quickly, but the danger comes not necessarily from the water's speed. As we have noted, even the fastest rivers move at only 20 miles an hour; in comparison, avalanches race at 100-plus miles an hour. Most often, the greatest danger in a flash flood comes from being trapped in a steep, unclimbable passage while the water rises.

"An entire oak tree paused at the margin of a falls, then tumbled, branches prying away as I watched," Craig Childs wrote of one flood he witnessed from a perch on a canyon ledge. The water crashed against rocks and sent them flying into the air, "pelting anything nearby, breaking into shrapnel. I could not count fast enough. Maybe ten rocks a second, six hundred rocks a minute, into the air. The tree was pulverized at the bottom. Any jag or protrusion was beaten down. If I were to stick my hand in there, my bones would splinter."

Childs continued to stare, hypnotized. "The flood bounded over immovable boulders, churning into whirlpools. It had a texture like a rapid boil, sending debris up, sucking it down, each roil a barrel rising to the surface. Nothing stayed on top for more than a couple of seconds. The trunk of a cottonwood tree showed through. Smaller boulders, three feet across, stabbed and rolled against the larger ones."

Such was the creature at Powell's heels. He scrambled down a side canyon, hoping to outpace his pursuer before it gathered its strength. "I find a thousand streams rolling down the cliffs on every side, carrying with them red sand; and these all unite in the cañon below, in one great stream of red mud."

But the rain had never reached the canyon's lower reaches, and the racing water vanished into the dry, sandy ground. "Although it comes in waves, several feet high and fifteen or twenty feet in width, the sands soak it up, and it is lost. But wave follows wave, and rolls along, and is swallowed up; and still the floods come on from above." Powell managed to stay ahead of the floodwaters. "I hasten to camp, and tell the men there is a river coming down the cañon. We carry our camp equipage hastily from the bank, to where we think it will be above the water. Then we stand by, and see the river roll on to join the Colorado."

•   •   •

The returning hero met a sullen reception in camp. Hectic and danger-filled as the day had been, it had brought the expedition only a mile and a half nearer its destination. The men's grousing, which to this point had reflected little more than the understandable weariness of the overworked and underfed, began to take on a sharper edge. "Another day wasted foolishly," Bradley complained to his diary. ". . . Major wished to land and climb the mountains so five of us started on a wild-goose chase after pitch but it was so hot we all backed out except the Major who says he climbed the cliff, but I have my doubts."

Bradley's explicit criticism of Powell marked an escalation in the tensions swirling around the expedition. The distrust of Powell's claim was all the more telltale in that it was unfounded—of the five men who had set out to reach the top of the cliff, only Powell had made it. Sumner did acknowledge Powell's accomplishment, but he was grudging, at best, in his praise of the one-armed climber who accomplished what none of the able-bodied men could match. "5 of the men tried to climb the cliff to get some rosin from the pine trees at the top," he wrote, "but all failed but the Professor, he being lucky enough to get about 2 lbs."

The mood improved considerably the next day, although not until late afternoon. The day began with a portage of two hundred yards, a marathon for men carrying boats and supplies on their backs, over boulders, in the desert heat. It continued through a string of "very bad" rapids, which called for still more lining. At the same time, the canyon was narrowing ominously. "In many places [the walls] meet the water on both sides," Bradley wrote, "so that if we meet an impassable rapid we shall have to run it with all the risk, or abandon the expedition."

The men had seen such fearsome spots before, but they had tried to convince themselves that they had made it through the worst of the tight squeezes. As recently as three days before, Bradley had reassured himself that there was "no place in the Colorado where we could not land on either side of the river." Now that hope, like so many others, had been yanked away.

But the good news, which temporarily banished any worries, was that the hunters managed to shoot two bighorn sheep. The men rounded a bend, saw the sheep in the rocks, and raced to shore. While one anxious faction pulled the boats against the rocks, out of sight, some of the men hurried off in pursuit of dinner. They bagged one sheep at once and lost the others, but then the flock turned and ran straight toward the boats. When the sheep were within twenty yards, one of the men took up his rifle and fired.

Everyone was overjoyed. "In the present reduced state of our ration," Bradley wrote, the killing of two sheep was "hailed as the greatest event of the trip." Sumner hailed the bounty as "a Godsend," for "sour bread and rotten bacon is poor diet for as hard work as we have to do." (The bacon had been salted and smoked to preserve it, but bacon was best stored in a cool, dry pantry, not dunked in a river and exposed to the desert sun. After two months, the bacon fat had turned rancid.) The men happily set to work dressing the kill and then lashed the prizes to the deck of one of the boats.

Downstream a bit, when they had found a good camp, they would prepare dinner. "But fresh meat is too tempting for us," Powell wrote, "and we stop early to have a feast. And a feast it is! Two fine, young sheep. We care not for bread, or beans, or dried apples tonight; coffee and mutton is all we ask."

Revived by their feast, Powell and the crew set out early the next morning. At once, the river did its best to dispel any lingering good cheer. The morning was largely taken up with two hard portages, one of them the longest on the Colorado thus far. By midday, the expedition had advanced only two discouraging miles.

The pace picked up after lunch, when the river seemed to turn less nasty. Quickly it became clear that the pace was *too* fast. At the foot of one steep stretch, the river made a sharp right, and the water piled up against the cliff that stood in its way. Out of control, the men pulled with all their might to stay off the cliff, but plunged straight toward it. Before the boats smashed against the rock wall, though, the waves rebounded off the rock and against the boats, sending them careening downstream helpless and full of water but safe.

Almost at once, as if to show just how broad its repertoire of tricks could be, the river played an equally malicious but quite different prank. Powell and the men had just learned what it was to see disaster looming directly in front of them and be helpless to avoid it. Now they would run along half blind, sensing danger just ahead but unable to make it out. Like Columbus's sailors, but with better reason, Powell and his men believed that at any moment they might plunge over the edge of the world.

They had been pinned between sheer cliffs before, but never for so long and never with so limited a downstream view. "The walls suddenly close in," Powell wrote, "so that the cañon is narrower than we have ever known it. The water fills it from wall to wall, giving us no landing place at the foot of the cliff; the river is very swift, the cañon is very tortuous, so that we can see but a few hundred yards ahead; the walls tower over us, often overhanging so as to almost shut out the light."

Powell stood on deck, gazing downstream with "intense anxiety," watching for boulders in the channel or the spray of whitecaps or, the most dreaded sight of all, the clean horizontal line that signified a waterfall. The men at the oars sped backward to a fate they could only guess at. For a mile and a half, the river swept along between high rock walls that never permitted a long view.

Finally the narrow gorge opened into a more open, broken section of canyon. By three o'clock, the expedition had emerged, safe, from the dreaded Cataract Canyon.

The ordeal had lasted perhaps fifteen minutes, a short enough time in ordinary circumstances. For Powell and the crew, the time seemed endless, as if they had been sentenced to race down a booby-trapped stairwell in the dark knowing that at any moment they could find themselves stepping helplessly into empty air.

That single excruciating episode captured, in a way, the great burden that Powell and his men confronted. Uncertainty sucked out their strength and resolve more surely than any other hazard, even if, in the end, nothing happened. "Now that it is past," Powell wrote, with magnificent understatement, "it seems a very simple thing indeed to run through such a place, but the fear of what might be ahead made a deep impression on us."

Dirty Devil Creek

Glen Canyon

Paria River

Crossing of the Fathers

Glen Canyon

San Juan River

UTAH

ARIZONA

CHAPTER EIGHTEEN
TO THE TAJ MAHAL

It had been two months since the expedition set out from Green River Station. While Powell and his men struggled through rapids and whirlpools, like Ulysses and his crew in some modern retelling of the *Odyssey*, the rest of the world had been left with no clue to their fate but a handful of out-of-date letters.

A conversation between two strangers in a stagecoach revealed how completely Powell and his men had vanished. Richard Townshend was a blue-eyed, curly-haired young Englishman in search of experiences beyond those available at Trinity College, Cambridge. In 1869, he had set out to see the American West. (Frank Goodman, who had left the Powell expedition in the Uinta Valley, was not the only Englishman smitten with the romance of the Wild West.) Townshend seemed bound for trouble—his nickname was "Cherub"—but he thrived among his new gold-seeking, gun-packing acquaintances. On a stagecoach, he happened to meet one of the West's most colorful characters, William Gilpin. A biography of Gilpin would make a fat and lively book, and the scene changes would give a reader whiplash. Raised a Quaker, Gilpin became a renowned soldier. Born in the East, he became one of the great spokesmen for Manifest Destiny and the vital role of the American West. Along the way, he explored Oregon with Frémont, served as Colorado's first territorial governor (appointed by Abraham Lincoln), and made a fortune in land speculation.

An acclaimed orator, Gilpin had an overblown rhetorical style that was excessive even by the standards of the day. In full bray, he sounded more than a bit like a nineteenth-century W. C. Fields. "The *untransacted* destiny of the American people is to subdue the continent," he thundered, "to rush over this vast field to the Pacific Ocean . . . to stir up the sleep of a hundred centuries—to teach old nations a new civilization—to confirm the destiny of the human race—to carry the career of mankind to its culminating point . . . to dissolve the spell of tyranny and exalt charity—to absolve the curse that weighs down humanity, and to shed blessings round the world!"

On the day Townshend met him, Gilpin was preoccupied with Powell and the Grand Canyon. "The greatest gash on the earth's surface, Mr. Townshend," Gilpin declared. "It is 500 miles in length and a mile in vertical depth. . . . Never has it been traversed by mortal man. Dead bodies of Indian braves have been washed down it, bodies of Utes or Navajos or Apaches slain in their tribal wars and given to the current, but never yet has any living human being passed that way."

Now, "that grand explorer, Major Powell" and his "brave companions" had set out to do what had never been done. Gilpin had his doubts about whether they could succeed. "The torrent may have swallowed them up; the Indians may have destroyed them; their boats may have been wrecked with the loss of all their provisions; and they may consequently have been starved to death; we know nothing." He begged Townshend to keep his ears open for any news as he continued his roaming.

Despite the bluster, Gilpin had made a good guess about Powell's whereabouts. The expedition had not yet reached the Grand Canyon, but it was only a few days away. Better yet, the furies of Cataract Canyon seemed to have been left upstream. On July 28, for the first time in many days, Powell and his men had a long, almost routine, run that stretched a dozen miles. They passed a stream that joined the Colorado from the west and was not marked on any of their maps. "Is it a trout-stream?" Powell recalled someone calling out hopefully, but it was muddy, not clear, and strewn with flood-borne, rotting vegetation. Dunn, in the lead boat, shouted back disgustedly that it was a "dirty devil."

"The water is about as filthy as the washing from the sewers of some large, dirty city," Sumner complained, and later he amended that description to make it still more repellent. The Dirty Devil, as they named it, "stinks bad enough to be the sewer from Sodom and Gomorrah, or even hell. I thought I had smelt some pretty bad odors on the battle field two days after action, but they were not up to the standard of that miserable little stream."

In light of what was soon to come, the vehemence of those descriptions is eye-catching. Hawkins would later dispute Powell's innocuous anecdote about the origin of the name Dirty Devil. As Hawkins told the story, it was Powell, not Dunn, who coined the name. More to the point, in Hawkins's version Powell chose "Dirty Devil" as an allusion to Dunn and a mark of contempt for him. On a trip as long and hard as this one, sulking and grousing were almost inevitable. Outright feuding, on the other hand, could be deadly.

Bradley mentioned the naming of the muddy river, too, but only in a single, hard-to-interpret sentence. "Major named the new stream 'Dirty Devil's Creek,'" he wrote, "and as we are the only white men who have seen it I for one feel quite highly complimented by the name, yet it is in keeping with his whole character which needs only a short study to be read like a book." Bradley's meaning is unclear, though the critical tone comes through plainly enough. And if we pair this remark with Bradley's criticisms of Powell from only three days before, for frittering away precious time and telling tall tales, there seems little doubt that the mood in camp was deteriorating.

The criticism is especially telling because Bradley was a hard man to provoke. Powell moved him to occasional fits of sputtering frustration (though he nearly always kept his fuming confined to his diary), but, to a remarkable extent, Bradley managed to ignore what he could not change. He traveled week after week in the same boat with the sullen, moody, half-demented Walter Powell, for example, and virtually never mentioned him in his diary.

At the same time that the men were growing ever more tense, the river seemed to be feeling almost mellow, at least in comparison with what it had been. "Run 20 miles with ease," Bradley wrote on July 29, one of the few references in two weeks to anything at all coming easily. "Found many small rapids or what we call small ones now but which would pass for full-grown cataracts in the States."

Someone spotted the stone ruins of an Indian dwelling alongside the river and about two hundred feet above it. For once the men were nearly as willing as Powell to linger and explore. They spent two hours looking at rock etchings, gathering arrowheads, and picking through pottery shards. Bradley took the trouble to measure the dimensions of each room and pocketed a few especially fine pottery bits as souvenirs.

Neither Powell nor the men believed that a place was only discovered when the first white men arrived. But even if they *had* felt inclined to belittle the Indians who had once lived here, the signs all around them of what had been routine, ongoing life would have raised troubling questions. To men who were only a few bags of moldy flour away from starvation, the Indians seemed like figures to puzzle over and admire, not to deride. "How they contrived to live is a mystery to me," Sumner marveled, for the country was "as destitute of vegetation . . . as the paper I write on." (Animals were nearly as scarce as plants. Dunn shot a coyote on July 29, and the half-starved creature was the only animal anyone had seen in three days.)

Returning to the river and heading downstream again, the men soon found another group of ruins. They made camp, and as usual Powell set out to climb the cliffs so he could scan the countryside. Quite near the top, he found himself stymied, unable to climb higher. Looking around, he eventually discovered steps cut into the rock and, nearby, an ancient, rickety ladder. For Powell, a man fully as romantic as any poet contemplating mortality in a country churchyard, this was a powerfully poignant moment. He climbed to the cliff top in the steps of his anonymous predecessors. "Here I stand," he wrote, "where these now lost people stood centuries ago, and look over this strange country."

And there he stayed, gazing out at the distant mountains until it was too dark to see. This made for a difficult return to reality, and to camp. "It is no easy task to find my way down the wall in the darkness," Powell wrote, "and I clamber about until it is nearly midnight, before I arrive."

They had reached a lyrically beautiful spot they would soon name Glen Canyon. "Tranquility" was the last word anyone would associate with Cataract Canyon—"violence" would be a likelier choice—but Glen Canyon was, in many ways, a gentle place. In comparison with the Grand Canyon, just down-stream, it was almost cozy, a mere thousand feet high in comparison with the Grand Canyon's five thousand.

The sense of imminent disaster that marked Cataract was absent here. Glen Canyon was marked more by sweeping curves than by sharp angles; the river that had crashed in fury through Cataract Canyon now meandered aimlessly along. Here and there the boats darted through riffles, but there were almost no rapids worthy of the name. Glen Canyon inspired dreamy reveries rather than nightmares.

"The curves are gentle," Powell wrote, "and often the river sweeps by an arc of vertical wall, smooth and unbroken, and then by a curve that is varie-gated by royal arches, mossy alcoves, deep, beautiful glens, and painted grottos." The sandstone cliffs rose from the water in great, swooping walls of orange and burnt red and purple-black. Enormous swatches of "desert varnish," black or tan stains formed of dissolved manganese and iron, draped the walls like tapes-tries. Here and there, springs refreshed the rock, and mosses and ferns grew on the moistened sandstone. Where bigger springs burst from the cliffs, cotton-woods and willows and oaks took root.

Shortly after lunch on July 31, the expedition reached the mouth of the San Juan River, which joined the Colorado from the east. The temperature remained at 100 degrees and the rocks were "almost hissing hot," but river junctions were always significant milestones and the men stopped to explore. Beautiful as it was, the canyon now revealed a harsh aspect that it had previously concealed. "The remainder of the afternoon is given to hunting some way by which we can climb out of the cañon," Powell wrote, "but it ends in failure."

There was no pressing need to climb out (except that a view from the rim would help in mapping the region), but this was a worrying sign even so. Later on, if the men lost another boat or found their way blocked by a waterfall, they would be hunting in earnest for such an escape route. Failure then could mean death.

Bradley was preoccupied not with beauty but with hunger. On the morn-ing of August 1, the men saw three bighorn sheep but failed to shoot any. "They seem to have no fear of falling," Bradley noted with melancholy admiration, "but will leap from rock to rock, never stumbling nor slipping though they will be a thousand ft. above us and a single mis-step would dash them to atoms."

The talk of food took on a sharper edge. The endless monotony of the meals was no longer the chief complaint. Survival, not variety, was fast becoming the issue. "[The sheep] are very good eating and we need meat very much, not having over 15 lbs. of bacon in the whole outfit," Bradley wrote. "We are short of everything but flour, coffee and dried apples and in a few days our rations will be reduced to that." Those would be starvation rations even for idle men, let alone for men sentenced to long days of hard labor.

Powell made matters worse by showing no inclination to hurry. Through all the changes in the men's mood, and the river's, the one constant was Powell's fascination with geology. On August 1 the men pulled to shore to explore an alcove they could see from the river. On entering a little grove of box elder and cottonwood trees, they saw around a corner a vast chamber cut into the rock, two hundred feet high by five hundred long by two hundred wide. "And this," Powell marveled, "is all carved out by a little stream, which only runs during the few showers that fall now and then in this arid country."

Powell was face-to-face here with the great intellectual challenge posed by the seven hundred miles of canyon he had passed through and by the side canyons and caverns along the way. The challenge was to accept the dizzying lesson the rock landscape proclaimed—in the immensity of time, water prevails over stone and shapes it as it pleases. It is no great feat to mouth the words, but *believing* them is another matter. To try to grasp the unfathomable stretches of time required for a trickling stream to carve a cathedral-sized cavern is to risk a kind of intellectual vertigo. Geologists today call this time-induced dizziness rock-shock.

The shock was all the more profound in an era when it was still commonly believed that the earth was a mere six thousand years old. But for Powell, the rebellious son of a minister who believed in the Bible's literal truth, the notion of limitless time was a liberation rather than a consternation. It would become a central theme of his intellectual life and the great lesson he was to draw from the Grand Canyon.

The men camped on August 1 in the stone chamber that had astounded Powell. The Howland brothers and Bill Dunn scratched their names onto a smooth space on the wall. Walter Powell entertained the others with a Civil War ballad called "Old Shady"—he had sung it so often before that the men called *him* "Old Shady"—and Powell named their temporary quarters Music Temple in Walter's honor. Despite its bucolic-sounding title, this was a bitter, vengeful song, perhaps a good match for bad-tempered, war-damaged Walter Powell. "Old Shady" was a newly freed slave, and his song is his mocking farewell to his ex-master and to the South. "Good bye, Massa Jeff! Good-bye, Missus Stevens," the singer calls to Jeff Davis and Davis's vice-president, Alexander Stephens, and he warns them that "Pretty soon, you'll see Uncle Abram's comin', comin'! Hail, mighty day."

All the following day was spent in the same camp because Powell was intent on climbing the cliffs. After a false start or two, he reached "a point of commanding view" where he could see mountains in the distance on three sides and long stretches of the San Juan River and the Colorado. By this time, Bradley's journal included almost daily passages lambasting Powell for dawdling. "In the same camp," Bradley complained while Powell explored, "doomed to be here another day." Powell's journal entry recorded the same information in a tone of perfect contentment. "We sleep again in Music Temple," Powell wrote, as cheerily as if he had finagled an extra day's vacation at a favorite weekend retreat.

Bradley fretted that perhaps they might be stuck in place even longer than a day. "Major has been taking observations ever since we came here and seems no nearer done now than when he began. He ought to get the latitude and longitude of every mouth of a river not before known, and we are willing to face starvation if necessary to do it but further than that he should not ask us to wait and he must go on soon or the consequences will be different from what he anticipates."

Bradley's reference to "consequences" is ambiguous—was he threatening mutiny or merely forecasting bad times ahead?—but there is no missing his frustration with Powell. Indeed, this first explicit mention of starvation reflected dismay that passed far beyond frustration. "If we could get game or fish we should be all right," Bradley went on, "but we have not caught a single mess of fish since we left the junction."

In camp that evening, August 2, Bradley or one of the others confronted Powell about the urgency of getting under way. "Major has agreed to move on in the morning," Bradley noted happily, "so we feel in good spirits tonight."

The following day brought more good news. In an "easy run," the men advanced thirty-three miles downriver. Better still, Sumner managed to kill one of the elusive bighorn sheep. "The one we got today is quite fat and will weigh about 80 lbs. dressed," Bradley exclaimed. The last sheep the men had killed had spoiled in the heat before they could eat their fill. This time they took the trouble to dry some of the meat.

They saw countless fish as they rowed downstream, but the fish refused to bite. "Where the water is still we could see them catching small flies that the river seems covered with," Bradley noted. Nowadays the Colorado often runs cold and clear, because it is bottled up by dams that catch its sediment and form deep, frigid reservoirs. A fly fisherman can stand in the cold river in the hot desert and catch fish that are native to mountain streams. But in an era when the Colorado ran warm and muddy, most of the fish eluding Bradley were chub and sucker and squawfish, not trout. Squawfish, in particular, were prizes worth winning. Before the dams changed the Colorado, the biggest squawfish could reach six feet and eighty pounds. Old photographs often display one of these giants dangling from a hook, and next to it, for scale, a child no taller than the fish. Not this day. With fish all around them, Bradley and the others came up empty-handed.

Glen Canyon's splendors continued unabated. "Past these towering monuments," Powell wrote, "past these mounded billows of orange sandstone, past these oak set glens, past these fern decked alcoves, past these mural curves, we glide hour after hour, stopping now and then, as our attention is arrested by some new wonder." (The descriptions in Powell's river diary, written that day rather than years later, were more spartan. "Many huge Mts. seen near river and back some distance, low walls giving a view." Bradley dispensed with the scenery even more quickly than that, and Sumner found nothing noteworthy "excepting daily duckings and continuous fasting.")

Late in the day on August 3, the men reached the Crossing of the Fathers, near the present-day Utah-Arizona border. Here, in 1776, a fourteen-man team of explorers led by Father Silvestre Vélez de Escalante and Father Francisco Domínguez had forded the Colorado on their return to Santa Fe, having failed to find a new overland route from New Mexico to California. (The Powell expedition had crossed the Escalante party's path once before, hundreds of miles upstream in the Uinta Valley, where the Spanish explorers had forded the Green on the outbound leg of their journey.) The two missionaries became the first white men to describe great stretches of Colorado and Utah, but the experience almost cost them their lives. At one point in their six months' trek, they lost their way in the desert and nearly died of thirst; at another, they grew so hungry that they killed their mules for food; still later, they barely survived a blizzard.

The glories of Glen Canyon that Powell celebrated are vanished now, drowned under the lake formed by the immense Glen Canyon Dam. Like the dam, the lake is colossal—at 186 miles long and with a twisting shoreline longer than that between Seattle and San Diego, it is the second biggest man-made lake in the United States. (The biggest, Lake Mead, sits at the downstream end of the Grand Canyon, behind Hoover Dam.)

The dam took what had been a muddy, seasonal river that flooded in the spring snowmelt and trickled in the fall and transformed it into a dazzling blue lake shimmering in a red desert. Where once only a handful of rowboats and rafts ventured down a broad and winding river, today three and a half million visitors a year bring jet skis and motorboats and canoes and houseboats to a mirror-still lake.

Construction of the dam, a 710-foot-tall plug of white concrete, began in 1960. It is what is known as a "cash register dam," because one of its several roles is to bring in revenue by selling electricity to Phoenix and other Southwestern cities. Glen Canyon is a curious setting for a massive dam—the Bureau of Reclamation had originally opposed the site because a dam would be anchored in Navajo sandstone, which is porous and crumbly. "The dark red rock," one dam historian notes, "is actually solidified sand dunes."

In the end, Glen Canyon Dam was built as a kind of massive afterthought. The great dam-building battles of the 1950s were fought farther north, in Dinosaur National Monument, which stretches across the Utah-Colorado border. The Bureau of Reclamation had proposed two dams for Dinosaur. In one of the major environmental clashes of the modern era, conservationists desperately fought the proposal. They opposed the dams in their own right, and they vehemently opposed putting dams in national parks or monuments.

After a tumultuous battle, the plans for Dinosaur's dams were shelved in

1956. The two sides agreed to a compromise that was seen at the time as a triumph for environmentalists—in return for not building dams in Dinosaur, the Bureau of Reclamation could continue unopposed with its plans for a dam at a remote and almost unexplored spot in the desert called Glen Canyon.

The lake behind Glen Canyon Dam began to fill in 1963. Eventually the rising waters drowned the last thirty-five miles of Cataract Canyon, hundreds of Anasazi ruins, and countless grottoes, stone bridges, and natural auditoriums like Marble Temple, where Walter Powell sang "Old Shady" for Powell's crew. The few environmentalists who had seen Glen Canyon cursed the dam. "To grasp the nature of the crime that was committed," wrote Edward Abbey, "imagine the Taj Mahal or Chartres Cathedral buried in mud until only the spires remain visible."

It happened, perhaps, because Glen Canyon had no constituency. Only a tiny number of people ever saw what Powell and his men had seen. Eliot Porter was one of the last. The Sierra Club published a book of his photographs as a kind of canyon epitaph; the book bore the apt title *The Place No One Knew*.

The much-praised, much-reviled lake formed by the dam is named Lake Powell. It is perhaps a tribute to Powell's tangled character that his admirers cannot agree on whether this is an honor or a desecration. Those admirers, moreover, have nothing but Powell in common; though they share a hero, they can barely refer to one another without spitting in contempt. Edward Abbey, for instance, placed Powell high in his pantheon and advocated the *bombing* of Glen Canyon Dam. "Probably no man-made artifact in all of human history has been hated so much, by so many, for so long as Glen Canyon Dam," Abbey wrote. But the Bureau of Reclamation, which built the dam, bestowed Powell's name on the lake as the highest tribute it could offer.

Easy as the river seemed as their boats passed through Glen Canyon, the men were jumpy about the rapids still to come. Bradley suspected trouble ahead, although he tried to convince himself that his fears were unfounded. For seventy-five miles upstream of Cataract Canyon, he noted, the river had wandered "through a dark calm cañon which a child might sail in perfect safety." Then had come the madness of Cataract, but that had been followed by the long, gentle stretch of river through Glen Canyon. Perhaps Cataract had been a malevolent fluke?

The expedition continued on its way, hoping for the best and fearing the worst. "Today the walls grow higher, and the cañon much narrower," Powell noted on August 4. (This narrow spot would become the site of Glen Canyon Dam.) On they went, through "a perfect tornado with lightning and rain," past canyon walls of creamy orange and bright vermilion and purple and choco-late brown and green and yellow. They advanced thirty-eight miles, a fine total for a single day, and made camp near the mouth of the Paria River.

The Paria marked the end of one canyon and the beginning of another.

No one knew it, but the expedition was poised on the brink of the Grand Canyon. "With some feeling of anxiety, we enter a new cañon this morning," Powell wrote on August 5. "We have learned to closely observe the texture of the rock. In softer strata, we have a quiet river; in harder, we find rapids and falls. Below us are the limestones and hard sandstones, which we found in Cataract Cañon. This bodes toil and danger."

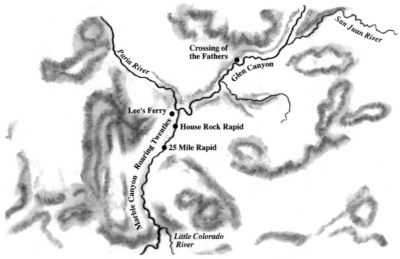

Crossing of the Fathers

San Juan River

Paria River

Glen Canyon

Lee's Ferry

House Rock Rapid

Marble Canyon

Roaring Twenties

25 Mile Rapid

Little Colorado River

CHAPTER NINETEEN
GRAND CANYON

They had camped just upstream of the Paria River, at a spot where the Mormons maintained a ferry. Indians had long crossed the river here as well, and Powell's men were dismayed to see fresh moccasin prints. "We were in no condition for a fight or a foot race," Sumner admitted.

The crossing is known today as Lee's Ferry, for John D. Lee, who was banished to this remote spot by Brigham Young in 1872. In his heyday a well-regarded member of the Mormon community—Lee was prosperous, handsome, a husband to nineteen wives, and an adopted son of Brigham Young himself—Lee ended up an outcast who met his death before a firing squad. In September 1857, Mormons and Indians had attacked a wagon train of "Gentiles" crossing southern Utah on their way to California. In what became known as the Mountain Meadows Massacre, some 120 men, women, and children were slaughtered. (The exact number of victims is unknown.) Only children too young to talk were spared. The Mormon leadership did its best to pin all the blame on the Indians, but when the truth began to seep out, Brigham Young assigned the role of scapegoat to his adopted son.* Lee, one of many who had participated in the massacre, was excommunicated from the church, assigned sole blame for the killings, and sent off to the barren hideaway that his wife named Lonely Dell.

Lee's Ferry is now the put-in spot for every river trip through the Grand Canyon. Rapids and other landmarks along the Colorado are labeled by their mileage below Lee's Ferry, which itself is designated as Mile 0. "It is desolate enough to suit a lovesick poet," Sumner remarked, and when the governor of Arizona Territory saw Lonely Dell, he remarked that he, too, would take up polygamy if banished to such a place. It is still desolate, but no one in search of solitude nowadays would venture to Lee's Ferry on an August morning. Easy access to the Colorado is a rare and valuable thing, and on summer days the boat landing at Lee's Ferry looks like a parking lot.

The boats—most of them inflatable rafts with an outfitter's name or logo brandished across a huge pontoon—sit at the water's edge, jostling one another like hippos. The boatmen, who will have worked through a blistering day and late into the night rigging their boats and checking their gear, do their best to deflect questions ("Are we gonna flip?") and concentrate on last-minute chores. Vans shuttling back and forth from nearby motels unload groups of eager, nervous passengers. They step out of the air-conditioning and into the sunlight, blinking, and take a first, wary look at the river they will be living on for the next week or two. When all the passengers have arrived and packed their clothes and sleeping bags away into waterproof rubber bags, the boatmen introduce themselves and give a lesson on adjusting a life jacket. It must be cinched *tight*, so that surging, whirling currents cannot yank it over a helpless swimmer's head. The trip leaders inject a small joke or two into their manda-tory safety talks, and everyone laughs a little too loudly.

At nearly the same spot a century and a third ago, Powell's men were jumpy, too. "Just below our camp a fine rapid commences that is roaring pretty loud and I can see the white foam for quite half a mile," Bradley wrote on August 4, 1869.

---

*New evidence now makes it seem that the Paiutes may have been framed. Recently, while digging a foundation for a monument to the victims, a backhoe operator unearthed thou-sands of bone fragments from the murdered emigrants. According to a page-one story in the *Salt Lake Tribune* on February 21, 2001, forensic anthropologists studying the bones found extensive evidence of gunshot wounds but no sign of injuries inflicted by knives or hatchets, the weapons supposedly wielded by the Paiutes.

"We have all learned to like mild rapids better than we do still water," he went on. "But some of the party want them very mild."

Bradley had fretted about the river and the weather almost from the day the trip began, and he had grown increasingly disdainful of Powell. But, until now, except for some muttering directed toward the hunters, he had never directed any criticism toward his fellow crewmen. Even here, he named no names. But Bradley's charge was clear enough—after seven hundred miles and uncounted rapids, some of the men were losing their nerve.

The next day he recanted a bit. "I said yesterday that we had learned to like rapids, but we came to two of them today that *suit us too well*. They are furious cataracts." The men had risen early and run one long rapid and then a dozen more in the next eight miles. Forced to portage around a rapid with a fifteen-foot drop, they lost their grip on the *Emma Dean* and stove a hole in her side while trying to carry her over the rocks. They made emergency repairs, then another brief run, and then another portage that even Powell acknowledged to be "long and difficult." By day's end, they had advanced twelve hard-won miles.

The men were beat. "Very hard work," Sumner scribbled in his journal that evening, and night brought no relief. Rain poured down and the wind howled. "Am very tired tonight," Bradley wrote. "Hope a good sleep will do me good but this constant wetting in fresh water and exposure to a parching sun begins to tell on all of us." Bradley's discouragement was especially noteworthy because he had become the expedition's most enthusiastic white-water boatman.

August 6 was a short day but a grueling one. The day's first task was repairing the *Emma Dean* properly. Then came what seemed like endless rapids and endless portaging. By day's end, Sumner, who often enlivened his journal with sharp observations and wry asides, had only enough strength for a bare-bones summary. "Made 3 portages and ran 10 bad rapids in 10½ miles. Walls of cañon 2000 ft. and increasing as we go. River about 50 yds. wide, rapids and whirlpools all the way." Bradley expressed his by-now-customary hope. "Three times today we have had to carry everything around rapids," he wrote, "but the last few miles we came tonight we found the rapids less furious and I hope we are out of the worst of this series."

In one stretch where the river ran between vertical cliffs, the expedition found itself forced to repeat the difficult and dangerous "leapfrog" maneuver they had devised a few weeks earlier. The three boats formed a precarious line, the first anchored to a rock, the second farther downstream and hanging on a line attached to the first, and the third still farther downstream and hanging on a line attached to the second. Then the last boat found a spot where it could anchor itself, and the upstream boats cast themselves free. Like the "catcher" in a trapeze act, whose job is to grab the daring young men somersaulting through the air to him, the downstream boat had to reel in its runaway companions and pull them to safety. It was a maneuver whose only merit was that the alternatives were worse, but it worked. Powell described the procedure at length, but Bradley, as if eager to put the whole desperate episode behind him, cut more quickly to the bottom line. "We succeeded in landing all our boats safely," he noted, and left it at that.

For decades to come, the rapids that had provided Powell and his men such a hostile welcome to the Grand Canyon would prove dangerous, even deadly. The next man after Powell to lead an expedition down the Green and the Colorado was one Frank M. Brown, a large, cheery, impatient, and recklessly optimistic Denver businessman. In 1889, Brown organized the grandly named Denver, Colorado Canyon, and Pacific Railroad Company and named himself president. Brown had looked at Colorado's coal, and at southern California's growing population, and seen a fortune beckoning. He planned to build a railroad line to carry that coal to market; the tracks would follow the course of the Colorado (thus, efficiently, running downhill all the way) as it cut its way through Cataract Canyon, Glen Canyon, and Grand Canyon on its way from the Rockies to the Gulf of California. A final section of track would cut across southern California toward San Diego.

Brown organized a team of sixteen men, with himself in charge. Robert Brewster Stanton, a railroad engineer, was second in command. Brown purchased five sleek but unsuitable wooden boats, each fifteen feet long, just over three feet wide, and round-bottomed. Long, skinny, and round, the boats were nearly as unstable as floating logs. At only 150 pounds apiece, they were featherweight. Knowing how Powell's men had struggled to carry their massive boats, Brown had opted for boats that would be easy to portage. He overdid it. The boats were "thin, light, red cedar," and so fragile that two of them had cracked end to end *while still on the train* to the Green River (Utah—not to be confused with Green River Station, Wyoming, Powell's starting point). When he first laid eyes on the boats, Stanton wrote later, "my heart sank within me."

All the crew were novices. Worse, many were cocky novices. Brown seemed to regard the whole expedition as a lark, and it fell to Stanton to press the case for prudence. But Stanton was a bit of a dry stick and more than a bit of a pedant, and the convivial Brown tended to brush his warnings aside. Stanton had proposed that the party include four experienced boatmen. Brown took Stanton's list and crossed off the boatmen's names, in favor of two Denver lawyers and men about town he wanted to impress. "I, with Reynolds and Hughes, who are to be *guests* of the expedition, will take their places and keep the boats up with the survey," Brown announced. A "thunderstruck" Stanton could only gulp and give in. Stanton had requested life jackets, too, and Brown vetoed that request as well. (Curiously, Stanton, like Powell, had only one good arm. Though he "could not swim a stroke," he followed Brown's orders and canceled the life jackets.)

They planned to pack the gear and much of the food in watertight, zinc-lined boxes three feet long, one for each boat. But the fully loaded boxes nearly swamped the boats, so they lashed them together into a clumsy raft. Stanton's two black servants, who had been dragooned into this makeshift navy as cooks, were assigned to drag the raft behind their boat.

On May 25, 1889, the expedition set out from Green River. Almost at once they had to stop to seal a leak in the cooks' boat. Five miles downstream, the first rapid tore three holes in another boat. Still, this was easy water, and by May 30, the Brown-Stanton expedition had reached Cataract Canyon. There the trouble began in earnest. "It would be a great relief, if it were possible," Stanton wrote later, "for me to blot out all remembrance of the two weeks following this evening of May 31st."

The supply raft was the first casualty. Just upstream of the first rapid in Cataract Canyon, the cooks' boat tried to pull to shore. Caught in the swift current and further hampered by the ungainly raft tied to them, the *Brown Betty* swept helplessly toward the brink of the fall. In desperation, the two men cut the rope to the supply raft and pulled for shore. They made it; the raft did not. The zinc boxes smashed into the rocks, flew apart from one another, and ricocheted wildly through the rapid. Some were eventually found downstream (with the food inside still intact), turning circles in eddies like toy boats in a pond.

From here on, one man recalled, it would be "disaster every day." On June 3, one boat hit a rock and sank; the crew swam to safety and even managed to rescue the boat, though nearly all its contents were lost. On June 4, Hughes and

Reynolds, the two "guests," capsized. They grabbed their runaway boat and clung to it for half a mile while it dragged them through a rapid. Finally they righted the boat and climbed in, but the boat was full of water and uncontrollable. For another half mile, the two would-be investors sped and spun along until a merciful eddy swung them to shore. On the same day, the same rapid snatched another boat that the men had been lining and spat it downstream. And *still* on the same day, the *Brown Betty* hit a rock and sank while the men were trying to line it past a rapid. Nearly all the pots and pans and much of the rest of the cooking gear went to the bottom, along with most of the food. "The matter of supplies . . . ," Stanton noted, "begins to look serious."

The expedition struggled on, making a mere mile or two a day. Each day brought new smashups and desperate swims and runaway boats. Even Stanton, a proud stoic, seemed nearly overwhelmed. "Such work as we had in Cataract Canyon, with our frail boats, being thrown into the water bodily every day, and working in water almost up to one's armpits for days at a time, guiding boats through the whirlpools and eddies, and, when not thus engaged, carrying sacks of flour and greasy bacon on one's back, over boulders half as high as a house, is not the most pleasant class of engineering work to contemplate."

Soon nearly all the food was lost or too spoiled to eat. For six days, breakfast and dinner consisted of a bit of bread washed down by coffee and a little condensed milk. Lunch was three lumps of sugar and all the river water you could drink. Morale, already low, fell even further when the men found a human skeleton in a pile of driftwood. "Ghastly suggestion of what might be our fate," Stanton shuddered. On June 16, the men took the most damaged boat, the *Mary* (named for Brown's wife), and chopped it apart to get desperately needed wood for repairs on the other boats. Brown watched and wept.

The Grand Canyon, when they finally reached it, made Cataract Canyon seem like a holiday jaunt. Food was no longer a concern, because it was easy to leave the river near Lee's Ferry, and Brown had ridden ninety miles to the town of Kanab and bought supplies. (This option was unavailable to Powell. There were no horses or mules to borrow in 1869, because what would become Lee's Ferry was then only another unnamed, uninhabited spot in the desert.) Half Brown's original crew, including both "guests," took this opportunity to return home. One man, a trapper and miner named Harry McDonald, joined the expedition.

But even with starvation no longer a threat, the Colorado itself presented more than enough danger. Brown's expedition entered the Grand Canyon on July 9, 1889. They made ten miles that first day, without mishap, and portaged the two biggest rapids, Badger and Soap Creek.

Brown woke up jumpy on July 10. He had slept poorly, troubled by dreams of rapids. This was uncharacteristic, for Brown's unrelenting good cheer was his most marked trait. Odder still, the only rapid in sight was Soap Creek, which was *up*stream and no longer a problem. Brown and Harry McDonald were on the river shortly after six o'clock. Stanton followed close behind, at 6:23.

Brown's boat turned sideways in what is now called Salt Water Riffle. It looked like nothing more than a series of small, harmless waves, but when the boat crossed an eddy line—the always treacherous boundary between the main current, headed downstream, and the eddy, moving upstream—it flipped in an instant. Brown and McDonald were thrown overboard. Brown found himself in a whirlpool along the eddy line. McDonald was thrown into the fast-moving water in midstream. Brown surfaced first. McDonald spotted him and called to him. "Alright," Brown hollered back.

The river swept McDonald downstream, where he saw his runaway boat speeding away, about fifty feet in front of him and upside down. Two hundred yards farther on, he crawled ashore. He climbed a rock, spotted Brown still struggling, and ran back upstream. Brown had been flung nearer to shore than McDonald but was in fact farther from safety. Even if he had been able to break free from the whirlpool and had swum toward the riverbank, he would have had to fight his way across the strong eddy current. There was another possibility—swimming *away* from shore and into the main current in hopes of washing ashore farther downstream, as McDonald had—but that, too, would have meant breaking out of the whirlpool.

McDonald ran near the spot where he had seen Brown. By this time, Stanton's boat had reached the scene. Stanton saw McDonald waving and screaming. After a few moments, he deciphered the frantic shout: "Mr. Brown is in there." Stanton fought his way out of the rapid. Brown had vanished, but suddenly his pocket notebook shot high above the surface. Peter Hansbrough, a crewman in Stanton's boat, plucked it from the river.

The men scanned the churning water. Without a life jacket, and weighed down by shirt, pants, and coat (and, perhaps, boots), Brown had not lasted long. From the moment Brown's boat flipped to the instant he disappeared beneath the waves, perhaps ninety seconds had gone by. For the rest of the day, the men carried out a fruitless search of the riverbank for miles below the scene of the accident, hoping at least to find a body they could bury. Hansbrough cut a careful inscription, complete with ornate capitals, into the rock near the bit of shore that Brown had tried to reach. "F. M. Brown Pres. D.C.C. & P. RR Co was drowned July 10 1889 opposite this point."

The survivors could easily have left the canyon by hiking up Salt Water Wash and then walking to Lee's Ferry, but they chose to stay with the river instead. The next morning, having found Brown's boat safe in an eddy, the men set out again. For three days, they struggled along, running and portaging and lining the rapids known today (for their distance from Lee's Ferry) as the Roaring Twenties. Then came Sunday, a day of rest, and then Monday, July 15.

Henry Richards, one of Stanton's black servants, and Peter Hansbrough, the man who had carved Brown's epitaph, pushed out into 25 Mile Rapid. In the lower section, the force of the current beating sharply against a steep limestone wall had cut an overhanging shelf a few feet above the water. Hansbrough and Richards seemed in fine shape, sweeping along in midstream, when suddenly the current grabbed them and pushed them left, toward the cliff and the overhang. In a moment, they were against the cliff, caught in swirling water.

The two men shipped their oars and tried to push off the rock, to keep from being shoved under the shelf. Richards shoved the bow of the boat clear of the cliff; Hansbrough, in the stern, had been caught under the shelf but managed to push himself clear as well. Stanton, watching from shore, relaxed. "They are all right now," he said, but at precisely that moment, their boat flipped and the two men were flung into the water.

McDonald and another man jumped into a boat and sped toward Richards, but he disappeared before they could reach him. Hansbrough had vanished beneath the waves the moment the boat capsized. Both men drowned. "I then realized fully what it meant to be without life preservers, in such work on such a river," Stanton lamented.

With the crew near mutiny and, in any event, too few men left to portage the boats, Stanton decided to abandon the expedition.* The disheartened

---

*He would return later in the year, determined to finish his railroad survey. The second Stanton expedition (in sturdier boats and with life jackets) would become the first since Powell's to make it through the Grand Canyon, though nothing ever came of the railroad scheme.

group continued downstream only far enough to find a place where they could hike out of what they now thought of as "death's canyon." On July 17, as they climbed a side canyon to safety, Stanton glanced back at the Colorado and saw "something like a large bundle floating down the river." It was Brown—the men recognized his coat—but the current carried his body out of view long before the two men who had rushed after him could get anywhere near.

John Wesley Powell had inspired Brown's expedition. Twenty years before Brown's death, on August 6, 1869, Powell ordered his crew to stop early "at a place where it seems possible to climb out." (The innocuous-sounding remark had a macabre flip side, for the unspoken message was that at many places there was no way out.) The next day, the almanac said, would bring an eclipse of the sun. By measuring the exact moment the sun disappeared and comparing that figure with the time forecast in the almanac, Powell would be able to determine his longitude precisely.

Powell had been looking forward to this chance for weeks; he anticipated the eclipse with all the excitement of a five-year-old trying to fall asleep the night before his birthday. Countless Americans shared that eagerness. "The line of totality was almost one continuous observatory, from the Pacific to the Atlantic," one scientific journal reported in November 1869. Powell's men felt no such zeal. "Tomorrow is the eclipse," Bradley grumbled, "so we have to stop and let Major climb the mountain to observe it."

Early on the morning of August 7, Powell and his brother Walter set out on a climb to the canyon rim, lugging their measuring gear with them. After four hard hours, they reached the summit and stacked some rocks into a viewing platform. Then they sat down to wait for the great moment to arrive, "but clouds come on, and rain falls, and sun and moon are obscured."

Miserably disappointed, the one-armed Major and his half-mad brother set out on the return climb to camp. Night overtook them, and the two men inched their way along for two or three hours, feeling their way in the dark. "At last we lose our way, and dare proceed no farther. The rain comes down in torrents, and we can find no shelter. We can neither climb up nor go down, and in the darkness dare not move about, but sit and 'weather out' the night."

Stuck in place on the cliff, like gargoyles on a cathedral, the brothers passed an endless night in the pouring rain. They had spent the previous day fervently hoping to see the sun disappear; now their only prayer was for it to appear quickly. Dawn finally arrived. "Daylight comes, after a long, oh! how long a night," Powell moaned, "and we soon reach camp."

While Powell and Walter had been climbing and waiting, the men in camp had been repairing the boats. Bradley had replaced four of his boat's ribs and recaulked her seams and noted proudly that "she is tight again as a cup." Even so, the boats as well as the men were nearing the limits of their strength. "Constant banging against rocks has begun to tell sadly on them," Bradley wrote, "and they are growing old faster if possible than we are."

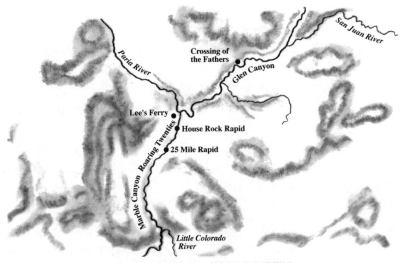

CHAPTER TWENTY
TIME'S ABYSS

"Though it is Sunday," Bradley wrote on August 8, "it brings no rest for us." Indeed, the work the men were now called on to perform made a mockery of the notion of a day of rest. (Bradley's weekly complaint about Powell's impiety had become their only Sunday ritual.) "Pulled out early and did a terrible hard work," Sumner wrote. Even Powell painted a somber picture. "It is with very great labor that we make progress," he wrote, "meeting with many obstructions, running rapids, letting down our boats with lines, from rock to rock, and sometimes carrying boats and cargoes around bad places." They portaged five rapids—their record for a single day—and advanced only three and a half miles.

Late in the day they made one last portage and stumbled exhausted into camp. As the rain pelted down, Powell and the crew retreated beneath a ledge. It took a hard search to find even a few sticks of driftwood, only enough for a tiny fire and barely adequate even for making coffee. The expedition, never a grand affair, had taken on a decidedly shabby look. "We begin to be a ragged looking set," Bradley wrote, "for our clothing is wearing out with such rough

labor and we wear scarce enough to cover our nakedness for it is very warm
with a sun pouring down between sand-stone walls 2000 ft. high."

Hawkins cut an especially striking figure. "I had a pair of buckskin
breeches," he wrote. "They were so wet all the time that they kept stretching
and I kept cutting off the lower ends till I had nothing left but the waist band.
When this was gone I was left with a pair of pants and two shirts. I took the
pants and one shirt and put them in the boat's locker [for safekeeping] . . . I
cut holes in my shirt tail and tied the loose ends around my legs so they would
not bother me in the water."

With the exception of the men themselves, the scenery was dazzling. Weary
as everyone was, they all gazed around in wonder at walls that gleamed gray and
pink and cream and purple. "The limestone is coming up again and there are
some of the most beautiful marbles I ever saw," Bradley wrote, "not excepting
those in the Cap. at Washington. They are polished by the waves, many of them,
and look very fine." These walls were formed of immense, sheer blocks and
looked unclimbable, as if built for the ramparts of some mythic castle. Bradley
briefly considered hunting for a broken chunk of stone to take home but
decided that "the uncertainty of adequate transportation" ruled out that idea.

Powell, naturally, was more effusive. "And now, the scenery is on a grand
scale. The walls of the cañon, 2,500 feet high, are of marble, of many beautiful
colors, and often polished below by the waves, or far up the sides, where showers have washed the sands over the cliffs." (Powell and the men may have
known that the "marble" was in fact limestone, though perhaps "marble" had
a more poetic ring.)

"At one place I have a walk," Powell went on, "for more than a mile, on a
marble pavement, all polished and fretted with strange devices, and embossed
in a thousand fantastic patterns. Through a cleft in the wall the sun shines on
this pavement, which gleams in iridescent beauty."

Even Sumner was bowled over. "There is marble enough of all kinds to
build forty Babylons, walls and all, with enough to spare to build forty other
cities before it would be missed. If geology is true—and it certainly is, if anything is—what vast ages the little insects must have worked to furnish the material for two thousand feet of marble! And what a length of time it took to form
the miles of lime and sandstone that overlie the marble! . . . And then, how long
did it require the Colorado River to cut its channel through all the sedimentary rocks and twelve hundred feet and more into the Archean formation? Who

knows? The testimony of the rocks cannot be impeached. I think Moses must be mistaken in his chronology as recorded in Biblical history."

To modern ears, "geology" is dry as dust, its subject matter lifeless in every sense. In the nineteenth century, geology was the stuff of drama and controversy. Powell was fascinated, and so were countless others. Though no one would have used the word, geology was sexy.

The key was that geology was also theology. The common belief among educated people (though, by Powell's day, no longer among scientists) was that God had created the earth a mere six thousand years before. The poet who described Petra as "a rose-red city half as old as time" meant his comparison literally. Then came the geologists brandishing pickaxes and fossils and spouting theories that contradicted Genesis. In a culture steeped in biblical doctrine, this was subversive. Geology might appear to be just another academic discipline, but both sides grasped the real issue. Sumner had put his finger on it. To study the age of rocks was to question the Rock of Ages.

The great, dizzying discovery of the nineteenth century was that time stretched backward farther than the human mind could grasp. At the dawn of the century, man had presided at the center of a cozy new homestead built especially with him in mind. Now he found himself wandering lost and forlorn in an ancient, rambling labyrinth with an absentee landlord. The earth's age could not be measured in thousands of years, the geologists declared, but in millions, or tens of millions, or hundreds of millions. The notion left Victorians clutching their heads in bewilderment and dismay. Einstein would befuddle later generations with his paradoxical insights into the nature of time, but that would be an affair largely for intellectuals. *This* shock was visceral.

Even a mere one million years is unimaginable in human terms. One million years is equivalent to fifty thousand human generations. Most people can, with considerable effort, imagine the world of their great-grandparents, *three* generations ago. One hundred generations bring us back only to the birth of Jesus, a hundred and sixty only as far as King Tut. To try to look back tens of thousands of generations brings the kind of queasy disorientation one gets from gazing down into a void. "It has taken several years of mountain climbing to cool my nerves, so that I can sit, with my feet over the edge, and calmly look down a precipice 2,000 feet," Powell once wrote. Even so, he could not bear to watch anyone else perch so precariously. "I must either bid him come away, or turn my head." Powell was referring to the view into a bottomless canyon, but he might have been describing anyone gathering his nerve to peer into time's abyss.

The reason the view back in time proved so unsettling had to do not with the age of the earth itself but with what the idea of unbounded time implied about mankind's place in the scheme of creation. The one point that both devout believers and hardheaded scientists agreed on was that humans were newcomers on earth. That was fine if the earth was young. But if the earth was ancient and man was the pinnacle of God's creation, why had He left the stage empty for so long? Mark Twain, a proud nonbeliever whose atheism was part intellectual exercise and part personal feud with God, put the challenge most starkly. "If the Eiffel Tower were now representing the world's age, the skin of paint on the pinnacle-knob at its summit would represent man's share of that age; and anybody would perceive that that skin was what the tower was built for."

Powell believed with all his heart that science was the only path to truth and that religion was a tangle of myth and obfuscation. He was far too prudent a man to express himself with Twain's gleeful sarcasm, but the explorer and the writer (who were almost exact contemporaries) were of one mind on the question of man's place in the cosmos.

In 1869, no topic was more fraught. The bombshell that was Darwin's *Origin of Species* had exploded only ten years before. Evolution was more controversial than geology because its assault on religion and human dignity was even clearer. But the step from geology to evolution was a small one. Endless time made room for endless change. If small changes accumulating over endless eons could level mountains and dig canyons, then it became comparatively easy to imagine that life itself could slowly and inexorably change as the eons unrolled.

For Powell, the Grand Canyon was the culmination of his expedition. Shutting his ears to his men's pleas to get moving, ignoring the urgent message of the dwindling food supplies, striving to forget the rapids lying in wait, he reveled in the geological wonderland that rose around him on all sides.

This was, finally, the land of fable, and Powell had penetrated to its core. Other scientists had seen it from above or even climbed down a side canyon to the river's edge, but their views had been mere peeks in comparison with Powell's detailed examination. No one had ever done anything to compare with this, and despite all the pressing reasons to stick to business, Powell made time to luxuriate in his surroundings.

No man ever took more delight in the glories of geology than Powell. The sheer scale of the Grand Canyon made other men feel insignificant—even Powell wrote occasionally of man's puniness in comparison with nature's majesty—but in truth he seemed more exalted than humbled. "In the Grand Canyon," he exclaimed, "there are thousands of gorges like that below Niagara Falls, and there are a thousand Yosemites. . . . Pluck up Mt. Washington by the roots to the level of the sea and drop it headfirst into the Grand Canyon, and the dam will not force its waters over the walls. Pluck up the Blue Ridge and hurl it into the Grand Canyon, and it will not fill it." Here, truly, was "the most sublime spectacle on earth."

The canyon packed an intellectual punch that fully matched its visual power. Powell found running rapids thrilling, but toppling outmoded doctrines was even better sport. Truth, he insisted, was written in the rocks and not in the Bible. Reading those rocks, more than reading the river, was his great ambition. Scientists had explained long before that to look out in space is to look back in time, for the starlight that seems to have reached us instantaneously has in fact been inching across the heavens like a postcard in a mailman's pouch. Today's light carries yesterday's news. Geology, Powell explained, teaches a parallel lesson. To scan a thousand-foot cliff in the Grand Canyon is to look back a hundred million years.

The Grand Canyon was a "library of the gods," Powell wrote, in which "ten thousand dark, gloomy alcoves" served as reading rooms. "The shelves are not for books, but form the stony leaves of one great book. He who would read the language of the universe may dig out letters here and there, and with them spell the words, and read, in a slow and imperfect way, but still so as to understand a little, the story of creation."

The difficulty in grasping this new story of creation reflected human limitations, not any flaw in the story itself. Looking at the buttes so common throughout the West, for example, Powell wrote that every one was "so regular and beautiful that you can hardly cast aside the belief that they are works of Titanic art. . . . But no human hand has placed a block in all those wonderful structures. The rain drops of unreckoned ages have cut them all from the solid rock."

The raindrops are easy to picture, but coming to terms with those "unreckoned ages" is the great hurdle that confronts all aspiring geologists. Perhaps a Swedish folktale comes as close to making the mystery palpable as any geological text. "High up in the North in the land called Svithjod, there stands a rock. It is a hundred miles high and a hundred miles wide. Once every thousand years a little bird comes to this rock to sharpen its beak. When the rock has thus been worn away, then a single day of eternity will have gone by."

Powell was sometimes hailed as a great geologist, though in truth he was not. But if he was not a brilliant innovator, he *was* a born teacher, with a knack for explaining complex ideas in simple language. Powell was not a detail man, but his boldness and his dramatic flair were well suited to geology's sweep. When he looked at the landscape, he pictured mountains rising and crumbling and rivers slicing their way to the sea. He had taught himself to think—and to see—in geologic time.

Seen over spans of tens of millions of years, the world is transformed. The fixed and solid-seeming surface of the earth rises and falls like the ever-changing face of the sea. Rivers carry mountains to the sea, and the seafloor rises to form new mountains.* If we could see on a geologic time scale, John McPhee writes, "continents would crawl like amoebae, rivers would arrive and disappear like rainstreaks down an umbrella, lakes would go away like puddles after rain."

The changes are real but too slow for us to grasp. Though we are looking at a movie, we see only a single frozen frame. Our lives are simply too brief. A firefly, which lives a few short weeks, might as well be expected to comprehend that the four-year-old chasing after it with a glass jar will someday grow to resemble her arthritic grandmother watching from the porch.

By geologic standards, ten million years is quick. *One* million years is the blink of an eye. Geologists are slow blinkers. Powell knew in his bones that, in a sense, a mountain is as fleeting and impermanent a feature of the world as a sand castle. Mountains rise up and fall down. The life span of a mountain range is on the order of sixty million years. The Rockies *will* crumble, Gibraltar *will* tumble.

It was a sermon Powell loved to preach. Nature achieved its grandest effects, he wrote, "not by an extravagant and violent use of power, but by the slow agencies which may be observed generally throughout the world, still acting in the

---

*Seashells had been observed on mountaintops at least as early as Leonardo da Vinci's day, but theologians had happily cited them as proof of the reality of Noah's flood.

same slow, patient manner." Like the claim that the earth was ancient, the claim that change was slow and uniform rather than sudden and violent had a double appeal for Powell. None of his nineteenth-century readers would have missed the unstated half of his argument—since the world is shaped and reshaped one grain of sand at a time, there is no need to talk of global floods and miraculously parting seas and the like. To the dustbin with all such superstitions.

Powell did not invent the notion that the sweep of time is nearly endless. Before him, Lyell and Darwin and countless others had painted the same picture.

"Water-drops have worn the stones of Troy and blind oblivion swallowed cities up," Shakespeare had written nearly four centuries before. But the Grand Canyon served as an unmatchable emblem of that almost ungraspable abstraction.

Still, even if Powell had never emerged from the Grand Canyon to spread the news about earth's antiquity, the message was in the air. Nineteenth-century Americans recognized that the story of the arid West is an epic tale of erosion, and slowness is erosion's defining trait. Where wind and rain and rivers have carved mesas and buttes and canyons, we can be sure the work has been going on through long ages. If we see a deep well, and then see that the man digging it is equipped with only a teaspoon, we can be sure that he has been digging for a good long while.

Dinosaurs provided the public with further proof that the earth had a long history, and the first decades after the Civil War were the great age of dinosaur discovery. Scientists roaming the West stumbled over dinosaur bones at every turn. At one site in Wyoming, bones of bygone creatures lay so thick on the ground that a sheepherder used them to build a cabin. During a brief stop at an isolated train station in Nebraska in 1868, a Yale scientist took a moment to examine a jumble of bones unearthed by someone digging a well. He recognized them as coming from ancient ancestors of the modern horse. A discovery made while an impatient conductor held the train proved to be one of the great finds in the history of science.

But, on a geologic scale, the first dinosaurs stumbled to their feet only the other day. The oldest rocks in the Grand Canyon, in contrast, are nearly two billion years old, not quite half as old as the earth. On a timeline where the history of the earth is compressed to a single year, dinosaurs show up only in mid-December and vanish on December 26. (Humans make their appearance on the evening of December 31. Columbus discovers America at three seconds before midnight.)

The first white men to peep over the rim of the Grand Canyon were Spanish conquistadors, in 1540, led by Hopi guides. (For more than two hundred years afterward, no whites returned.) The Spaniards could make no sense of what they were seeing. The conquistadors' "discovery" of the Grand Canyon, one early writer scoffed, was "akin to a dog's discovery of the Moon." A few agile conquistadors tried to climb down to the river, but they gave up after making it about one-third of the way. "From the top they could make out, apart from the canyon, some small boulders which seemed to be as high as a man," Pedro de Castañeda reported. "Those who went down and who reached them swore that they were taller than the great tower of Seville."

It is hard to place ourselves in the position of those who encountered the Grand Canyon unprepared by countless descriptions and pictures. For the modern traveler, familiarity breeds, if not contempt, then complacency. No early visitor tiptoeing to a canyon rim responded with anything but awe. Consider, for example, this description of the Grand Canyon of the Yellowstone, a petite version of the Grand Canyon of the Colorado. The writer was Nathaniel Pitt Langford, a brave and adventurous explorer and one of the first white men to describe what is now Yellowstone National Park. He wrote in 1870, a year after Powell's exploration of the Grand Canyon.

"Standing there or rather lying there for greater safety," Langford recalled, ". . . I realized my own littleness, my helplessness, my dread exposure to destruction, my inability to cope with or even comprehend the mighty architecture of nature. . . . A sense of danger, lest the rock should crumble away, almost overpowered me. My knees trembled, and I experienced the terror which causes men to turn pale and their countenances to blanch with fear, and I recoiled from the vision I had seen, glad to feel the solid earth beneath me." Langford was clinging to a spot of ground tourists now know as Inspiration Point.

Canyons stunned their first eyewitnesses partly because their size is so far beyond human scale. The Grand Canyon especially, at a mile deep, ten miles across, and nearly three hundred miles long, dwarfs all who would take its measure. But size is only part of the story. Unexpectedness plays just as large a role. Mountains are miles high and the ocean is seemingly infinite, but they announce their presence from a distance. The sight of a snowy peak or the tang of an ocean breeze beckons the traveler.

From an airplane, where even the Grand Canyon looks like a gash ripped in the earth, the view makes a kind of sense. From river level, as Powell and his men had seen, a canyon emerges gradually, unfolding over the course of days and weeks. But approached on foot, a canyon is a shock. Until the last moment, there is no hint of anything special in the neighborhood—park rangers at the Grand Canyon are asked, over and over again, where the canyon is—and then, suddenly, the ground disappears from before one's feet.

The Grand Canyon proclaims, with the force of a slap to the face, "This is *very* old." Curiously, that message is only half right. The rocks that the canyon reveals are billions of years old, but the canyon itself is young. This is no paradox. A tree might be hundreds of years old, but the axe cut that revealed its rings might have been made only yesterday.

The Grand Canyon is somewhere between four and six million years old. Geologically speaking, to be a mere five million years old is to be barely out of diapers. (In that short time, the river has carried away all the rock that once filled the canyon rim to rim. Before the dams were built, the Colorado carried an average of five hundred thousand tons of sand and silt—enough to fill a line of pickup trucks stretching the full width of the United States—past a single point *every day*. It is no wonder that the river ran red and muddy.)

"The most emphatic lesson that the canyon teaches," observed the geologist William M. Davis, one of Powell's disciples, "is that it is not an old feature of the earth's surface but a very modern one . . . It is properly described as a young valley." If the Grand Canyon were truly old, Davis explained, it would hardly be worth going to see. In time, as erosion continues on its relentless course, "the main canyon shall slowly wear deeper and deeper, and the side canyons, increasing in length and width, shall consume the entire plateau, hundreds of miles north and south of the river, and thus the whole region shall be reduced to a featureless lowland but little above sea level."

Most of the Grand Canyon is an enormous layer cake (or, more precisely, a hole cutting through an enormous layer cake). The individual layers are hundreds of feet thick and have formed over tens of millions of years. The deeper the layer, the older, like the clothes on a teenager's bedroom floor. The changes from one layer to another testify to a landscape that has changed again and again through the ages.

The Coconino Sandstone, for instance, is near the rim and therefore young. A mere 200-odd million years old, it is a remnant of a former desert, an eternity of blowing sand dunes now glued in place. But what was once a desert has also been a coastal plain and the site of a warm, clear sea. The Hermit Shale, which sits beneath the Coconino Sandstone, was laid down by rivers dropping their loads of silt and mud at what was once a seacoast. The five-hundred-foot-thick Redwall Limestone is formed, in part, of countless tiny seashells that rained gently down onto the ocean floor three hundred million or four hundred million years ago.

More is gone than remains. If the Grand Canyon is a journal recording earth's history, vandals have torn out over half the pages. At one juncture that Powell named "the Great Unconformity," layers of rock that differ in age by 1.2 billion years lie one atop the other. All the intervening years have eroded away without a trace. It was another of Powell's pet themes. "Beds hundreds of feet in thickness and hundreds of thousands of square miles in extent, beds of granite and beds of schist, beds of marble and beds of sandstone, crumbling shales and adamantine lavas have slowly yielded to the silent and unseen powers of the air, and crumbled into dust and been washed away by the rains and carried into the sea by the rivers." Someone left the cake out in the rain.

We can see those layers and deduce their history because the Colorado River has sliced its way downward inch by inch through thousands of feet of rock. "The finest workers in stone are not copper or steel tools," wrote Thoreau, "but the gentle touches of air and water working at their leisure with a liberal allowance of time."

Even more so when those touches aren't so gentle. The Colorado cuts down as it flows not simply because the water is moving but also because the river carries a swirl of rocks and pebbles and grit that help it abrade its way. (For similar reasons, stonecutters in quarries use abrasive-coated wire blades to cut blocks of stone.) The cutting does not proceed at a steady pace. For much of the year, the rocks and sand may sit in place on the river bottom. Come the spring thaw or a sudden flood, though, and the river grinds those rocks against the channel bottom. Multiply by ten thousand centuries and you have a canyon.

The river's sawing is only part of the story. The Grand Canyon is so deep because it is part of an immense geological formation called the Colorado Plateau, which has been soaring upward while the river has been cutting its channel. The entire plateau, which straddles the Four Corners region where Utah, Colorado, New Mexico, and Arizona meet, has risen one thousand feet in just the past million years. Picture, Powell said, a buzz saw spinning in place while the log it was cutting rose up: "The river was the saw which cut the mountains in two." (And the canyon is so wide—rather than merely as wide as the river itself—because erosion bites into its exposed sides.)

For Powell and his fellow geologists, all this was irresistibly exciting. In the nineteenth century, the West was a geologic boomtown. Geology in Europe and back East was far more difficult because the landforms there had been squeezed and twisted into almost unfathomable shapes and then hidden by trees and bushes and grass. What was elsewhere concealed by vegetation was here stripped bare. Better yet, Grand Canyon country was a region of neat layers resting tidily one atop another, color-coded, hundreds of feet thick, and extending for hundreds of miles. Best of all, the landscape was neatly sliced open by canyons that served geologists as dissecting kits served anatomists.

Old World geologists could scarcely conceal their envy of their counterparts in the American West. "The history of the rocks there is so simple," one British geologist complained when he learned of Powell's findings, "that a child could read it."

This was partly true and partly sour grapes, for the challenge of the Grand Canyon is as much psychological as intellectual. The river is barely visible from the rim. Where the Colorado does appear, glinting in the desert sun, it seems tiny and insignificant. (The Spanish conquistadors had estimated that it was six feet wide.) "It looks about large enough to turn a village grist-mill," observed Clarence Dutton, the great nineteenth-century geologist, "yet we know it is a stream three or four hundred feet wide. Its surface looks as motionless as a lake seen from a distant mountain-top." How could this placid stream cleave five thousand feet of rock?

For Powell and his weary men, who were battling the river on a daily basis rather than squinting to make it out in the distance, it did not take a leap of imagination to believe in the Colorado's power.

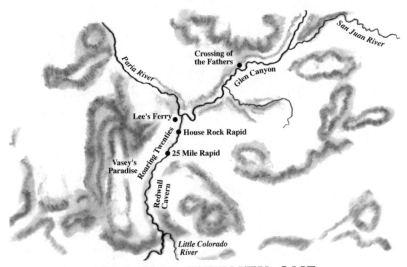

CHAPTER TWENTY-ONE

THE GREAT UNKNOWN

Late in the afternoon of August 8, after an exhausting day of portaging, Powell's men dragged themselves into camp and collapsed. The roar of a rapid just downstream promised that the next day would be another ordeal. A second rapid loomed about five hundred yards beyond the first one, "but we can't get down to look at it," Bradley wrote, "for the walls come down to the water and it is too deep and swift to wade past it." The trap was squeezing shut.

Powell was happily studying the scenery, but Bradley had no time for such distractions. Restless and jumpy, he found himself veering back and forth between frank confessions of dread and hearty proclamations of confidence. "We are interested now only in how we shall get through the cañon and once more to civilization though we are more than ever sanguine of success. Still our slow progress and wasting rations admonish us that we have something to do. Fortunately we are a happy-go-lucky set of fellows and look more to our present comfort than our future danger."

That "comfort" didn't amount to much. They were all stuck "in a cave of the earth," Bradley complained, trying to look forward with anticipation to a breakfast of beans that Hawkins had promised for the next morning. The "future danger," on the other hand, was all too real. "Hard at work early," Sumner wrote the next day, August 9. "Made 4 portages and ran 27 bad rapids in 13 miles." The rapids were "furious," Bradley wrote, though he recorded his customary hopes that "this series of heavy rapids is about ended."

Not all the sights were ominous. "The river turns sharply to the east," Powell wrote, "and seems enclosed by a wall, set with a million brilliant gems. What can it mean? Every eye is engaged, every one wonders. On coming nearer, we find fountains bursting from the rock, high overhead, and the spray in the sunshine forms the gems which bedeck the wall. The rocks below the fountain are covered with mosses, and ferns, and many beautiful flowering plants." (And poison ivy, as modern tourists have learned.)

It was, Sumner agreed, "the prettiest spring I ever saw." Powell named the spot Vasey's Paradise, in honor of a botanist who had traveled in the Rocky Mountains with him the year before. Struck by the welcome burst of greenery against "the unending barrenness of the cañon," Bradley lauded it as "the prettiest sight of the whole trip." Sumner, characteristically, downplayed his own enthusiasm. "The white water over the blue marble made a pretty show," he conceded. "I would not advise anybody to go there to see it."

Only two miles downstream they came to another spectacular site, now called Redwall Cavern. "The water sweeps rapidly in this elbow of river, and has cut its way under the rock, excavating a vast half circular chamber," Powell wrote. The space, "if used for a theater, would give sitting to fifty thousand people."

No theater could have a more spectacular setting, though Powell exaggerated its size perhaps tenfold. Today, with the Colorado dammed, the undercutting that formed the cavern is at an end. (In the flood year of 1983, though, river runners floated deep inside the cavern and tied up to its ceiling.) For boatmen and passengers on the Colorado today, Redwall Cavern provides a welcome chance to leave the cramped confines of the boats and run and play on the vast sandy beach. No sooner has the first boat tied up than the Frisbees start flying. Where Powell and his hungry, fearful crew gawked in wonder, tourists in cutoff dungarees and Day-Glo bathing suits roam the beach and slather suntan lotion on their peeling skin and decide if it's late enough in the day to crack open a beer.

Redwall Cavern itself remains exactly the immense stone clamshell that Powell saw. In some ways, the Grand Canyon has changed considerably since Powell's day. Because the dams catch the Colorado's load of mud and silt, for example, the river often runs green rather than red. Other changes show the hand of man more directly. Navajo Bridge soars overhead at Mile 4, for instance. A rustic lodge called Phantom Ranch sits at Mile 88. Test holes drilled in the cliffs at Mile 40 mark the spot where the Bureau of Reclamation intended to site Marble Canyon Dam.

Against a backdrop as enormous as the Grand Canyon, though, such signs of human intrusion are rare. For nearly all its 277 miles, the Grand Canyon looks from river level just as it did a hundred years ago, or a thousand, or a hundred thousand. Photographers have traveled throughout the Grand Canyon rephotographing the earliest pictures of it, taken shortly after Powell's first expedition. In nearly every case, trying to tell which photo was taken when Ulysses S. Grant was president and which taken only the other day is nearly impossible, a harder variant of the children's puzzles that challenge the reader to "find six differences between these two drawings." Is that the same cottonwood tree in both photographs? What happened to that piece of driftwood? In some cases, where the river flows happened to match, photographs of the same rapid taken one hundred years apart show the identical holes in the identical places.

If Powell and his men were captured by a time machine and sent to the Grand Canyon today (at a time when the river happened to be muddy), they would see nothing for miles and miles to indicate that it was not 1869. The illusion would last until a jet passed high overhead or a raft of tourists rounded a corner. Up above, at Grand Canyon Village, the canyon's South Rim is thick with tour buses and restaurants. Down on the river, there are no buildings, no signs, no maps or trash cans, no more than a handful of human alterations to the landscape. Nearly everywhere it is many hours or days to the nearest electrical outlet. There are no camps as such, simply beaches where the river has deposited some sand.

The camping spots are pristine, scrupulously maintained by hikers and boaters who would no sooner leave a gum wrapper in the Grand Canyon than they would a Big Mac in the Sistine Chapel. Twenty thousand visitors a year travel the river and leave scarcely a sign. Commercial boatmen, in their shorts and T-shirts, are more punctilious than the starchiest English butlers; let a single blob of jelly from a sandwich fall onto the sand and someone will scoop it up and carry it off before it stops trembling. The National Park Service's river rangers scour campsites inch by inch looking for "microtrash." A cigarette stub or a bit of candy wrapper a quarter-inch square is a major find. A river ranger will cross a churning river to retrieve a beer can he has spotted bobbing in an eddy, an act roughly akin to a trooper's darting across a six-lane highway to pick up a plastic bag snagged on the guardrail.

If rangers find signs of a campfire (fires are forbidden), they clear away the evidence to avoid giving ideas to future arrivals. Half-burnt logs are flung into the river, blackened rocks scattered, even ashes lugged away in a bucket. As in Powell's day, anyone venturing into the canyon must bring his supplies with him. (Fishing is permitted.) And just as everything must be brought in, so must everything be carried out. Every river trip carries its own "toilets"—metal boxes with the top lid temporarily removed in favor of a toilet seat. Many a river passenger pays $200 a day to ride downstream in the company of an ammo can or two of human waste.

Powell and his men had grimmer concerns than the tidiness of their campsites. They had just passed the rapids that, twenty years later, would prove disastrous for the Stanton expedition. Frank Brown, the would-be railroad tycoon, would drown just downstream of Soap Creek Rapid, at Mile 12, below Lee's Ferry. Hansbrough and Richards would drown at 25 Mile Rapid. By the time Powell and his men reached Redwall Cavern, at Mile 33, they had made it by Badger Creek Rapid and Soap Creek and Sheer Wall and House Rock and the nine or ten rapids that make up the Roaring Twenties.

There was no letup, and there would be none, and, as if in passing, Powell made the key observation that explained why. "A creek comes in from the left," he wrote, "and just below, the channel is choked with boulders, which have washed down this lateral cañon and formed a dam, over which there is a fall of thirty or forty feet." Wherever side canyons join the Grand Canyon, that is, sudden floods can dislodge boulders and unstable chunks of cliff and fling them into the main channel where they squeeze the river into abrupt, angry rapids.

Powell was the first to figure it out, but Stanton, in the ill-fated expedition of 1889, was the first eyewitness. Stanton and his exhausted men climbed out of the Grand Canyon just thirty-two miles downstream of Lee's Ferry. Already, they had endured three drownings and countless near misses. Still to come was one final fright, a kind of parting gift from the Grand Canyon. The day before, the survivors had watched helplessly as their leader's corpse floated downstream. Now, as they hiked out of South Canyon to what they hoped was safety, the skies opened and the rain poured down in what seemed like solid sheets of water. Stanton described the scene:

> As the rain increased, I heard some rock tumbling down behind us, and, looking up, I saw one of the grandest and most exciting scenes of the crumbling and falling of what we so falsely call the everlasting hills. As the water began to pour over from the plateau above, it seemed as if the whole upper edge of the Canyon had begun to move. Little streams, rapidly growing into torrents, came over the hard top stratum from every crevice and fell on the softer slopes below. In a moment they changed into streams of mud, and, as they came farther down, again changed into streams of water, mud and rock, undermining huge loose blocks of the harder strata, and starting them they plunged ahead. In a few moments, it seemed as though the slopes on both sides of the whole Canyon, as far as we could see, were moving down upon us, first with a rumbling noise, then an awful roar. As the larger blocks of rock plunged ahead of the streams, they crashed against other blocks, lodged on the slopes, and, bursting with an explosion like dynamite, broke into pieces, while the fragments flew into the air in every direction, hundreds of feet above our heads, and as the whole conglomerate mass of water, mud, and flying rocks came down the slopes nearer to where we were, it looked as if nothing could prevent us from being buried in an avalanche of rock and mud.

This was a debris flow. Powell and his men had never seen such a thing, but they would come to recognize its signature all too well, for in the Grand Canyon it is debris flows that make rapids. A debris flow is a kind of thick mud river, like an enormous, slow-motion avalanche made of wet concrete. Because it is so viscous and so powerful, a debris flow can dislodge and carry boulders that a flash flood could never budge. Boulders and trees slide and tumble and spin along helplessly as the flow snags them and drags them downhill. Some accounts describe "car-sized objects" riding atop the flow, but reality is more dramatic. Cars generally weigh under two tons; in 1990, a debris flow in the Grand Canyon swept a 280-ton boulder into the river.

John McPhee described a recent debris flow in Los Angeles. A house sat in its path:

> The parents' bedroom was on the far side of the house. Bob Genofile was in there kicking through white satin draperies at the panelled glass, smashing it to provide an outlet for water, when the three others ran in to join him. The walls of the house neither moved nor shook. As a general contractor, Bob had built dams, department stores, hospitals, six schools, seven churches, and this house. It was made of concrete block with steel reinforcement, sixteen inches on center. His wife had said it was stronger than any dam in California. His crew had called it "the fort." In those days, twenty years before, the Genofiles' acre was close by the edge of the mountain brush, but a developer had come along since then and knocked down thousands of trees and put Pine Cone Road up the slope. Now Bob Genofile was thinking, I hope the roof holds. I hope the roof is strong enough to hold. Debris was flowing over it. He told Scott to shut the bedroom door. No sooner was the door closed than it was battered down and fell into the room. Mud, rock, water poured in. It pushed everybody against the far wall. "Jump on the bed," Bob said. The bed began to rise. Kneeling on it—on a gold velvet spread—they could soon press their palms against the ceiling. The bed also moved toward the glass wall. The two teen-agers got off, to try to control the motion, and were pinned between the bed's brass railing and the wall. Boulders went up against the railing, pressed it into their legs, and held them fast. Bob dived into the muck to try to move the boulders, but he failed. The debris flow, entering through windows as well as through doors, continued to rise. Escape was still possible for the parents but not for the children. The parents looked at each other and did not stir. Each reached for

and held one of the children. Their mother felt suddenly resigned, sure that her son and daughter would die and she and her husband would quickly follow. The house became buried to the eaves. Boulders sat on the roof. Thirteen automobiles were packed around the building, including five in the pool. A din of rocks kept banging against them. The stuck horn of a buried car was blaring. The family in the darkness in their fixed tableau watched one another by the light of a directional signal, endlessly blinking. The house had filled up in six minutes, and the mud stopped rising near the children's chins.

Debris flows are rare because they require, simultaneously, torrential rains and cliff sides poised to collapse. But even rare events happen sooner or later, and the Grand Canyon has all the time in the world. Start with that abundance of time, add the canyon's endless cliffs, stir in its dramatic rainstorms, and you have a recipe for debris flows. And, accordingly, nowhere else in North America have debris flows shaped the landscape as they have in the Grand Canyon.

The first lesson for geologists is that the classic picture of erosion nibbling the canyon walls away grain by grain is wrong. The true picture is not one of crumbling rocks but of crashing boulders, as rare but violent storms take great bites from the canyon's flanks. "Debris flows move entire slopes," writes the geologist Robert Webb, "not just individual grains of sand." A second lesson is that the Grand Canyon is, for now, in equilibrium, a battleground for two stalemated opponents. Tiny sidestreams, aided by debris flows, throw boulders into the main channel. Even the mighty Colorado has only a certain amount of energy to expend, and the river is using its energy not on cutting a deeper channel for itself but on eroding and removing the boulders choking its path. When the climate changes again, one warring side or the other may gain an advantage, but for now the canyon's depth is unchanging.

For river runners, those geological insights pale next to a more practical one—debris flows make rapids, and the Grand Canyon is ideal spawning ground for debris flows. Powell had seen that rapids *could* form in this way, but the truth was far worse than he realized. The Grand Canyon is joined by countless side canyons, every one of them guaranteed, sooner or later, to fling its river-choking cargo into the Colorado's path. For Powell and his weary men, that meant an endless series of rapids.

•　　•　　•

On August 10, the men advanced only fourteen miles but encountered thirty-five rapids. "We run them all though some of them were bad ones," Bradley wrote. "One was the largest we have run in the Colorado for we have gone more cautiously in it than we did in the Green." Bradley had complained throughout the trip that Powell insisted too often on hauling the boats around rapids, and he had not changed his mind. At first he had argued that portaging and lining were difficult and dangerous and unnecessary. Later he complained that they wasted precious time (and therefore precious food). Those objections still held, and now Bradley added one more. "I think we have had too much caution and made portages where to run would be quite as safe and much less injurious to the boats."

At two o'clock in the afternoon, the expedition reached the Little Colorado, a tributary of the Colorado that the men also called the Chiquito or the Flax. They were not impressed. "It is a lothesome little stream," Bradley wrote, "so filthy and muddy that it fairly stinks. It is only 30 to 50 yds. wide now and in many places a man can cross it on the rocks without going in to his knees." Sumner's judgment made Bradley's sound like a love poem. The Little Colorado was "as disgusting a stream as there is on the continent . . . half of its volume and ⅔ of its weight is mud and silt." It was little but "slime and salt," "a miserably lonely place indeed, with no signs of life but lizards, bats, and scorpions. It seemed like the first gates of hell. One almost expected to see Cerberus poke his ugly head out of some dismal hole and growl his disapproval of all who had not Charon's pass."

The mud that so revolted Bradley and Sumner was runoff from summer thunderstorms. In dry weather, the Little Colorado runs a brilliant, tropical blue (because of calcium and sodium and other dissolved minerals). The green Colorado and the blue Little Colorado run side by side for a time, independent ribbons of color like stripes on a flag.

In rainy weather, the Little Colorado runs muddy. Unlike Powell's men, modern tourists tend to revel in the mud, as if the river were a mud bath at a posh spa. They splash in the goop like puppies, reveling in the warmth after the Colorado's iciness. (The water released into the Colorado at Glen Canyon Dam comes from two hundred feet below the surface of Lake Powell and is a painful forty-eight degrees. Just splashing water on your face to wash up takes an effort of will. Holding your head underwater long enough to wash your hair actually hurts. Hypothermia can become severe in five minutes. For those who fall off their boat, it is a greater risk than drowning. Often the two work in tandem, for a person plunged into cold water immediately begins to hyper-ventilate, and the desperate, reflexive attempts to catch one's breath may lead instead to swallowing great mouthfuls of water.)

But the Little Colorado is enticingly warm. Commercial trips all pull over at the river junction, like mini-cruise ships at a Caribbean harbor, and spill their passengers ashore. If the water level permits, the more adventurous—and less fastidious—attach their life jackets around their waists like giant diapers (to cushion collisions), lie on their backs bobsledder style, and shoot and slither their way down the muddy river through a mini-rapid or two. It is like bob-bing in an enormous glass of chocolate milk while a titanic six-year-old stirs it with a giant spoon.

Powell planned to stop for a few days to measure the latitude and longitude of the mouth of the Little Colorado. The men were not pleased, but they knew this was part of the bargain. "We are sorry to be delayed as we have had no meat for several days and not one sixth of a ration for more than a month," Bradley wrote, "yet we are willing to do all that we can to make the trip a success."

On August 11, their second day in place, Sumner and Walter Powell set out to take measurements. The walls, they found, were three thousand feet high, more than half a mile. "The ascent is made, not by a slope such as is usually found in climbing a mountain," Powell wrote, "but is much more abrupt—often vertical for many hundreds of feet—so that the impression is that we are at great depths; and we look up to see but a little patch of sky."

Powell hiked upstream a few miles along the Little Colorado to explore. He went, as usual, without a weapon. "As we were on the edge of the Apache and Havasupai Indian country, his act was foolish, to say the least," Sumner com-plained, but Powell saw nothing more dangerous than a couple of rattlesnakes. The most striking sight was the river's "tumbling down over many falls."

Had Powell ventured eighty or a hundred miles upstream, and had the recent rains been heavy enough, he would have had a jaw-dropping surprise. There he would have encountered the muddy waterfall now known as the Grand Falls of the Little Colorado. In dry weather, there is almost nothing to see. But in full flow, the cataract is 185 feet in height, slightly higher than Niagara. An exploring party coming down the Little Colorado above the fall would have almost no warning that they were about to plunge to their deaths. The country just above the waterfall is fairly flat, and the river's descent is gradual. Not until about a quarter mile above the fall does the pace begin to pick up. By then it would likely be too late.

A geologist who had grown up near Niagara Falls once asked Powell why he assumed there were no such falls on the Colorado. "Have you never seen the river?" Powell shot back. "It is the muddiest river you ever saw . . . I was convinced that the canyon was old enough, and the muddy water swift enough and gritty enough to have worn down all the falls to mere rapids."

This was more good-natured swaggering than a serious response. In the long run, Powell was quite right—waterfalls eventually create the conditions for their own demise, like termites who consume the home they live in. The problem, which Powell understood perfectly well, is that we live in the short run. His "argument" amounted to the claim that, because ice cream eventually melts in the heat, no one should expect to see an ice cream cone in July. Powell had made up his mind to tackle the Grand Canyon, risks be damned. The indisputable existence of the Grand Falls of the Little Colorado, in a region geologically identical to the one he was traversing, highlights how much of a gambler he was.

While Sumner and the Powell brothers ventured off exploring and taking measurements, Bradley stayed in camp and sulked. He had good reason. Like a soldier stuck in the trenches for months and no longer able to imagine any luxury that could compare with a dry pair of socks, Bradley was beset by mundane miseries. Camp was "filthy with dust and alive with insects," and Bradley could not escape "for I have nothing to wear on my feet but an old pair of boots in which I cannot climb the mountains and which are my only reliance for making portages." Modern boatmen wear rubber sandals (a Grand Canyon boatman invented Tevas), but Bradley and the others were condemned to leather boots that fell apart in the water and tore open on the rocks.

Bradley went barefoot in his boat and in camp when he could, and he had a pair of moccasins to wear when the sand was too hot or the rocks too sharp. But his spirits were as low as they had been all trip. "I have given away my clothing until I am reduced to the same condition of those who lost by the shipwreck of our boats. I cannot see a man of the party more destitute than I am. Thank God the trip is nearly ended."

Without maps, this was little more than a hopeful guess, but Bradley tended to round off nearly all his gloomy thoughts with an optimistic note, the better to keep himself going. The others shared his melancholy and frustration. Only Powell, still reveling in the delights of geology, seemed cheerful, and the men found their leader's good humor more infuriating than inspiring. "If this is a specimen of Arrazona a very little of it will do for me," Bradley wrote. "The men are uneasy and discontented and anxious to move on. If Major does not do something soon I fear the consequences, but he is contented and seems to think that biscuit made of sour and musty flour and a few dried apples is ample to sustain a laboring man. If he can only study geology he will be happy without food or shelter but the rest of us are not afflicted with it to an alarming extent."

On August 12, *still* in camp, the men were growing frantic. Bradley had copied his notes, but he had finished with even that make-work project. The men had pinned down their position, so the mapmaking duties were discharged. "There remains nothing more to be done that is absolutely necessary for lat. and long. are sufficient and we ought to be away in the morning," wrote Bradley. "Don't know whether we shall or not."

It was Powell who had measured their latitude. The finding had a practical significance that none of the men could miss. The expedition had already made it as far south as they needed to go. They were now at the latitude of Callville, Nevada, the tiny Mormon settlement that signified a safe escape from the Grand Canyon. From now on, every mile west was a mile toward home, and every mile in any other direction was a life-threatening detour.

In the meantime, the weather did its bit to add to everyone's misery. "I am surprised to find it raining nearly every night in a country where they say rain seldom falls," lamented Bradley. After long, wet days battling rapids on the river, tourists hit by monsoons today echo the same words as they cower in their tents in their soaked clothes and listen to the rumbling thunder and worry about where they left their life jacket and whether the whipping wind will find it and toss it into the river.

For no clear reason, Powell seemed to consider the junction of the Colorado and the Little Colorado to be the true beginning of the Grand Canyon. By modern reckoning, the expedition had entered the Grand Canyon some sixty miles before. Powell called that first stretch of canyon, from Lee's Ferry to the Little Colorado, Marble Canyon. Today the name "Marble Canyon" is still used, but it designates a part of the Grand Canyon rather than a separate entity.

On their last day camped at the junction of the two rivers, Powell scrawled a note in his river diary. "Take obs. Capt. climbed Mt.," it read in its entirety. Years later, in Washington, D.C., Powell composed a journal entry to mark the same milestone. The words are the most famous ever written about the Grand Canyon.

We are now ready to start on our way down the Great Unknown. Our boats, tied to a common stake, are chafing each other, as they are tossed by the fretful river. They ride high and buoyant, for their loads are lighter than we could desire. We have but a month's rations remaining. The flour has been resifted through the mosquito net sieve; the spoiled bacon has been dried, and the worst of it boiled; the few pounds of dried apples have been spread in the sun, and reshrunken to their normal bulk; the sugar has all melted, and gone on its way down the river; but we have a large sack of coffee. The lighting of the boats has this advantage: they will ride the waves better, and we shall have but little to carry when we make a portage.

We are three quarters of a mile in the depths of the earth, and the great river shrinks into insignificance, as it dashes its angry waves against the walls and cliffs, that rise to the world above; they are but puny ripples, and we but pigmies, running up and down the sands, or lost among the boulders.

We have an unknown distance yet to run; an unknown river yet to explore. What falls there are, we know not; what rocks beset the channel, we know not; what walls rise over the river, we know not; Ah, well! we may conjecture many things. The men talk as cheerfully as ever; jests are bandied about freely this morning; but to me the cheer is somber and the jests are ghastly.

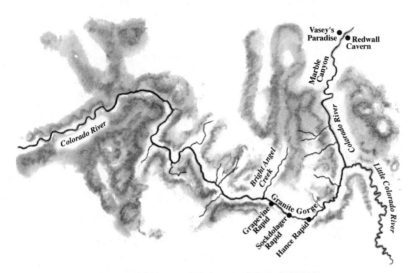

CHAPTER TWENTY-TWO

SOCKDOLAGER

Powell's remarks were striking and not just for their eloquence. *His* men talked "as cheerfully as ever," at the same time that Bradley described himself and the rest of the crew as "uneasy and discontented and anxious." Was Powell oblivious to his men's mood? Or did he recognize their dismay but choose, on literary grounds, to portray the men as a hardy band of adventurers joking in the face of death?

Powell's description of himself as oppressed by "somber" and "ghastly" thoughts strikes a curious note as well, for Powell was no angst-ridden prince of Denmark. His remark a few sentences before, that the boats were now so light that they were easy to carry, rings truer. It evokes Henry V's startling argument to his troops at Agincourt that it was a *good* thing to be outnumbered four to one. "The fewer men, the greater share of honor," Shakespeare's Henry declared, and now Powell, with astonishing but characteristic optimism, insisted that to be nearly out of food was in truth a happy turn of circumstance.

At eight in the morning on August 13, Powell and his fretful crew untied their battered boats and pushed out into the unknown. "With some eagerness, and some anxiety, and some misgiving," Powell wrote, "we enter the cañon below, and are carried along by the swift water through walls which rise from its very edge."

Immediately they found themselves fighting through a nest of rapids. "The rapids are almost innumerable," Bradley wrote, "some of them very heavy ones full of treacherous rocks." In the course of a long day, they ran too many rapids to count, lined three others, and advanced a total of fifteen miles.

They camped that night at the head of a rapid they could scarcely bear to look at. It was "the worst rapid we have found today and the longest we have seen on the Colorado," wrote Bradley. "The rocks are seen nearly all over it for half a mile or more—indeed the river runs through a vast pile of rocks." Hance Rapid, as it is known today, is the rockiest and steepest rapid in Grand Canyon. In a mere half mile, the river drops thirty feet.

It is named for John Hance, a prospector who found asbestos near here in the early 1880s. The asbestos was valuable (primarily for fireproof theater curtains, in an era when footlights were open flames), and Hance constructed two mule trails from the rim to the river to haul it out. Eventually the tourists who began coaxing Hance to lead them into the Grand Canyon proved more lucrative than the asbestos. Hance abandoned mining for tourism and became the first white resident of the South Rim.

Like the tourists Hance led down his trails, Sumner could only gape at the rapid in front of him. He reckoned that it was "about 1 mile long with a fall of 50 or 60 ft. that has about 100 rocks in the upper half of it. How anyone can ride that on a raft is more than I can see. Mr. White may have done so but I can't believe it."

Powell and all his crew knew James White's story, but until now they had never had occasion to ponder it in detail. On September 8, 1867, a naked, incoherent, starving prospector was found clinging to a raft about one hundred miles below the Grand Canyon. "i was ten days With out pants or boos or hat," James White wrote to his brother as soon as he recovered, and "i Was soon bornt so i Cold hadly Wolk."

White and two companions had been searching for gold somewhere in the vicinity of the San Juan River, which joins the Colorado a short distance north of the Grand Canyon. They were attacked by a band of fifteen or twenty

Indians, and one of the three prospectors was killed. White and the other man escaped. When they reached the Colorado, the pair found some driftwood, lashed together three cottonwood logs each about ten feet long and eight inches in diameter, and, still afraid for their lives, set out down the river. Four days later, White's partner washed overboard and drowned. White continued alone. For seven full days, he had nothing at all to eat except for two rawhide knife scabbards. To keep from drowning, he tied himself to the raft with a rope around his waist and plunged helplessly along as the raft tumbled downstream. By the reckoning of his rescuers—White himself contributed almost no specifics to his story, except that he had been afloat fourteen days—White had traveled 550 miles and was the first man to have traversed the Grand Canyon. "i see the hardes time that eny man ever did in the World," he wrote his brother, "but thank god that i got thrught saft."

White's story was widely believed at first, but nearly all modern writers dismiss it. Five hundred miles was a tremendous distance to float in only two weeks. Rather than speed downstream, a raft drifting willy-nilly would likely break up on a rock or fly apart in a rapid or hang up in an eddy, and probably sooner rather than later. Just as puzzling, White said that he had encountered only one major rapid; and that the cliffs above him were only three or four hundred feet tall; and that the walls were not variously red and black and purple and tan but uniformly whitish-yellow. In what should have been the white-water nightmare of Cataract Canyon, if White's geography was correct, "the water was so smooth that George and I sat on the raft with our feet in the water."

In truth, most white-water historians believe, White was simply lost. He had told the truth as best he knew, and the wild claims about the Grand Canyon represented neither lies nor a hoax on White's part but overexcitement on the part of his rescuers. White's descriptions did not fit the Grand Canyon—but *did* fit a stretch of river below the Grand Canyon—because White was already beyond the Grand Canyon before his raft ever touched the water. He had traveled a total of 60 miles, not 550.

True or not, White's story may have inspired Powell's Grand Canyon expedition. What a single, unprepared man clutching a log raft could blunder through, an organized expedition in newly made boats might pull off in style. Powell claimed that the idea of exploring the Colorado in boats was his. (Sumner made the same claim.) But, as a result of White's strange adventure, such talk was in the air in 1868 and 1869, before Powell set out. The *Rocky*

Mountain News, for example, not only reported White's story but even reprinted White's letter to his brother. Oramel Howland worked at the *News* and Sumner's brother-in-law was the editor.

White's story also appeared in General William Palmer's railroad survey of 1869 and in a paper presented to the Academy of Science of St. Louis. The effect was not only to publicize the tale but also to imbue it with authority. The key passage in the railroad report drew on White's experience. "The absence of any distinct cataracts or perpendicular falls would seem to warrant the conclusion that in times of high water, by proper appliances in the form of boats, good resolute men, and provisions secured in water-proof bags, the same passage may be safely made, and the actual course of the river mapped out, and its peculiar geographical features properly examined." For John Wesley Powell, such prose would have seemed an invitation and a challenge.

Whether or not Powell had been lured to the Colorado by White's story, he and the crew concluded as soon as they saw Hance Rapid that no one had come this way on a log raft. In any case, they had more urgent problems to deal with. "At the lower end of the rapid the granite rises for the first time," Sumner wrote. "There is no granite whatever (except boulders) from Green River City to the head of this rapid." Powell took a moment to explain why that news was so ominous. "Heretofore, hard rocks have given us bad river; soft rocks, smooth water; and a series of rocks harder than any we have experienced sets in. The river enters the granite!"

Powell was right to be afraid of hard rocks, but his reasoning was wrong. In the Grand Canyon, hard rock *does* mean bad rapids but not because the river bottom forms a sharp ledge, as it does in a waterfall. Instead, hard rock makes for steep side canyons that fling enormous boulders into the river and block the channel. But Powell did have the big picture right—this was trouble to dwarf anything they had yet encountered.

Bradley was worried, too. "Major has just come in and says the granite is coming up less than a mile down the river," he wrote. This was bad news, and the news had seemed bad enough already. "One thing is pretty certain," Bradley noted somberly. "No rocks ever made can make much worse rapids than we now have."

They were entering a dark, frightening canyon-within-a-canyon now called the Upper Granite Gorge. To a geologist, this gorge marks the onset of metamorphic rocks—ones transformed by heat and pressure deep within the earth—as opposed to the sedimentary formations, born of wind and water, that had grown so familiar. At two billion years old, these are the oldest rocks in the canyon and among the oldest exposed rocks on earth. To the layman, the transition is more basic. The "granite," which is in fact mostly Vishnu schist, *looks* hostile and dangerous, with sharp, swooping curves and nasty, jagged edges. If Edward Gorey had designed the Bat Cave, it might look something like the Upper Granite Gorge.

It gleams coal black in the sun, though the dark, tortured rock is crossed by scars of red and pink and cream formed of Zoroaster granite or quartz. The gorge squeezes the river against its will, and the sound of roaring, protesting rapids carries upstream. Powell and his men strained to make out what lay ahead. "We can see but a little way into the granite gorge," wrote Powell, "but it looks threatening." Emery and Ellsworth Kolb, seeking to duplicate Powell's trip in 1911, echoed that judgment, but comforted themselves with the knowledge that others had come this way. "The granite gorge seemed to us to be the one place of all others that we had seen on this trip that would cause one to hesitate a long time before entering," Ellsworth Kolb wrote.

For Powell and his men, there could be no hesitating. After breakfast on August 14, they set out into the gorge. "At the very introduction," Powell wrote, "it inspires awe. The cañon is narrower than we have ever before seen it; the water is swifter; there are but few broken rocks in the channel; but the walls are set, on either side, with pinnacles and crags; and sharp, angular buttresses, bristling with wind- and wave-polished spires, extend far out into the river. Ledges of rocks jut into the stream, their tops sometimes just below the surface, sometimes rising few or many feet above; and island ledges, and island pinnacles, and island towers break the swift course of the stream into chutes, and eddies, and whirlpools."

They portaged one rapid and ran the next two. The walls of their stone cage rose steadily higher. At about eleven o'clock, they heard the familiar roar of ground-level thunder. "The sound grows louder and louder as we run," Powell wrote, "and at last we find ourselves above a long, broken fall, with ledges and pinnacles of rock obstructing the river. There is a descent of, perhaps, seventy-five or eighty feet in a third of a mile, and the rushing waters break into great waves on the rocks, and lash themselves into a mad, white foam." The seventy-five-foot descent was an exaggeration for dramatic effect—in his river diary Powell wrote, "Fall 30 ft. probably, huge waves"—but the danger was real.

The men stopped to consider their options. It didn't take long. With cliffs that extended to the water's edge, portaging and lining were impossible. The men could perhaps climb to the summit a thousand feet above them and then descend on the far side of the rapid, but it was inconceivable to do so while carrying the boats. "We must run the rapid," Powell concluded grimly, "or abandon the river."

This was Sockdolager, nineteenth-century slang for a knockout punch (pronounced Sock-DOLL-a-jur).* The early river runners who described it sounded as intimidated as if they had encountered a one-eyed, man-eating Cyclops or snake-haired Medusa herself. "Just as the dinner hour was near and the threatening black granite had risen to one thousand feet above the water," one of these early explorers wrote, "we heard a deep, sullen roar ahead and from the boats the whole river seemed to vanish instantly from earth. At once we ran in on the right to a small area of great broken rocks that protruded above the water at the foot of the wall, and stepping out on these we could look down on one of the most fearful places I ever saw or ever hope to see under like circumstances—a place that might have been the Gate to Hell."

At the same vantage point, even Sumner, as proud and feisty a man as ever lived, conceded that he had met his match. "We finally encountered a stretch of water and canyon that made my hair curl," he wrote. "I don't know how it affected the other boys. The walls were close on both sides, with a fall of probably thirty feet in six hundred yards, a white foam as far down as we could see, with a line of waves in the middle, fifteen feet high." Beyond that, the canyon veered left, cutting off the view.

---

*The word was one of the last that Lincoln ever heard. John Wilkes Booth lurked outside the presidential box at Ford's Theatre on April 14, 1865, while the audience enjoyed a frothy comedy called *Our American Cousin*. Booth stood silently, waiting for the line that would unleash a burst of laughter from the audience and help cover the sound of his gunshot: "Don't know the manners of good society, eh? Well, I guess I know enough to turn you inside out, old gal—you sockdologizing old man-trap."

The rapid, as bad in fact as in appearance, had the power to reduce grown men to cowering wrecks. In 1903, a prospector and skilled boatman named Hum Woolley decided to take a boat through the Grand Canyon. He recruited two cousins, John King and Arthur Sanger, as crew. Neither cousin had any idea what he was in for. They quickly learned. Not far below Lee's Ferry, Sanger already sounded desperate. "Thank God we are still alive, it is impossible to describe what we went through today. Only the wonderful river knowledge and oarsmanship of Hum Woolley saved us from the vortex . . . I am scared."

In Sockdolager, the two cousins pressed themselves flat against the floor of the boat while Woolley battled the waves. "I thought this was the end as we rushed up a great wave, then another until I was almost sick and dizzy with fear," Sanger wrote. "Wolly would yank the boat this way and that, until one of the great waves or whirlpools suddenly swung us broad-side to a great wave coming up the river and towering at least 20 ft high, curling right over us . . . it came down and almost filled the boat again, the next one did fill it John and myself was clear under water Wolly shouted to lie still and paddled the boat over to the shore."

Woolley's boat was well designed for white water, and he ran the Colorado facing forward, as today's boatmen do. Powell's men had no such advantages. By this time, they had gained a degree of river-running skill, but it was nothing compared to what they needed. Crueler still, the caution that had helped them get this far now threatened to undo them. By portaging and lining every rapid that looked dangerous, the men had avoided calamity. But they had also missed the chance to hone their skills in more or less manageable rapids. Now, with virtually no experience in big rapids, they had come to "a perfect hell of waves." This time there was no avoiding it. They were troops on the eve of a great battle.

From here on, the Colorado's ferocity would far outdo anything Powell and his crew had seen. With their fledgling skills made irrelevant by the river's power, the men were no longer navigating but simply hoping to survive. Once they had tried to plot the best course downstream. Now they were in the position of a hapless pedestrian who has slammed his jacket in a car door and is being dragged down the street.

"I decided to run it," wrote Sumner, "though there was a queer feeling in my craw, as I could see plainly enough a certain swamping for all the boats. But what was around the curve below out of our sight?" If there was a waterfall lurking just out of range, everyone understood, they were about to speed to their deaths. But, with no options, Sumner announced that he was ready to start. "Who follows?" he cried. Hawkins and Hall, the two youngest members of the expedition, one the none-too-expert cook and the other the ex-mule driver who had once complained that his boat would neither gee nor haw, answered first. "Pull out!" they yelled. "We'll follow you to tidewater or hell."

"We step into our boats, push off, and away we go," wrote Powell, "first on smooth but swift water, then we strike a glassy wave, and ride to its top, down again into the trough, up again on a higher wave, and down and up on waves higher and still higher, until we strike one just as it curls back, and a breaker rolls over our little boat."

It nearly did them in. "The *Emma Dean* had not made a hundred yards," wrote Sumner, "before an especially heavy wave struck her and drove her completely under water. Though it did not capsize her or knock any one out, the wave rendered her completely unmanageable. Dunn and I laid out all our surplus strength to keep her off the rocks, while Major Powell worked like a Trojan to bail her out a little."

Powell picked up the story. "Still, on we speed, shooting past projected rocks, till the little boat is caught in a whirlpool, and spun around several times. At last we pull out again into the stream, and now the other boats have passed us. The open compartment of the *Emma Dean* is filled with water, and every breaker rolls over us. Hurled back from a rock, now on this side, now on that, we are carried into an eddy, in which we struggle for a few minutes, and are then out again, the breakers still rolling over us. Our boat is unmanageable"— Sumner and Bradley chose the identical word—"but she cannot sink, and we drift down another hundred yards, through breakers; how, we scarcely know."

Finally, the *Emma Dean* caught up with the freight boats, safely tucked into a quiet cove. The rest of the crew sent up a cheer as Powell, Sumner, and Dunn came into view. For Andy Hall, who viewed the most terrifying misadventures with the glee of a small boy at a Fourth of July parade, it was all great fun. He only regretted that he didn't have a photograph of the floundering *Emma Dean*.

Sumner was not as quick to recover his nerve. "I have been in a cavalry charge, charged the batteries, and stood by the guns to repel a charge," he wrote. "But never before did my sand run so low. In fact, it all ran out, but as I had to have some more grit, I borrowed it from the other boys."

•   •   •

If the Colorado is indeed a monster—a "liquid predator," in the words of one modern-day river runner—it is a monster with decidedly odd habits. For one thing, since virtually all its rapids are formed by side canyons, the Colorado, unlike some smaller rivers, is *not* a continuous string of rapids. Instead, the Colorado is a "pool and drop" river, with long, placid stretches punctuated by bursts of danger, like a restful sleep broken by nightmares. Rapids form only 10 percent of the Colorado's length through the Grand Canyon, yet they account for 50 percent of the river's drop.

On Grand Canyon trips today, the calm interludes provide an opportunity for tourists to sprawl out and work on their tans, and for boatmen weary of rowing to play the good host and let someone else take a turn at the oars, and for restless teenagers to ask, "How long till the next one?" For Powell and his men, the frequent breaks were a cruel tease, the river behaving like a taunting bully who lets his victim up, pretends to leave, and then returns to the attack.

On average, because its pools are so long, the Colorado drops only eight feet a mile through the Grand Canyon. But averages can obscure useful information. On average, snakes and centipedes each have fifty legs. If we consider only its twenty-odd miles of rapids, the Colorado falls a far more daunting forty feet per mile. The overall drop seems steeper than it really is, partly because the rapids are so dramatic and partly because the Colorado Plateau, as it rose into the air, rose unevenly. Lee's Ferry is three thousand feet above sea level, for example, and the North Rim of the Grand Canyon, ninety miles downstream, is at eight thousand feet. In that stretch, the river drops six hundred feet, and the canyon rim *rises* a mile. The effect is to exaggerate the river's fall, which seems hardly necessary. This is like outfitting Shaquille O'Neal with elevator shoes, dressing Dolly Parton in a Wonderbra.

Even a forty-foot-per-mile drop is only middling by white-water standards. Many rivers are far steeper and more rock-studded, which means that they demand quicker, more precise moves from the boatman. The steepest section of North Carolina's Green River, a mile-long run that is one of the most difficult in North America, drops over 350 feet.

What sets the Colorado apart from many other rivers is its volume and the sheer difficulty of coping with so many tons of angry, surging water. The flow in the North Carolina Green's most dangerous stretch is usually less than 250 cubic feet a second. On the Colorado, that would be barely a trickle. A flow of less than ten thousand cubic feet per second is low. Thirty thousand is a good flow but hardly remarkable. As recently as 1983, when spring floods filled Lake Powell so high that Glen Canyon Dam was in danger, the Colorado's flow reached *one hundred thousand cubic feet per second.*

"This is *big* water," says Zeke Lauck, a professional boatman with decades of experience on some eighty rivers around the world. "In some rivers, you see a line of current running left to right and you can just cross it. On the Colorado, if *it's* moving right, *you're* moving right." Running rapids in the Grand Canyon, Lauck says, is like wrestling an overwhelmingly strong but not especially subtle giant. The challenge is not in responding to tricky spins and reverses and feints, as in a kung fu contest, but in coping with an opponent who is unrelenting, overpowering, and unforgiving.

Roughly speaking, the danger in most rivers is dodging the rocks, whereas the danger in the Colorado is coping with the force of tons of water forming towering waves and holes ten feet deep and whirlpools like light-devouring black holes. Whirlpools form at the junction of strong currents moving in different directions, and Powell and his men marveled at their power. "In high water there are a lot of very dangerous whirlpools that will catch a boat and whirl it round till it makes the occupants dizzy," Sumner wrote. "If the boat is drawn under the vortex of the whirlpool," he added, with forgivable exaggeration, "there is not one chance in a hundred of its ever rising again."

A swimmer who falls into a whirlpool can be in deadly trouble even if she is wearing a life jacket. Just above a black rock at the bottom-right-hand side of Lava Falls in the Grand Canyon, for instance, sits a small, fast whirlpool that boatmen call the side pocket. Its terrifying feature is that much of the water that forms it makes its escape by racing toward the river bottom and then squeezing between the black rock and the shore. The deep-diving current glues any swimmer to that underwater sieve.

Whirlpools can make trouble even without dragging their victims underwater. "The *Emma Dean* was caught in so severe a whirlpool that Dunn and I could not pull out of it to save our lives," Sumner recalled. "It spun us round like a roulette wheel. I thought I saw a chance to get out as the boat spun round past a rock about thirty feet away. Seizing the rope in my teeth, I made a desperate plunge. Good lively swimming, combined with the momentum of my jump, enabled me to make the rock by a scratch. Seizing it with a death grip, I was able to pull the bow of the boat out of the swirl, whereupon it shot ahead like a scared rabbit."

Eddy lines mark the boundary between the main downstream current and water moving upstream in an eddy. The colliding currents make for whirlpools and what boatmen call squirrelly water. In smaller rivers, the eddy line truly is a line, or perhaps a narrow band. In the Colorado, the "line" can be a zone eight or ten feet across. If the water in the eddy is moving fast, the eddy sits *lower* than the main river—picture water in a glass that someone is stirring furiously—and a boatman can be trapped by an unclimbable "eddy fence." Even a low fence can block a boat as effectively as a curb stymies a wheelchair. On smaller rivers, eddy fences six inches high have trapped boats for hours. In the 1983 flood in the Grand Canyon, the eddy fence at Mile 24½ Rapid was *four feet* tall.

Boils form when powerful currents deflect off the river bottom and head back to the surface. They can come out of nowhere, as if a fountain has suddenly been turned on, and can rise several feet above the river surface. In the Grand Canyon, they range from three feet across to forty feet. Boils can swamp a boat, if they happen to come up under one side and push the opposite side underwater, or they can lift it up and fling it aside.

Holes are even more formidable in a river as big as the Colorado. "At thirty-two thousand cubic feet per second," the veteran boatman Larry Stevens calculates, "one thousand tons of water are moving through the river channel every second. If an average elephant weighs about five tons, this means that the flow of the river is equal to two hundred elephants skipping by every second. A hole in the river may take up about a third of the channel, so the hydraulic dynamics in that hole are about the same as sixty-seven elephants jumping up and down on your boat."

Such hazards make river running a high-stakes game of pinball, with the caveat that the boatman is both the player and the ball.★ The pros make it look easy. Clair Quist, one of the best, takes a wry delight in pretending to think there's nothing to it. "Just remember," he tells a novice about to run his first Grand

---

★River runners use the word "boatman" for both men and women. The first woman to row the Colorado through the Grand Canyon was the legendary Georgie White, in 1952, but it took until the 1970s for the commercial companies in the Grand Canyon to hire women in any numbers. Today everyone agrees that many of the best boatmen are female.

Canyon rapid, "take the waves head-on." As advice, this is essential but gravely inadequate, akin to telling a first-time tightrope walker to remember to keep his balance.

The art in river running consists not in trying to outmuscle the river—no human stands a chance in that test of strength—but in identifying a path that leads smoothly downstream. Quist and his fellow veterans would have you believe there's nothing to it. All you have to do, says the boatman John Running, is find a stretch of river that's going where you want to go and then hop aboard. Easy does it. "The important thing to remember is that you're dancing with the river"—Running pauses for a theatrical beat—"and *you're* not leading."

With the current running upstream here, downstream there, diagonally or even downward in still other places, and all at different speeds—imagine a criss-crossing series of airport-style moving sidewalks designed by a sadist whose aim was to knock passengers off their feet—it's *not* easy. The trickiest zones are at the junctions where two "moving sidewalks" merge or collide or interact in some still stranger way. The catch is that often the boatman must seek out these junctions, intentionally moving into harm's way, in order to maneuver downstream. Rocks mean trouble, for example, but a boatman might need to slow down and spin the boat around in preparation for a move farther downstream. Often the best way is to duck partway behind a rock so that one end of the boat is in water pushing upstream while the other end is still in water moving downstream.

In small, steep, fast, rock-strewn rivers, such moves have to be just so. In the Grand Canyon, the margin of error is generally bigger. "You can say, 'Oh, I really wanted to be three or four feet farther right,' and most times you'll get away with it," says the boatman Zeke Lauck. "On the other hand, because of the force that's hammering on you, skills that work other places don't work here. Boatmen think, 'Gee, I've got my boat sideways, what I would usually do is put this oar in the water and—' Well, in the Grand Canyon, if you've got your oar in the wrong place, you're in trouble."

By now, Powell and his men were perpetually in trouble. "Down in these grand, gloomy depths we glide," Powell wrote, "ever listening, for the mad waters keep up their roar; ever watching, ever peering ahead, for the narrow cañon is winding, and the river is closed in so that we can see but a few hundred yards, and what there may be below we know not."

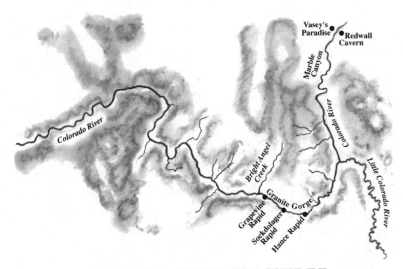

CHAPTER TWENTY-THREE

FIGHT

Sockdolager marked the first in a forty-mile-long series of rapids now nicknamed Adrenaline Alley. The waters churn and roil, and huge waves collide with one another as the river crashes in fury against the black walls that would presume to hem it in. The options of portaging or lining, which Powell had embraced so many times, often did not exist in this devil's gorge.

"The walls now, are more than a mile in height—a vertical distance difficult to appreciate . . ." Powell wrote. "A thousand feet of this is up through granite crags, then steep slopes and perpendicular cliffs rise, one above another, to the summit. The gorge is black and narrow below, red and gray and flaring above, with crags and angular projections on the walls."

Even a single day in the Granite Gorge had left everyone keyed up (with the possible exception of the unshakable Andy Hall), and no one had any idea how long the granite continued. "This is emphatically the wildest day of the trip so far . . . ," Bradley wrote on the evening of August 14, after losing an oar in Sockdolager but otherwise escaping intact. "The waves were frightful beyond anything we have yet met and it seemed for a time that our chance to save the boats was very slim but we are a lusty set and our good luck did not go back on us then." Then, just below Sockdolager, came "a rapid that cannot be run by any boat, half a mile long, 75 yds. wide, fall of 50 ft. and full of rocks."

They struggled through this horror as best they could, clinging to the cliff face and inching their way downstream by snubbing ropes around protrusions in the rocks. From two o'clock until sundown, the men labored, and still they had made it only halfway down what is now called Grapevine Rapid. Like a boy trying to crack open nuts by hurling them to the ground, the river repeatedly flung the boats against the granite. Bradley yearned to take a chance and run the rapid, but he kept his thoughts to himself. Exhausted and discouraged, Powell and the men knocked off early. They sat down to a meager dinner of bread and coffee and ate to the accompaniment of the rapid's roar.

The bare rock offered few places to sleep. Sumner and Powell found a spot big enough for their bedrolls, though Bradley half-expected to hear one of them fall into the river in the night. Bradley wedged himself into a seam in the rocks, and Hall found a tiny perch at the water's edge. The rest of the men were "tucked around [the cliff] like eve-swallows."

The next day was August 15, another Sunday without rest. The first order of business was getting through the tumultuous rapid they had begun battling the day before. After a near disaster with the *Emma Dean*—it looked for a moment as if the rocks had bashed a hole in her side—Bradley volunteered to run the rapid in the *Maid*. Powell granted his wish, and Bradley swept his way downstream. He was shoved this way and that "with terrible force" and "whirled round and round" in whirlpools, but the run was a success. Then, momentarily safe, Bradley relaxed a bit. Like every boatman delighted to find himself upright and intact at the bottom of a fearsome rapid, he settled in to watch the others try. "Rowed into the eddy and laid on our oars to have the fun of seeing [Hawkins] run it," Bradley wrote. Hawkins provided some entertainment by nearly swamping and breaking an oar, but he, too, made it downstream.

Somewhere in the chaos, Oramel Howland lost his notes and his map of the canyon from the Little Colorado on. This was not life-threatening—whether or not they had compass directions, there was no choice but to go where the river dictated—but it was a severe blow. Powell saw himself as head of a scientific expedition charting unknown territory. He had no interest in leading a pack of adventurers.

•   •   •

It was hard to know which was worse, the slow agony of lowering the boats with ropes or the sudden danger of running the rapids. Even Powell acknowledged that it was "not easy to describe the labor" involved in lining the boats through the Granite Gorge. The men clambered along the sharp, steep cliffs while holding a rope attached to a bucking boat, struggling to find a place to secure the rope. Where the river dropped off too quickly, they had to let go of the boat for a moment and hope to haul it in below before it sank or split open on the rocks. Sometimes there seemed no route downstream, and someone would throw bits of driftwood into the maelstrom to see which path they took, to get some idea of where to try to coax the boat. "And so we hold, and let go, and pull, and lift, and ward, among rocks, around rocks, and over rocks."

Running the boats through these angry, twisting rapids was usually just as hard. (Once or twice, the fickle river smiled on the men. In one rapid with an eight-foot drop spanning the full width of the river, Sumner gulped hard and braced for trouble "and jumped it like jumping a hurdle with a bucking horse—and didn't ship enough water to moisten a postage stamp.")

Far more often, the river showed its angry side. "The river is very deep, the cañon very narrow, and still obstructed, so that there is no steady flow of the stream," Powell wrote, "but the waters wheel, and roll, and boil, and we are scarcely able to determine where we can go. Now, the boat is carried to the right, perhaps close to the wall; again, she is shot into the stream, and perhaps is dragged over to the other side, where, caught in a whirlpool, she spins about. We can neither land nor run as we please. The boats are entirely unmanageable; no order in their running can be preserved; now one, now another, is ahead, each crew laboring for its own preservation."

Only one shaft of light cut through the gloom. On the afternoon of August 15 the men came to a swift, beautiful creek coming in from the north. It was "clear as crystal," Bradley exclaimed, and he could see fish swimming in it. Powell named it Silver Creek (but later changed the name to Bright Angel, to contrast with the stinking, muddy Dirty Devil they had encountered two weeks before.)★ A large willow tree growing from a sandbar just above the creek provided welcome shelter and an ideal spot to camp. "Stretched our

---

★Tourists today know Bright Angel Creek as the site of Phantom Ranch, a small group of buildings that offer the only food and lodging on the floor of the Grand Canyon.

weary bodies on the sand under the willow and rested the remainder of the day," wrote Sumner.

The respite was brief. An eddy had grabbed the *Emma Dean* the previous morning and thrown her against the rocks, and she needed emergency repairs. (Bradley's boat had escaped the same fate by two feet.) The men also needed to find wood suitable for cutting several new oars. It was a hard search, Sumner noted, for the river was "so terrific it seems to smash everything into pieces, leaving nothing large enough to make an oar." Half a mile up Silver Creek, they discovered a huge pine log that a flash flood had tossed aside and began the arduous task of rolling it down to the beach on skids.

Everyone's nerves were frayed, rubbed raw by a routine that Sumner summarized as "rapids, daily duckings, and 'heap hungry' all the time." The weather seemed to know only two extremes: Either rain flooded down, or the sun beat down without mercy. On August 16, the canyon was a blistering 115 degrees, and making the oars was "considerable of a task." Bradley was in Eeyore mode, feeling grumpy and put upon. "They have come to think that my boat should carry all the rations, go into all dangerous places first and get along with least," he sulked. "So be it." He consoled himself with the hope that "the trip is nearly ended."

Turning the tension up another notch, Hawkins managed to worsen the already dire food situation. He had laid out the food on the riverbank to dry in the sun and had turned his attention to making oars. One of the boats swung around in the current, and its rope knocked the baking soda into the river. No more bread, then, except sodden lumps of dough. From here on, Sumner noted, meals would consist of "rotten flour mixed with Colorado River water."

This was hardly an exaggeration. "Our rations are still spoiling," Powell wrote. "The bacon is so badly injured that we are compelled to throw it away. . . . We now have only musty flour sufficient for ten days, a few dried apples, but plenty of coffee." Even the fish in Silver Creek proved impossible to catch. Like the most famous desert wanderers of all, thousands of years before, the Powell expedition would from this point on subsist on little more than unleavened bread.

"To add to our troubles," Sumner lamented, "there was a nearly continuous rain and a great rise in the river that created such a current and turmoil that it tried our strength to the limit. We were weakened by hardships and ceaseless toil for twenty out of twenty-four hours of the day. Starvation stared us in the face. I felt like Job: it would be a good scheme to curse God and die."

•   •   •

As if to mark their misery, a festering feud between Powell and Dunn came to a head. "At noon one day when the boats were being let over a bad place," Hawkins wrote, "Dunn was down by the water's edge with a barometer, taking the altitude." Somehow a rope fastened to one of the boats caught him under the arms and knocked him into the water. Dunn had nearly drowned, but he had managed to catch hold of a rope and drag himself to safety. Though Dunn was unhurt, he had been carrying a watch that belonged to Powell—the man measuring altitude was assigned to note the exact time—and the watch was ruined. Powell told Dunn that he would have to pay $30 for the watch (about ten days' pay for a skilled workman) or leave. "Dunn told him a bird could not get out of that place, thinking the Major was joking," Hawkins went on, "but all of us were very quickly convinced that every word the Major said was meant."

Dunn "really should have been a little more careful" on the day of the dunking, Sumner conceded, but "the Major evidently wanted to impress his military standing upon Dunn, and proceeded to give him a tongue-lashing that roused his ire to such a pitch that I think only the fact that the Major had but one arm saved him from a broken head, if nothing worse." The crew had sided with Dunn, and the sniping grew nasty. When Sumner remarked that Dunn had come close to drowning, Walter Powell muttered that it would have been no great loss.

Hawkins had fixed dinner. He poured each man a cup of coffee, and they began eating. Powell sat a few yards from the others, as he often did. The routine called for Hawkins to serve Powell. Not this night. If Powell wanted to eat, Hawkins snarled, "he would have to come and get his grub like the rest of the boys." Walter took his brother dinner.

The immediate crisis had passed like a fast-traveling thunderstorm, but the tension lingered, like an electric charge in the air. Like it or not, the men were utterly dependent on one another and stuck with each other's company. They had not seen another human being in nearly six weeks, since the Uinta Indian Agency (and Bradley, Sumner, the Howland brothers, and Dunn had been stuck in camp and missed even that slight diversion). Two weeks is a long trip in remote country, a month an eternity. Powell and his men had been on the river more than eighty days. Cabin fever is a hazard of every prolonged trip. For these men, condemned to rowboats careening down a wild river, in perpetual danger of drowning and on the edge of starvation, the homeliest cabin would have seemed like a castle. Dunn, Sumner, and Powell in particular spent all day together, every day, fearing for their lives and crammed in the sixteen-foot-long *Emma Dean*. It was not a setup to soothe ruffled tempers.

On August 16, a day spent stuck in camp repairing the boats and cutting oars, Powell tore into Oramel Howland for losing the maps. Howland and Powell had been quarreling over the maps since Disaster Falls, when Howland had lost his first set of maps, his gear, and nearly his life. Powell had chewed Howland out more than once since then, and Sumner had tried to intervene on his friend's behalf. Now Powell started in on Howland again, and Sumner stepped in again. The war was over, he reminded Powell, and "he couldn't come any damned military there."

Then Powell turned to Dunn, his other whipping boy. Powell snarled at Dunn, and Dunn snapped back. Powell cursed Dunn out, and Dunn said that if Powell weren't a cripple, he would make him take it back. Half-mad, ox-strong Walter Powell leaped in, shouting that *he* was no cripple. Eyes flashing in fury, he charged Dunn, screaming that he would kill him.

Hawkins grabbed Walter by the hair and shoved his head underwater. "For God's Sake, Bill, you will drown him!" Dunn cried, and Howland and Dunn together dragged Walter from the water and threw him down on a sandbar. Walter coughed and choked and, once he had caught his breath, shouted that Hawkins was a coward and a lowlife, a "Missouri puke." Walter stormed off to get his gun, vowing to shoot both Dunn and Hawkins. As Walter bent over to unlash the gun from the deck of his boat, Andy Hall came up behind him and slugged him on the side of the head. Walter whipped around. Hall stood, gun in hand, shouting at Walter to back off before Hall took his head off. Walter backed off.

The next day, with a fragile peace in place, the leaky boats and the feuding men "pulled out again for more of the Great Unknown." The contrast with the jaunty party that had left Green River Station not quite three months before was hard to miss. "There was not much talk indulged in by the grim squad of half-starved men with faces wearing that peculiar stern look always noticed on the faces of men forming for a charge in battle," Sumner wrote.

Nearly all the men had indeed seen battle. Bad as their predicament was now, the war had been worse. The Civil War had been a kind of horrific canyon journey of its own, a marathon that wended through unknown and unspeakably perilous territory, with long, dull stretches punctuated by episodes of sudden, shocking violence, and with no good way out but straight ahead, come what may.

They could not help thinking of what they had endured before, but even the memory of hardships overcome seemed little consolation now. "This part of the canyon is probably the worst hole in America, if not in the world," Sumner wrote. "The gloomy black rocks . . . drive all the spirit out of a man. And the excessive drenching and hard work drive all the strength out of him and leave him in a bad fix indeed."

Bad fix or not, they had no choices left. "We had to move on or starve," Sumner wrote, and even Powell agreed. "We must make all haste possible," he wrote. "If we meet with difficulties, as we have done in the cañon above, we may be compelled to give up the expedition, and try to reach the Mormon settlements to the north." The precious barometers had broken, too, so there was no longer any way to tell how far the river still had to fall.

They had begun as a scientific expedition, more or less, but the time for science had long passed. Angry, hungry, exhausted, scared, Powell and his eight companions were in a race for their lives. They had ten days' rations remaining, and they could only guess how much farther they had to go.

CHAPTER TWENTY-FOUR

MISERY

They needed to race, but they had to crawl. "Although very anxious to advance," Powell wrote, "we are determined to run with great caution, lest, by another accident, we lose all our supplies. How precious that little flour has become! We divide it among the boats, and carefully store it away, so that it can be lost only by the loss of the boat itself."

On August 17, Powell and the men set out at eight in the morning. They sped along for three miles but ground to a halt when they came to a rapid too big to run. As they lowered the boats on ropes, the *Maid* leaped ahead on her line while the men tried futilely to rein her in, and smashed into a rock. Rain fell in intermittent bursts that proved more maddening than a steady downpour would have. "Have been thoroughly drenched and chilled," Powell wrote, "but between showers the sun shines with great power, and the mercury in our thermometers stands at 115°, so that we have rapid changes from great extremes, which are very disagreeable."

They stopped to repair the *Maid*, finally returned to the river, and promptly came to another impassable rapid. They lined it, too, only to run into still another rapid they could not run. For the third time in a day, they took on the dreaded task of lowering the boats by rope. This last time was the worst. It took all afternoon, and, when it was finally over, the men found still another "very bad" rapid that they would have to line first thing the next morning.

Bruised and exhausted by the day's labor, the men sat down to a dinner of unleavened bread. When the sun sank, the temperature plummeted and the rain picked up. The weather in the Grand Canyon is as extreme as the scenery. Clear days can be not just sunny but scorching, as if the sunlight has passed through the lens of a magnifying glass. The most nondescript hat seems like a priceless boon. When it rains, Powell noted, it seems as if "some vast spout ran from the clouds." Few sights are as magnificent as a thunderstorm sweeping through the Grand Canyon, few experiences as miserable as being caught in a downpour at night without a tent or a ledge to hide under, trying against all odds to fall asleep.

"It is especially cold in the rain to-night," Powell wrote. "The little canvas we have is rotten and useless. The rubber ponchos, with which we started from Green River City, have all been lost; more than half the party is without hats, and not one of us has an entire suit of clothes, and we have not a blanket apiece. So we gather drift wood, and build a fire; but after supper the rain, coming down in torrents, extinguishes it, and we sit up all night, on the rocks, shivering, and are more exhausted by the night's discomfort than by the day's toil."

When the long night finally ended, the wet, wretched men set out downstream again. The rapids were "very numerous and very large," Bradley wrote on August 18, and "the worst kind of a rapid because you can see rocks rising all over them with no channel in which to run them." A day spent portaging and lining earned them only four miles. "Hard work and little distance seems to be the characteristic of this cañon," Bradley wrote wearily, and meals of coffee and moldy flour were "not sufficient to anything more than just to sustain life."

While the crew manhandled the boats through the rapids, Powell set off alone to climb the cliffs, intent on geology. Both leader and crew may have welcomed a break from one another. ("Major Powell . . . was a nuisance in the work of portaging," Sumner wrote. "His imperious orders were not appreciated. We had troubles enough without them.") Up Powell climbed, past the black granite and the rusty sandstone and the greenish shale to the base of the red-stained limestone. "I climb so high that the men and boats are lost in the black depths below, and the dashing river is a rippling brook; and still there is more cañon above than below. All about me are interesting geological records. The book is open, and I can read as I run. All about me are grand views, for the clouds are playing again in the gorges. But somehow I think of the nine days' rations, and the bad river, and the lesson of the rocks and the glory of the scene is but half seen."

In the afternoon, the cavorting clouds gave way to storm clouds. "This P.M. we have had a terrible thunder-shower," Bradley wrote. "We had to fasten our boats to the rocks and seek shelter from the wind behind bowlders. The rain poared down in torrents and the thunder-peals echoed through the cañon from crag to crag making wild music for the lightning to dance to. After a shower it is grand to see the cascades leap from the cliffs and turn to vapor before they reach the rocks below. There are thousands of them of all sizes, pure and white as molten silver."

Such hymns to nature were more in Powell's line than Bradley's, but Powell's spirits were so low that he managed only to moan, "Still it rains." Bradley's "pure and white" cascades of rain seemed only to heighten the contrast with the swollen, muddy river, now so full of dirt and grit that the men could no longer drink it. (They could have filled a bucket or a big pot with river water and let the mud settle to the bottom overnight, but ever since the fire, when they lost so much of their mess kit, they had not even had enough cups to go around. Now they drank rainwater puddled up in the rocks.)

On his climb, Powell had peered downstream in the hope of seeing the river emerge from the dreaded granite. He strained to see some encouraging sign of change in the cliffs but could make out "only a labyrinth of deep gorges."

On August 19, it rained all day. "Still we are in our granite prison," Powell noted bleakly. They began the day with a rapid they could run, for once, but it nearly did them in. "The waves were frightful and had any of the boats shipped a sea it would have been her last for there was no still water below," Bradley wrote. "We run a wild race for about two miles, first pulling right— then left, now to avoid the waves and now to escape the bowlders, sometimes half full of water and as soon as a little could be thrown out it was replaced by double the quantity."

They paused for lunch, such as it was, and ate on a cliff side as the rain poured down. Back in the boats again, the men came at once to a "furious" rapid. It had no rocks, though, and Powell decided to run it. The *Emma Dean* went first but immediately swallowed a wave and flipped, flinging Powell, Sumner, and Dunn into the river. Helpless, the three men managed at least to cling to the overturned boat as it raced downstream. Bradley and Walter Powell, in the *Maid of the Cañon,* saw what had happened and did their best to come to the rescue, but "the whirlpools below caught us and our furious speed threw us against the rocks with terrible force." Reeling but still afloat, the *Maid* staggered toward the *Emma Dean.* "It seems a long time before they come to our relief," Powell wrote. "At last they do come; our boat is turned right side up, bailed out, . . . and on we go, without even landing."

Remarkably, they lost nothing but a pair of oars. The clouds cleared, and soon after the men found some driftwood lodged in the rocks and built a huge fire. The flames leaped up and stars twinkled overhead, but it was hardly paradise. The expedition had advanced not quite six miles in another gruesome day. Even with the fire, all the bedrolls were sopping wet, as they had been for a week. Come morning, the sun rose in a cloudless sky but the men were too miserable to push on. Much as they needed to hurry, they needed to rest and dry their gear even more. Powell and the crew spent the morning in camp.

Life had become an endless round of work, and the first men to explore one of the world's greatest natural wonders found themselves stuck in a colossal rut. Each day was, in Sumner's words, "a ceaseless grind of running or letting down rapids with lines, varied in places by making portages of boats and contents. The contents were a small item, but the boats, water-logged and very heavy, taxed our strength to the limit." John Henry, the ex-slave, supposedly died with a hammer in his hand, building the railroad. Powell's men were John Henry's contemporaries. If they did not drown or starve, it began to seem as if they, too, might work themselves to death.

Powell had lightened up a bit, though, at least according to Sumner. After the debacle in camp, "everything was as smooth as with two lovers after their first quarrel and make-up. Major Powell did not run the outfit in the same overbearing manner after that. At a portage or a bad let-down he took his geological hammer and kept out of the way."

The pace picked up. On August 20, the men advanced eight miles after their morning break. Bradley dared to hope that they were nearly home. "We must be getting near to where the Mormons run the river," he wrote, "for they have run it 65 miles above Callville and one would think we had run rapids enough already to be allowed a respite soon."

The river had other plans. The next day ranked "first for dashing wildness of any day we have seen or *will* see if I guess rightly," Bradley wrote. After six bad rapids in seven miles, they came to what Sumner called "a perfect hell—a rapid with a fall of 30 ft. in 300 yards." (This was not their first "perfect hell." Sumner glimpsed hell almost as often as Bradley spotted the worst rapid conceivable.) After a sharp bend to the left, the river swung sharply back to the right. The *Emma Dean* was flung broadside to the current, somehow not capsizing but uncontrollable, as helpless in the pounding waves as a bobbing cork. The river's twists cut off the view ahead, but there was no cutting off the sound of roaring rapids. Black granite walls rose to the sky on both sides of the channel, squeezing the river between its walls and amplifying its thunder. There was no place to pull to shore, no chance of portaging or landing, nothing to do but hang on and hope.

Powell stood on deck, clutching a strap tied across the boat, ducking the crashing waves. Sumner and Dunn struggled to row, their backs to the action. For ten miles the boat leaped and bounded and spun. "The excitement is so great that we forget the danger," Powell wrote, "until we hear the sound of a great fall below; then we back on our oars, and are carried slowly toward its head, and succeed in landing just above, and find that we have to make another portage."

That was melancholy news (although the thrilling run had at least broken the routine), but after the portage came a fabulous discovery. "Just here," Powell rejoiced, "we run out of the granite!"

"Ten miles in less than half a day, and limestone walls below," Powell exclaimed. "Good cheer returns. We forget the storms, and the gloom, and cloud covered cañons, and the black granite, and the raging river, and push our boats from shore in great glee."

The good news came just in time. With the crushing work and the nasty feuding and the pounding rain and the lack of food, no one had much in reserve. "I feel more unwell tonight than I have felt on the trip," Bradley wrote. "I have been wet so much lately that I am ripe for any disease and our scanty food has reduced me to poor condition, but I am still in good spirits." If he ever made it back to civilization, Bradley vowed, he would eat to bursting.

Then, as suddenly as the vanishing of a dream, the good times fled. The rain resumed and kept up all day. No sooner had the exultant men set out again than they found the river carrying them in precisely the wrong direction. "We wheel about a point again to the right, and turn, so as to head back in the direction from which we come, and see the granite again, with its narrow gorge and black crags."

The return to the granite was only half the bad news. The river that they fervently prayed would head west toward civilization and safety now turned almost due east. "What it means I don't know," Bradley wrote, "but if it keeps on in this way we shall be back where we started from."

Now that they were somewhere near their destination and near the end of their food supplies as well, the river's every twist and turn seemed a cruel mockery. The river giveth and the river taketh away. The mood in the boats veered wildly from hope to despair and back again a dozen times a day. Fast water raised the men's spirits. Rain and portaging and the sight of black rocks knocked them back down. Sumner acknowledged the river's treachery in a few terse words. "Ran the granite up and down again," he wrote on August 21, meaning that the boats had escaped from their granite dungeon and then the river had sneeringly returned them to it.

On August 22, Sumner recorded the same spirit-sapping message again. (On the same day, Powell noted forlornly that "a part of our flour has been soaked in the river again.") Then, on August 23, came a change. "Camped on the south side between perpendicular walls 2000 ft. high," Sumner wrote, "all marble." The last two words, though easy to miss, were in fact a cry of joy. Reporting the same news, Bradley sounded more relieved than joyful, like a patient who has recovered from a lingering illness. "This P.M. we got out of the granite rock and . . . the river has now got back to its propper direction again."

August 24 began with a bad portage, but it turned into another encouraging day. The highlight was a fifteen-mile run through a narrow gorge ("all marble," Sumner noted again, still thrilled). Despite pouring rain at night, Powell and the men lay snug and dry in an immense alcove cut into the stone. Sometime soon now, they reassured one another, they would come to the Grand Wash Cliffs, which mark the end of the Grand Canyon. "We cannot now be very far from it unless the river turns back again which it shows no sign of doing," Bradley noted hopefully.

The Mormons, who knew this area better than any other white men (which is to say, hardly at all), had guessed that the distance from the mouth of the Little Colorado to the Grand Wash Cliffs was seventy or eighty miles. Powell and the crew, zealously recording their mileage quarter mile by quarter mile, reckoned that they had already run more than 120 miles since the Little Colorado. How many times could this accursed river bend?

"It is curious how anxious we are to make up our reckoning every time we stop," Powell wrote, with remarkable restraint, "now that our diet is confined to plenty of coffee, very little spoiled flour, and very few dried apples. It has come to be a race for a dinner. Still, we make such fine progress, all hands are in good cheer, but not a moment of daylight is lost."

The boats were hardly fit for racing. All three were leaking so badly that the men had to caulk and recaulk them at nearly every stop. On August 25, they lined one of the freight boats down a rapid, holding a rope tied to an iron ringbolt in the bow. They had lined innumerable rapids before, but now the boats were near collapse. As the men strained at the rope, the bolt tore out like a tooth yanked from the diseased gums of a starving man. The boat leaped forward, but by good fortune, four of the crew were in the river with it, trying to help maneuver it downstream. Before the boat could make its getaway, the men managed to put a line through a ringbolt in the stern.

The landscape changed again. Lava cones from extinct volcanoes sat perched on the canyon rim and a forty-foot-tall shaft of jet-black basalt now called Vulcan's Anvil rose from the river itself. A million years ago, countless tons of molten lava spilled over the canyon rims here. That prehistoric dam rose 2,330 feet high in the Grand Canyon (Glen Canyon Dam is 710 feet) and transformed the ferocious Colorado into a placid puddle that may have extended the full 179 miles back upstream to Lee's Ferry. It took the relentless river a mere quarter of a million years to wear the dam away and restore the status quo.

For Powell, half-starved and lost on an endless river though he was, the signs of a geological battle on a grand scale conjured up a spectacle glorious to contemplate. "What a conflict of water and fire there must have been here! Just imagine a river of molten rock, running down into a river of melted snow. What a seething and boiling of the waters; what clouds of steam rolled into the heavens!"

The men spent three hours portaging Lava Falls, today one of the two or three most dreaded rapids in the Grand Canyon, and then ran another twenty miles of rapids. "Thirty five miles today. Hurrah!" Powell wrote, and Bradley noted optimistically that "the country begins to look a little more open and the river still improves." They were not home yet. Bradley concluded his journal entry for August 25 on a more somber note. "We commenced our last sack of flour tonight."

Late in the morning on August 26, the men made a discovery they considered far more exciting than any extinct volcano. They found a carefully tended vegetable garden, with corn and melons and squash and no sign of the Indians who had planted it. The corn and melons were too young to eat, but the squash were large and tempting. Powell and the others raced off with a dozen squash, hurrying in case the Indians returned. "What a kettle of squash sauce we make!" Powell cheered. "True, we have no salt with which to season it, but it makes a fine addition to our unleavened bread and coffee. Never was fruit so sweet as these stolen squashes."

Dinner precisely duplicated lunch. No one complained. "What a supper we make," Powell wrote, "unleavened bread, green squash sauce, and strong coffee. We have been for a few days on half rations, but we have no stint of roast squash."

The meals were the highlights of the day, but there was other good news. For the second day in a row, the weary party had advanced thirty-five miles. "A few days like this," Powell wrote, "and we are out of prison."

The river responded as maliciously as if Powell had whispered his hope aloud. The next day it swung in the wrong direction yet again, this time heading south. The dark granite, which the men had left behind, now loomed ominously in the distance once more. Toying with its victims like a Lothario trifling with his admirers' hearts, the river turned and twisted as if for sport. "Now and then the river turns to the west, and excites hopes that are soon destroyed by another turn to the south," Powell wrote.

Then came a sudden, cold end to the teasing. "About nine o'clock we come to the dreaded rock. It is with no little misgiving that we see the river enter these black, hard walls."

CHAPTER TWENTY-FIVE

SEPARATION RAPID

At noon on August 27, 1869, the expedition came to a rapid that outdid their worst forebodings. The men pulled to shore and stared in horror. "The water dashes against the left bank and then is thrown furiously back against the right," Bradley wrote. "The billows are huge and I fear our boats could not ride them [even] if we could keep them off the rocks. The spectacle is appalling to us."

This latest bit of bad news might have been easier to take if it had been just one more dreary note in a long dirge. But the day before, the discovery of the Indian garden had sent everyone's spirits soaring. Hope had begun stirring, whispering hints of a life outside prison walls. And then this.

With the boats tied up on the right shore, Powell and the crew set out up the cliff in search of a path around the rapid. They climbed up and across the granite for a mile or two but saw no place where they could line or portage and no path through the rapid. "To run it," Powell concluded bleakly, "would be sure destruction."

Everyone straggled back to camp, choked down lunch, and set out again, this time after crossing to the river's other shore. Up they climbed, to the top of the granite, only to find that the crags and pinnacles made it nearly impossible to see down to river level. They wasted another hour climbing high above the river on the left, in a futile search for a useful vantage point. Giving up, they recrossed the river to search the right bank a second time, as if in the hope that *this time* the iron bars of their cell would somehow give way. A close look only confirmed how much trouble they were in. To transport the heavy boats around the rapid would mean climbing eight hundred feet to the summit of the granite, and then back down, while somehow belaying or carrying the boats. By Powell's calculation, it would take ten days. Only five days' food remained.

"We appeared to be up against it sure," Sumner wrote. "There were two side canyons coming into the Colorado nearly opposite each other, the river not being over fifty yards wide and running like a race horse. The first part of the rapid is caused by the big rocks carried out of the side canyon coming in from the south. The second part of the rapid is formed by rocks from the northside canyon and a granite reef that reaches one-third of the way across the river, making a Z-shaped rapid. We spent the day trying to solve the problem."

Oramel Howland and Dunn confided to Sumner that they had found a solution of their own. They intended to leave the river and try to make it overland to one of the nearby Mormon settlements. The rest of the party should hike out, too. The alternative was an almost certain drowning. "I did what I could to knock such notions out of their heads," Sumner wrote, "but as I was not sure of my own side of the argument, I fear I did not make the case very strong."

Just two weeks into the trip, Howland had survived the wreck at Disaster Falls and emerged undaunted. Two weeks later, he had proclaimed that nothing was as abhorrent to him as "a calm, smooth stream" and nothing as appealing as the sight of "white foam" in a mad river. Now, after eight more weeks of rapids and half rations and rain and quarreling, this brave man had endured enough. "[Howland] had fully made up his mind to quit," Sumner wrote, "since the rapids had become a holy terror to him."

Bradley had his nerves under better control, but even his sober assessment of the situation was deeply discouraging. "We have only subsistence for about five days and have been trying half a day to get around this one rapid while there are three others in sight below," he wrote. "What they are we cannot tell only that they are huge ones. If we could get on the cliff about a hundred yards below the head of this one we could let our boats down to that point and then have foothold all the rest of the way, but we have tried all the P.M. without success."

Late in the afternoon, Powell climbed down from the cliff and announced his decision. The next morning they would lower the boats past the first waterfall, then run the rapid that extended to the head of the second fall, and then squirt down a chute Powell had spotted along the right side of that fall. Finally, they would pull for their lives across the river, to get to the left of the channel-blocking boulder before the angry river flung them into it headlong.

It was more a prayer than a plan. Even the first step, lining the upstream waterfall, was one they had already dismissed as impossible. After Powell broke the news, he and the crew took the boats across the river and camped, forlornly, by a creek at a side canyon on the river's left side.

They drank their black coffee and gulped down chunks of unleavened bread and tried to ignore the rapid's din. Scared and despondent, the men muttered to one another and tried to think things through. "There is discontent in camp tonight and I fear some of the party will take to the mountains but hope not," Bradley wrote. "This is decidedly the darkest day of the trip."

Just after dinner, Oramel Howland approached Powell and asked if they could talk in private. The two men walked a short distance along the creek and talked out of earshot of the others. To take on the rapid, Howland said, was suicide. They should all abandon the river and take their chances on hiking out to safety. But regardless of what the others decided, Howland announced, he, his brother, and Dunn had made up their minds to leave.

Powell might have tried ordering Howland to stay with the expedition, but instead he tried to persuade him not to leave. The canyon walls soared thousands of feet above their heads. They might be trapped, like children in the bottom of a well. They had hardly any food. If they *did* manage to climb out, could they make it across the desert to the nearest Mormon town? The best choice was to trust their lives to the river.

In the end, neither man could sway the other. Howland grabbed his bedroll and headed off to find a place to sleep. It was a starry night, and Powell took out the sextant to measure their position. The straight-line distance to the mouth of the Virgin River, he estimated, was perhaps forty-five miles. No more than twenty miles upstream on the Virgin River were Mormon settlements, and safety. Even if the forty-five miles turned out to be ninety, because of the river's meandering, they *had* to be nearly home. And the Colorado was known to be fairly tame for a long stretch above the mouth of the Virgin.

Powell woke Howland and made his case one more time, this time drawing on his new calculations. No sale. "We have another short talk about the morrow, and he lies down again; but for me there is no sleep," Powell wrote. "All night long, I pace up and down a little path, on a few yards of sand beach, along by the river. Is it wise to go on?"

Powell weighed and reweighed the possibilities through the night. He checked the rations again. The expedition had come down to this one judgment call, he knew, and decisive as he ordinarily was, Powell wavered. On the one hand, they had survived more rapids than they could count. On the other hand, they had never run a rapid that looked as bad as this one. On the one hand, this rapid could be a fluke, a one-time horror. On the other, it might be a preview of what the river had in store from here on. On the one hand, they might be able to climb out of the canyon. On the other, they would then have to survive a long trek across the desert. On the one hand, it had rained so much lately that, if they hiked out, they might find enough water to keep them alive. On the other, they might not.

"At one time, I almost conclude to leave the river," Powell wrote. "But for years I have been contemplating this trip. To leave the exploration unfinished, to say that there is a part of the cañon which I cannot explore, having already almost accomplished it, is more than I am willing to acknowledge, and I determine to go on."

Powell woke Walter and told him about Howland's decision to leave. Walter said he would stay with his brother. Powell talked with Hawkins and Hall. The two men, boat mates in the *Kitty Clyde's Sister*, were the expedition's youngest, boldest members. They had come a long way, in every sense, since the first day of the trip when they had not known how to pull to shore even in easy water. They proclaimed their determination to stay with the river.

Powell turned toward Bradley. Time after time, Bradley had favored running rapids that Powell had insisted on portaging. That choice had pitted a minute's danger against hours of crushing labor. Now the choice was between a minute or two of hell and days wandering in the desert. Bradley chose the river. "I shall be one to try to run it rather than take to the mountains," he had written the previous night just before falling asleep. "'Tis darkest just before the day' and I trust our day is about to dawn." With slightly more hesitation—he was worried that this looming rapid was only the first of a deadly series—Sumner made the same decision.

• • •

In the morning, no one had the heart to resume the debate. "At last daylight comes," Powell wrote, "and we have breakfast, without a word being said about the future. The meal is as solemn as a funeral. After breakfast, I ask the three men if they still think it best to leave us."

Oramel Howland did, and so did Dunn. Seneca Howland tried to convince the pair to stick with the others. Failing in his plea, Seneca threw in his lot with his older brother and Dunn. That made three who planned to trust to the desert's mercy, and six who "came to the determination," in Bradley's words, "to run the rapid or perish in the attempt."

The entire party crossed the river again. With only six men continuing downriver, two boats would do. The men unloaded the *Emma Dean* and left it behind, tied to the shore. ("Abandoned the small boat as useless property," Sumner wrote, with characteristic unsentimentality, of the boat he had piloted for nine hundred miles.) Left on shore, too, as if in mute acknowledgment that the trip was a scientific expedition no longer, were the broken barometers and the collections of fossils and minerals, as well as some beaver traps.

"With great labor," and with the help of the Howlands and Dunn—who were parting from the expedition but lingered to help with this last chore—the men managed to carry the two freight boats over a twenty-five-foot boulder. Once there, they snubbed a rope around a convenient pillar in the granite and lowered the boats into a little cove just barely big enough to hold them both.

Then it was time for farewells. The three climbers took two rifles and a shotgun, in the hope they would find game when they reached the plateau. Well aware of the plight of the men they left behind, they refused Powell's offer of a share of the rations but did take some lumps of dough that Hawkins had fixed. Powell gave Oramel Howland a letter to Emma, and Sumner gave him a watch to deliver to his sister as a keepsake in case he drowned. The trip records had been kept in duplicate, and Howland took one set. (In the haste and confusion, Powell inadvertently gave Howland two copies of some records and and no copies of others.)

Howland pleaded with the others, one last time, to reconsider. It was folly to challenge this rapid. Downstream, the river seemed to veer south once again, deeper into the granite. A few more miles would exhaust what few rations remained and then it would be too late to climb out. "Some tears are shed," Powell wrote. "It is rather a solemn parting; each party thinks the other is taking the dangerous course."

"Three men refused to go farther . . . and we had to let them take to the mountains," Bradley wrote somberly. "They left us with good feelings though we deeply regret their loss for they are as fine fellows as I ever had the good fortune to meet."

The *Kitty Clyde's Sister*, with Powell aboard, pushed out into the current. The *Maid* followed close behind. The Howland brothers and Dunn stood on a cliff, watching anxiously. "Dashed out into the boiling tide with all the courage we could muster," Bradley wrote. The *Sister* raced along the foot of the wall, scraped against a boulder, then plummeted over a fall, careened into the waves at its foot, and filled with water. A huge boulder loomed downstream, blocking the channel's right side. Caught in the torrent headed for the rock face, Hall and Hawkins pulled for their lives, desperate to get across the current and to the left of the boulder. The waves grew "too large to do anything but hold on to the boats." Somehow—it happened so fast they could not figure it out—they were safely by, bobbing in the choppy water at the rapid's downstream end.

The *Maid* made it, too. Both boats had nearly swamped, but they were right side up and everyone was safe and not even an oar had been lost. It had taken perhaps a minute.

"The men that were left sat on the cliffs and watched us go safely over," Powell wrote, "so we went into camp and waited two hours, hoping that they would join us with the boat left tied to the rocks above." Even without the *Emma Dean*, the Howlands and Dunn could have joined Powell and the others simply by climbing along the cliff past the rapid. That would have meant all nine men had to squeeze into two boats. Earlier in the trip, when there were countless miles still ahead and the boats were weighed down by tons of supplies, that would have been impossible. Now the boats were nearly empty, the trip was nearly done, and a human cargo that totaled perhaps five hundred pounds would have been no problem.

Sumner and the rest of the crew fired their guns into the air to signal that all was well and waved imploringly for their friends to come. To no avail. "The last thing we saw of them," Sumner wrote, "they were standing on the reef, motioning us to go on, which we finally did."

There stood Oramel Howland, his Lear-like beard now filthy and matted, and silent, loyal Seneca Howland by his brother's side, and Bill Dunn, a mountain man about to put his skills to the ultimate test. Rapid or no rapid, the three were determined to go their own way. They set out up what is now called Separation Canyon in somber procession.

Exhausted and emotionally drained, the six men turned to the river once again. It showed no mercy. "Ran 10 more rapids in 6 miles," Sumner wrote, "when we came to another hell." Hemmed in by high, narrow walls, Lava Cliff Rapid, as it is now called, was impossible to run. The men found they could climb the cliff along one side, though, and made a plan to line the boats through. Bradley stayed in his boat so he could fend off the rocks with an oar. The men climbed their way downstream while holding the *Maid* on the end of a 130-foot rope lashed to the cutwater, the long, curved piece of wood at the farthest forward point of the bow. (If a knife were held in cutting position, as if about to slice down into a loaf of bread, the forwardmost part of the blade would correspond to the cutwater.)

All went well for a few minutes, but the current grew fiercer and the layout of the cliffs forced the men to climb ever higher. Finally it became clear that without more rope to swing him free, Bradley and his boat were stuck. One man raced back along the shore for more rope. The others did their best to fight the current that was trying to grab Bradley for itself. Though only a rope's length apart, Bradley and the men on the cliffs could not hear each other over the rapid's roar. The cliff was so steep that the men could not *see* Bradley either, their view cut off by a point of rock. The *Maid* was trapped in the worst possible spot, where the current was at its strongest. Time and again, the river dragged the boat out from shore to the rope's fullest extension, like an archer pulling a bow to its ultimate tautness, and then shot it full force against the rocks.

With just four more feet of rope, Bradley figured, he would have enough slack to maneuver out of harm's way. He didn't have four feet, though, and the river seemed intent on pulverizing the *Maid*. Was he better off waiting for rescue or launching himself into the rapid while he still had a boat?

Bradley took out his knife and leaned down toward the line. "One look at the foaming cataract below kept me from cutting it and then I was suffering all the tortures of the rack." He held off and spent a moment scanning the rapid. Now two frantic scenes played out in parallel, out of sight of one another. On the cliff, the men finally had a second rope and had begun joining the two lines. In the *Maid*, Bradley still hesitated, knife in hand. Suddenly the river ripped the boat from the cliff once again, and the cutwater tore loose and flew thirty feet into the air, trailing the rope behind it.

The freed boat dashed into the rapid like a runaway horse. "On I went,"

Bradley wrote, "and sooner than I can write it was in the breakers but just as I always am, afraid while danger is approaching but cool in the midst of it, I could steer the boat as well as if the water was smooth. By putting an oar first on one side then on the other I could swing her around and guide her very well." Bradley raced downstream, now aligned with the current rather than broadside to it, while the men stared horror-struck from the cliff above.

"The boat is fairly turned," Powell wrote, "and she goes down almost beyond our sight, though we are more than a hundred feet above the river. Then she comes up again, on a great wave, and down and up, then around behind some great rocks, and is lost in the mad, white foam below. We stand frozen with fear, for we see no boat. Bradley is gone, so it seems."

Then a shape emerged from the waves. A boat, and Bradley standing on its deck, waving his hat back and forth to show that he was safe. "But he is in a whirlpool," Powell wrote. "We have the stem-post [cutwater] of his boat attached to the line. How badly she may be disabled we know not." Powell shouted to Sumner and Walter to see if they could grab a line and run along the cliffs to Bradley's rescue. At the same time, Powell, Hall, and Hawkins jumped into the other boat and sped into the rapid, in a frenzy to get to Bradley while there was still time.

They had lined and portaged countless rapids that seemed too dangerous to run. Now, without a plan, they had flung themselves into perhaps the worst rapid of all. "A wave rolls over us, and our boat is unmanageable," Powell wrote. "Another great wave strikes us, the boat rolls over, and tumbles and tosses, I know not how. All I know is that Bradley is picking us up."

Bradley had rescued his rescuers. When Sumner and Walter scrambled down the cliff a few minutes later, the party was intact again. It was an emotional reunion. "Major says nothing ever gave him more joy than to see me swing my hat," Bradley wrote proudly, "for they all thought that the boat and I too were gone to the 'Happy Hunting grounds' until then."★

Bradley had rated rapid after rapid as "the worst we ever run." Now he was finally ready to retire the title. The last rapid of the day on August 28, he declared once and for all, "stands A No. 1 of the trip."

In his river diary, Powell summarized all the mad adventures of the day in a bare handful of unpunctuated words. "28th," he wrote. "Boys left us ran rapids Bradey boat broke camp on left bank camp 44."

As usual, he misspelled Bradley's name.

★Later, on the lecture circuit, Powell would grow teary and the audience would burst into applause when Powell described Bradley waving his hat to signal "All's well!"

CHAPTER TWENTY-SIX
DELIVERANCE

The next day, August 29, was "the first Sunday that I have felt justified in running," Bradley wrote, "but it has now become a race for life. We have only enough flour to last us five days and we do not know how long we may be winding around among these hills." Everyone was exhausted now, staggering toward the finish line like first-time marathoners.

They were hopeful, though, for the country looked more open by the mile. By ten o'clock, they had made it out of the granite once more. They ran all the rapids now without even stopping to scout, "for we have no time to waste in looking." The farther they went, the easier the rapids became.

In two more hours, it was over. "At twelve o'clock," Powell wrote, "[we] emerge from the Grand Canyon of the Colorado." Free at last. "We came out to a low rolling desert," Sumner wrote, "and saw plainly that our work of danger was done—gave 3 cheers and pulled away steady strokes." The Howland brothers and Bill Dunn had abandoned the river and set out on foot just over twenty-four hours before.

For almost the first time in three months, the river view was not cut off by stone walls. "It was a strange and delightful sensation," Sumner wrote. "For my part, I felt as Dante probably felt as he crowded up from Hades." Rowing easily now, the men carried on like schoolboys. Walter Powell, "Old Shady," sang away at the top of his lungs, and Hall tried it, too. Hall's ventures into song never went well. This time Hawkins told him to stop before he drowned him. Lunch was another meal of "sinkers," the heavy lumps of flour they had grown to know too well, but the prospect of better meals just ahead helped a bit.

They camped that night on the river's left bank in a thicket of mesquite. As always in moments of high emotion, Powell turned to military images to convey his feelings. "The relief from danger, and the joy of success, are great. When he who has been chained by wounds to a hospital cot, until his canvas tent seems like a dungeon cell, until the groans of those who lie about, tortured with probe and knife, are piled up, a weight of horror on his ears that he cannot throw off, cannot forget, and until the stench of festering wounds and anesthetic drugs has filled the air with its loathsome burthen, at last goes out into the open field, what a world he sees! How beautiful the sky; how bright the sunshine; what 'floods of delirious music' pour from the throats of birds; how sweet the fragrance of earth, and tree, and blossom! The first hour of convalescent freedom seems rich recompense for all—pain, gloom, terror."

Powell and all his men were beside themselves with happiness and relief. "Ever before us has been an unknown danger, heavier than immediate peril," Powell wrote. Every waking hour in the Grand Canyon had been marked by danger and toil and the threatening growl of rumbling waters, all this endured in "gloomy depths" that transformed one's world into a stony tomb. "Now the danger is over; now the toil has ceased; now the gloom has disappeared; now the firmament is bounded only by the horizon; and what a vast expanse of constellations can be seen!"

Worn out though Powell and the other men were, they sat up late, too excited to sleep. "The river rolls by us in silent majesty; the quiet of the camp is sweet; our joy is almost ecstasy," Powell wrote. "We sit till long after midnight, talking of the Grand Cañon, talking of home, but chiefly talking of the three men who left us. Are they wandering in those depths, unable to find a way out? Are they searching over the desert lands above for water? Or are they nearing the settlements?"

•   •   •

As soon as the sun rose on August 30, Powell and the crew were back on the river. Now it was "nothing but smooth water and rolling desert." At ten o'clock, they caught sight of a band of Indians in the distance, their first glimpse of human beings in nearly two months (for the men who had not gone to the Indian agency, three months). By the time the boats approached, the Indians had fled. Two or three miles downstream, the boats rounded a corner and came upon more Indians. They seemed terrified, but Powell managed to convey, in the Indians' language, that they were friends. Powell asked a few questions about the nearest white people but got no response.

Taking to the boats again, the men continued downstream. By noon, they had covered twenty-six miles (they had run forty-two the previous day). They stopped quickly for lunch, and then set to work again, hoping with each pull of the oars that now, finally, they would reach the mouth of the Virgin River. After another seven miles, they saw figures in the distance. Sumner pulled out a spyglass and stared downstream. These were white men! He dove into the locker, pulled out the Stars and Stripes, and set it fluttering at the bow.

Three men and a boy stood fishing with a net in a large, muddy stream. The fishermen stared at the two battered, leaking boats and their crew of six half-starved, nearly naked, worn-out men. The boats drew near, and Powell and the men burst out with questions. Was this stream the Virgin? Were there Mormon settlements nearby?

The fishermen were Mormons helping to build a new town at the junction of the Colorado and the Virgin. They had been expecting Powell's crew—dead. Several weeks before, Brigham Young had sent orders from Salt Lake City to scan the river "for any fragments or relics of [Powell's] party that might drift down the stream."

No one had imagined that Powell and his crew could turn up alive. They had trouble believing it themselves. "We could hardly credit that all our trials were over," Bradley exclaimed. He took a few minutes to scribble a note "to assure Mother I was all right, but I was so intoxicated with joy at getting through so soon and so well that I don't know what I wrote to her."

Joseph Asey, their newfound friend, led the six men to his cabin and began cooking—fish and squash and whatever else he could lay his hands on—while Powell and his men tried to keep from tearing into it raw. Then "we laid our dignified manners aside and assumed the manner of so many hogs," Sumner wrote. "Ate as long as we could and went to sleep to wake hungry."

August 31 was a well-earned day of rest. Asey had sent a messenger to St. Thomas, a Mormon town about twenty miles upstream on the Virgin River, to deliver the news of Powell's arrival. A bishop from St. Thomas arrived in the evening, bringing mail, flour, cheese, bread, butter, and two or three dozen melons in his wagon. "We talked and ate melons till the morning star could be seen," Sumner wrote happily.

It was over. They had traveled for ninety-nine days and covered a thousand twisting, white-capped miles. Ten rowdy, hollering men had pushed out into the current at Green River Station, Wyoming. Six weary, frightened men had stumbled ashore below the Grand Canyon. One man had left nearly two months before, at the first opportunity. The fate of three more was unknown.

Sumner sat down and did his own summing up. "Rapids ran 414—Portages made 62—making 476 bad rapids . . . I never want to see it again anywhere. Near the Grand Cañon it will probably remain unvisited for many years again, as it has nothing to recommend it, but its general desolation as a study for the geologist."

The expedition was the making of Powell. Soon a nation would hail him as "the conqueror of the Colorado." Sumner, fully as proud and self-dramatizing a man as Powell but a far less sociable character, wanted only to put the whole experience behind him. "I find myself penniless and disgusted with the whole thing," he wrote, "sitting under a mesquite bush in the sand." He closed his journal with a flourish. "I shoulder my gun and bidding all adieu I go again to the wilderness."

Scarcely any supplies remained. The final tally, for six men, was ten pounds of flour, fifteen pounds of dried apples, and eighty pounds of coffee.

CHAPTER TWENTY-SEVEN
THE VANISHING

On September 1, the Colorado River Exploring Expedition disbanded, never to reunite. Powell and Walter left the river (the stretch that remained had been explored before) and rode off to St. George, Utah Territory, en route to Salt Lake City and, finally, a triumphant return back East.

Sumner and Bradley continued downstream in the *Maid of the Cañon* as far as Fort Yuma, near the Arizona-California border, for no reason much better than a desire to finish what they had started. Hawkins and Hall rowed the *Kitty Clyde's Sister* all the way to the ocean.

The visions of heaps of gold and stacks of beaver pelts had long since disappeared. Precisely what the crew did take away from the trip, beyond two badly worn boats, will probably never be known, but it wasn't much. Decades later, in 1902, Sumner would complain bitterly that "when Proff Powell left us at the mouth of the Virgin River he gave me 100 dollars in green backs worth then on the Coast 75.00 to Hall he gave the great sum of 20 dollars to Bradley a pleasant smile and a volume of thanks." In his old age, Hawkins complained in the same vein.

While his four ex–crew men proceeded downstream, Powell made his way north with Walter. At St. George, Powell asked anxiously if anyone knew anything about the Howlands and Dunn, but there had been no word. Powell's driver and mule team turned toward Salt Lake City, a week distant. On September 8, while Powell was on the road, the Salt Lake paper reported grim news. "Three of the Powell Expedition Killed by Indians," the *Deseret Evening News* announced, above a one-paragraph story:

> We have received a dispatch through the Deseret Telegraph Line from St. George of the murder of three of the men belonging to the Powell exploration expedition. It appears according to the report of a friendly Indian that about five days ago the men were found by peaceable Indians of the Shebett [Shivwits] tribe very hungry. The Shebetts fed them, and put them on the trail leading to Washington in Southern Utah. On their journey they saw a squaw gathering seed, and shot her; whereupon they were followed by three Shebetts and killed. A friendly Indian has been sent out to secure their papers. The telegraph does not give us the names of the men.

The Powell brothers reached Salt Lake City and heard the bad news on September 15. Powell deemed it not at all certain that the murdered men were the Howlands and Dunn, especially because the woman they had reportedly shot (and, by some accounts, raped) was unarmed. "I have known O. G. Howland personally for many years," Powell declared, "and I have no hesitation in pronouncing this part of the story as a libel. It was not in the man's faithful, genial nature to do such a thing."

Still, with his attention now turned back East, Powell let the matter drop. At seven o'clock on September 16, Powell delivered a lecture on the expedition. At nine the same evening, he and Walter caught a night stagecoach to Corinne, Utah. ("Major Powell, of the Powell Expedition who has been lost, drowned and resurrected a dozen times [on paper], arrived here last night from the south," the local paper reported, "in the best of health and spirits, a plain, unpretentious gentleman.") From Corinne, they hopped a Union Pacific train to Chicago. By September 20, Powell was ensconced at Chicago's Tremont Hotel, giving interviews to eager reporters and planning the stops on his lecture tour.

•   •   •

The 1869 expedition had been a first, but Powell had fallen dismayingly short of his scientific goals. In 1870, he announced a plan to do the whole horrendous trip a second time. The new expedition, he explained, would produce the careful maps and topographic data that the first one had not. Trading on his newfound reputation, Powell won a $12,000 appropriation from Congress—a significant sum in those days—"for completing the survey of the Colorado of the West and its tributaries." Success brought fame and fame brought money. Powell's days as an empty-pocketed freelancer had ended.

The key to this second expedition was a different approach to the problem of supplies. Rather than carry all the food from the outset, Powell intended to cache supplies ahead of time at various spots along the river. In the late summer of 1870, Powell was traveling in southern Utah in the expert company of Jacob Hamblin, the renowned Mormon scout, looking for manageable routes down to the Colorado. Hamblin, "the Leatherstocking of the Desert," was an ardent Mormon who served Brigham Young, in the words of one historian, "with a loyalty that amounted almost to worship."

Near Mount Trumbull, Powell and Hamblin met with a group of Shivwits. With Hamblin serving as interpreter, Powell explained what he was after. "I tell the Indians that I wish to spend some months in their country during the coming year, and that I would like them to treat me as a friend. I do not wish to trade; do not want their lands . . . I tell them of many Indian tribes, and where they live; of the European nations; of the Chinese, of Africans, and all the strange things about them that come to my mind. I tell them of the ocean, of great rivers and high mountains, of strange beasts and birds. At last I tell them I wish to learn about their cañons and mountains, and about themselves, to tell other men at home; and that I want to take pictures of everything, and show them to my friends."

Hamblin translated the chief's response. "We will be friends, and when you come we will be glad. We will tell the Indians who live on the other side of the great river that we have seen Ka-pu-rats ["Arm Off"], and he is the Indians' friend. . . . We have not much to give; you must not think us mean. You are wise; we have heard you tell strange things. We are ignorant. Last year we killed three white men. Bad men said they were our enemies. They told great lies. We thought them true. We were mad; it made us big fools. We are very sorry. Do not think of them, it is done; let us be friends."

Here was confirmation, then, and from the Indians' own mouths, of the story the newspaper had reported the year before. Powell seemed curiously unperturbed. "That night I slept in peace, although these murderers of my men, and their friends, the *U-in-ka-rets*, were sleeping not five hundred yards away." He never sought to punish anyone for the killings and almost never referred to the murdered men again.

It is a puzzling story, made all the more unsettling by Powell's reliance on Hamblin to interpret for him. The Shivwits knew perfectly well how swift and careless frontier justice could be. In 1866, at nearby Pipe Spring, for example, someone had killed two white ranchers. An armed posse arrived, accused a band of Paiutes of the murders, and left eight of them dead. Would Indians in 1870 freely confess to a white man that they had murdered three of his companions?

In camp below Separation Rapid on the night that the Howland brothers and Dunn hiked out, Sumner and the other men had tried to guess their friends' fate. The consensus, as Sumner put it, was that "the red bellies would surely get them." Sumner (though he was no friend of the Indians) disagreed. He feared that the three men "would not escape the double-dyed white devils that infested that part of the country. Grapevine reports convinced me later that that was their fate."

Sumner's suspicions had begun to take shape almost at once. On the second day at the Virgin River, while the men devoured melons and swapped stories with their Mormon greeting party, they tried to work out a plan to rescue the Howlands and Dunn. One of the listeners suddenly perked up when Powell mentioned the valuables the men had with them. "From one with a listless demeanor," Sumner wrote, "he instantly changed to a wide-awake, intensely interested listener, and his eyes snapped and burned like a rattlesnake's, particularly when Major Powell told him of an especially valuable chronometer for which he had paid six hundred and fifty dollars."

Soon after, Sumner heard the story of how his three friends had been killed by Indians intent on revenge. He dismissed it at once and never saw fit to reconsider. "I am positive I saw some years afterwards the silver watch that I had given Howland," he wrote in his old age. "I was with some men in a carousal. One of them had a watch and boasted how he came by it. I tried to get hold of it so as to identify it by a certain screw that I had made and put in myself, but it was spirited away, and I was never afterwards able to get sight of it. Such evidence is not conclusive, but all of it was enough to convince me that the Indians were not at the head of the murder, if they had anything to do with it."

There matters stood until 1980. Then Wesley Larsen, an amateur historian and a former dean of the college of science at Southern Utah University, found a letter that had lain undisturbed in a trunk with a jumble of other papers for ninety-seven years. The letter was written by one William Leany to John Steele and rediscovered in a trunk belonging to Steele's great-grandson.

Leany had been one of Brigham Young's bodyguards. Steele was a judge and a militia officer in the Blackhawk Indian War (and the father of the first white child born in Utah Territory). Both men were devout Mormons, but, in the most notorious incident in Mormon history, Leany had run afoul of his church. In 1857, as noted earlier, a group of Mormons (and, possibly, Indians) had perpetrated the Mountain Meadows Massacre, killing some 120 California-bound emigrants. Leany happened to see that ill-fated wagon train pass by as it wound its way across southern Utah headed in the direction of Mountain Meadows, and he recognized a man in the emigrant party. Years before, in Missouri, the man's father had saved Leany from an anti-Mormon mob that was threatening to kill him. Now the emigrants were in trouble themselves, for the Mormons had agreed to deny food or other assistance to the Gentile interlopers.

Seeing his rescuer's son, Leany defied the ban and gave the man a meal, a roof for the night, and some vegetables. Leany's fellow Mormons charged him with giving "aid and comfort to the enemy." To teach him a lesson, someone clubbed him over the head, fracturing his skull and leaving him for dead.

Leany survived. By 1883, he and Steele were old men. Steele evidently suggested to his good friend Leany that the time had come for them both to repent of their sins. Leany wanted no part of it. The church had blood on its hands, but *he* had nothing to repent. Like an Old Testament preacher, Leany thundered that "thieving whoredom murder and Suicide & like abominations" reigned in the land. Then came the sentence that, a century later, electrified Wes Larsen: "You are far from ignorant of those deeds of blood from the day the picket fence was broken on my head *to the day those three were murdered in our ward & the murderer killed to stop the shedding of more blood.*" [italics added]

Larsen (like the other players in the story, a good Mormon) embarked on a frenzied round of detective work. The reference to "our ward," a local Mormon district run by a bishop, was the first clue. Leany and Steele had lived in the same ward only once through the years, in 1869. And in that same fateful year, Larsen found, only one trio of men—Oramel Howland, Seneca Howland, and Bill Dunn—had been reported missing or killed in southern Utah.

Further, Larsen learned, only weeks before the Powell expedition reached Separation Rapid, Brigham Young had traveled throughout the region warning the faithful that the long-threatened invasion of Utah by Gentiles was imminent. When "war" came, Young warned his listeners, blood would rise "to their knees and even to their waist and to their horses' bridle bits." The Mormon leader ordered sentries posted at all the passes leading into southern Utah.

Then, at the worst possible moment, three white strangers wandered into no-man's-land spouting a cock-and-bull story about their trip down a river that everyone knew was impassable. The three men were dragged off and executed as spies, Larsen speculates, and the news of the unsanctioned executions triumphantly telegraphed to Salt Lake City. At virtually the same moment, church leaders in Salt Lake received word from the fishermen at the mouth of the Virgin River that John Wesley Powell had survived his trip down the Colorado. He had arrived alive, but he was asking after three of his men who had left the river and set out on foot across the desert.

In Larsen's scenario, the next step was an exact replay of the Mormon response to the 120 killings at Mountain Meadows. First came cover-up (thus the reference in Leany's letter to "the murderer killed to stop the shedding of more blood"), then a vow of silence on the part of those who knew the truth, and finally a finger of blame pinning the crime on the nearest Indians.

Three theories, then, and none of them good. The first assumes that Indians would nonchalantly reveal that they had killed three white men (and the Shivwits had a reputation as peaceable). The next bases an accusation of triple murder on a stranger's shifty eyes and Sumner's certainty that he recognized a favorite watch. The last takes a cryptic sentence in a private letter and spins a tale of conspiracy, blood oaths, and cold-eyed executions.

The only certainty is that Oramel Howland, Seneca Howland, and Bill

Dunn hiked up Separation Canyon on August 28, 1869, and were never seen alive again. They should have made it to the plateau, for the climb up from the river is rugged but doable and the heavy rains would have left water standing in puddles. Once at the rim, they would have headed north toward the Mormon settlements. It would have been only a short distance to a high point called Blue Knob that offers a view over the countryside. From there, the next conspicuous landmark to the north was Mount Dellenbaugh, as it is now called, about twenty miles distant. The summit is an easy hike and would have made a good spot to look for Indian trails or water.

Scratched into the rock near the mountain summit is an inscription, cut with a knife. "Dunn 1869," it reads, and "Water," above an arrow pointing north. If the inscription is a hoax, it is an old one. Ten years ago, a local historian tracked down a "hermit cowboy" who had lived near Mount Dellenbaugh since the early 1930s. He didn't know if the inscription was genuine or not, the old man said, but he knew that it had been there as long as he had.

The bodies of the Howland brothers and Dunn have never been found. The three men, along with their weapons, their papers, and the compasses and chronometers and other gear they carried, seem to have vanished into the air.

EPILOGUE

Frank Goodman, who left the expedition shortly after the wreck at Disaster Falls, settled in Utah and became the patriarch of a large family. ("He had the most sense of anybody in the whole expedition," one modern boatman half-jokes. "He walked out when he had the chance.") Goodman settled in Vernal, Utah, in the northeast corner of the state. Vernal happened to be the home-town of Nathaniel Galloway, the trapper who went on to invent the "face your danger" technique of river running. One white-water historian specu-lates that Galloway may have hit on the idea of taking rapids head-on, and the idea of building flat-bottomed, lightweight boats as well, by listening to Goodman's tales of misfortune in Powell's boats.

George Bradley spent the rest of his life near San Diego running a small fruit orchard. In 1885, in poor health, he returned to Massachusetts to be near his family. Bradley died a few weeks later, at fifty, in his sister's home. His death went unnoticed even in his hometown newspaper.

The irrepressible Andy Hall returned to his old trade, driving mule teams, in Arizona Territory. On an August afternoon in 1882, Hall and another man were driving a six-mule train for the Wells Fargo Express Company. They carried the U.S. mail and a miscellany of other goods, including a large box containing a $5,000 payroll in gold. Just outside Globe, Arizona, the mule train reached a small gully next to a giant boulder. A volley of rifle shots tore into the mules. One animal fell dead, and the others ran off in panic. Hall saw the gunmen chase down the mule carrying the gold, divide up the loot, and ride off. He set out to track the thieves while his partner rode to Globe to round up a posse.

The posse found a series of markers Hall had left. First were deep gouges he had scraped in the dirt with his boot heel, then (when the robbers moved into brush country) a trail of small, broken branches, and finally (when the route moved across bare rock) torn-up bits of handkerchief. In time, the posse found Hall himself, dead. He had been shot eight times.

The killers were caught two days later and hanged from a sycamore tree in Globe. The Chamber of Commerce now marks the site with a plaque that notes that "the culprits had a fair hearing" and goes on to add that "saloons were closed and it was an orderly lynching."

Jack Sumner and Billy Hawkins also lived well into the new century.* Sumner became a prospector who roamed across Utah and Colorado for another three decades without ever hitting pay dirt. (In 1902, he looked back at the great adventure of his young manhood and compressed the entire odyssey into a single sentence. "May 24th 1869," he wrote, "the Expedition pulled out into the swift current of Green River and Hell commenced and kept up for 111 days.") Hawkins, reluctant cook and possible fugitive from justice, took up farming and ranching near Eden, Arizona, and became a well-respected justice of the peace, a distinguished figure sporting a great, bushy mustache.

For a time, both men kept on good terms with Powell, but in the end the two men turned on their former leader. One great cause of the rift was a mistaken belief that Congress had appropriated $50,000 for the expedition and that Powell had kept back the men's share of the money.

Each man buttressed his case with further complaints. Hawkins focused on

---

*In their old age both men wrote about the 1869 expedition. I have drawn on those accounts but, for fear of jumping ahead of the story, avoided such expressions as, "Nearly four decades later, Sumner recalled . . ." in favor of "Sumner later recalled . . ." See the Epilogue notes for a discussion of these late-in-life recollections.

Powell's autocratic ways. "I can say one thing truthfully about the Major," he wrote in 1907, "that no man living was ever thought more of by his men up to the time he wanted to drive Bill Dunn from the party."

Proud, cantankerous Sumner came to feel that Powell had stolen not only the wages that were due him but the recognition he should have had as well. Bitter to the end, and past it, Sumner had the last word in an obituary that reads almost as if he had dictated it. The story focused on the 1869 expedition. "Without dwelling at length on the incidents of this thrilling and perilous journey—and there were many," the *Rocky Mountain News* declared, "it may be truthfully asserted that it was due almost wholly to Sumner's cool nerve, calm judgment and quick and ready resourcefulness in all circumstances and in the face of every trying situation that the party completed the journey." Sumner, not Powell, had truly been in command. "He was its real leader, and to him alone is due the fact that the entire party did not go down in the awful currents of the mighty cañon."

"Powell's Scurvy Trick," the *News* labeled one further outrage. "Powell left the six men who had accompanied him at Fort Yuma, on the lower river, and returned East to become famous as an explorer. To Sumner he gave sufficient money to return home, but left the other five penniless. Sumner promptly divided the money with his companions, and from that day on had no use for the man whose life he had twice saved, and who owed so much to him for the services he had rendered."

Sumner died in 1907, Hawkins in 1919.

Walter Powell lived a long, melancholy life. Plagued by depression and fierce headaches, he was unable to work after the early 1870s. One sister or another cared for him for many years, and when they could no longer cope, Walter was admitted to an asylum in Washington, D.C. "At one time, perhaps for two years he claimed to be a prophet," the Major recalled. "I never knew [Walter] to be dangerous to any one except once when I was afraid he would kill one of my men." Walter Powell died in 1915.

The Colorado River, too, deserves a eulogy of sorts. The river Powell knew is no more, for it is now interrupted by a series of great dams. In long stretches the river still runs fierce and formidable, but it is a penned beast, like a zoo lion. The Colorado is "the world's most regulated river," according to the

National Research Council, but it has worn away mightier dams than those that encumber it today, and, centuries from now, it will flow free again.

Separation Rapid is gone, drowned in 1938 as the rising waters of Lake Mead pooled up behind Hoover Dam.

Powell emerged from the Grand Canyon a hero and a celebrity, a kind of nineteenth-century astronaut. In a national series of lectures, the hardy explorer, "direct from scenes of his great exploits and discoveries," painted his audiences a series of word pictures of the West's greatest natural wonder.

Soon, though, Powell forsook the crowds and returned to the desert. On May 21, 1871, he set out from Green River Station a second time, intent on redoing the entire grueling trip, determined this time to bring back a scientific picture of canyon country. The crew was new and again included no skilled boatmen. Instead of feisty and hardheaded mountain men, Powell now opted for a group composed almost exclusively of his own friends and relatives. (Of the first expedition's crew, Powell invited only Sumner. He accepted, despite all his later condemnation of Powell, but heavy snows in the mountains kept him from joining the others.)

The second expedition endured all the hazards of the first—near-drownings, runaway boats, a fire, feuds, rations barely adequate to stave off starvation. Finally, only halfway through the Grand Canyon, Powell decided it was time to stop. (On top of everything else, Hamblin, the Mormon scout, had sent word that it was not a safe time for whites to venture farther into Indian territory.) Powell told the crew that he had decided to cut the trip short, and, as one man wrote, "Everybody felt like praising God."*

Powell's Grand Canyon trips were the most exciting and glamorous events of his life, but perhaps not the most important. Fittingly, Powell's place in American history is as marked by paradox as was his character. Powell was not only the first to explore America's greatest landscape but also its poet laureate, the first to convey the almost infinite depths of time and space the Grand Canyon represents.

---

*To the understandable dismay of his second crew, Powell never discussed this new expedition in print. Instead, he wrote as if there had been only one expedition (though he silently incorporated some material from journals kept by the second crew). In later years, the men of the second expedition often met with open disbelief when they insisted that they, too, had explored the Grand Canyon with Powell.

And yet at the same time as he was discoursing on infinity, Powell became the first great spokesman in American history for the notion of limits.

The lesson of the West, this astonishingly optimistic man declared, was that *not* all things are possible. The empty West was fundamentally different from the crowded East and always would be. The differences all had to do with water—the West was dry—and Powell insisted that every intelligent discussion of the settling of the West had to begin with that stark fact. Even if every river in the region were diverted from its banks, there would not be enough water to turn the West green. The Jeffersonian vision of an endless checkerboard of small farms reaching from shore to shore, the vision that had launched Lewis and Clark, was an impossibility.

Powell propounded those views in a variety of conspicuous forums. Like John Glenn a century later, he took his fame and reputation and founded a political career on it. In Powell's case, he followed up his second Grand Canyon trip with further surveys of the West and with studies of Indian languages and customs. In surprisingly short order, this formidably able and ambitious man came to know the West as deeply as anyone in the United States. At the peak of his career, in the early 1880s, Powell was simultaneously director of the U.S. Geological Survey and director of the Bureau of Ethnology (dedicated to the study of Indian cultures), an organization he had founded.

Powell's prominence and his insistence that the West could not be turned into a shimmering green oasis kept him permanently at the center of political controversy. For the legions of boosters, politicians, and land speculators chanting "Go west, young man!" and loudly insisting that "Rain follows the plow," Powell was a pariah. One of the loudest of those voices belonged to William Gilpin, the bombastic orator and spokesman for Manifest Destiny who had once worried aloud about whether "that grand explorer, Major Powell" would emerge alive from the Grand Canyon. Now Gilpin and Powell were archenemies. "In readiness to receive and ability to sustain in perpetuity a dense population," Gilpin thundered, "[the West] was more favored than Europe." Powell responded with statistics-laden reports and maps of rainfall patterns.

In the end, the boosters won, and the onetime titan of government science was driven out of town. The canyons of the Colorado had proved easier to negotiate than the corridors of power in Washington, D.C.

•   •   •

In 1889, after the Brown-Stanton expedition lost three men by drowning, the *New York Tribune* came to Powell for comment. How was it, the reporter asked, that Powell had not only succeeded in making it through the canyons but had succeeded on his first try?

"I was lucky," Powell replied.

This was indisputably true, but Powell and his crew were also brave, resourceful, and resolute in the face of danger and hardship that far surpassed anything they had foreseen. Though Powell alone rose to fame after the expedition, he and all his men were true American heroes.

In 1895, Powell published a new edition of *Exploration of the Colorado*. Brusque and autocratic though he could sometimes be, he struck a gentler note as he looked back almost three decades to his "noble and generous companions" on the first expedition. "Many years have passed since the exploration, and those who were boys with me in the enterprise are—ah, most of them are dead, and the living are gray with age. Their bronzed, hardy, brave faces come before me as they appeared in the vigor of life; their lithe but powerful forms seem to move around me; and the memory of the men and their heroic deeds, the men and their generous acts, overwhelms me with a joy that seems almost a grief, for it starts a fountain of tears. I was a maimed man; my right arm was gone; and these brave men, these good men, never forgot it. In every danger my safety was their first care, and in every waking hour some kind service was rendered me, and they transfigured my misfortune into a boon."

Powell died in 1902, at sixty-eight, at his summer home in Brooklin, Maine. In death as in life, he inspired strong and conflicting responses. The *Washington Post* ranked Powell with Columbus, Magellan, and Lewis and Clark. The *New York Times* made do with a few curt paragraphs ("Noted Ethnologist Dies") alongside news of a flower show that featured a "Large Display of Dahlias."

At first, as so often with public figures, the tributes tended toward the vague and grandiose. Powell was "a moral giant" and the "brightest exemplar of human knowledge." A few years would have to pass before this complicated man came into sharp focus.

In 1918, at a ceremony to honor Powell's memory, the secretary of the interior caught the essence of the man in a mere handful of words. "Major Powell, throughout his life, was the incarnation of the inquisitive and courageous spirit of the American. He wanted to know and he was willing to risk his life that he might know."

NOTES

This is a work of nonfiction. If the text says that the temperature was 115 degrees or the bacon was vile, one of the men made that observation. If thoughts are attributed to someone—Bradley thought this was the worst rapid yet—he said so.

Ten men set out on a journey through the desert. On their way they met hardly a soul. What they did not write, we will never know. I have chosen to leave the gaps rather than guess at how they might be filled. We lose the glossy finish of fiction but gain the tang and texture of reality.

The narrative is built almost exclusively from eyewitness accounts, and time has placed all the witnesses beyond the reach of cross-examination. Firsthand testimony is seldom straightforward, even in ordinary circumstances. Witnesses can be biased or careless or confused, and memory is a notoriously tricky tool. In the case of Powell and his men, all the generic difficulties are magnified. Some accounts were written on the day they describe, and some fifty years later. Some were written for private diaries, and some for publication. Powell wrote at a congressman's urging, in the hope of winning government support for further exploration. Sumner and Hawkins wrote in their old age, at the urging of a rival of Powell's, when Powell was famous and they were unknown.

In order to keep the Notes in bounds, I have given references for all quotations but, most often, not for facts cited in passing. In almost every case, a reader curious about the source of any such factual remark would find it in one of the journal entries cited in the same passage. (Occasional references to such things as the color of the cliffs may be based on my own observations.) If there are several quotations in a paragraph but only one is cited in these notes, the other quotations in that paragraph are from the same source.

All students of the Powell expedition are indebted to the *Utah Historical Quarterly* for making the expedition diaries and related materials so easily accessible. (This indispensable material is also available on an excellent but hard-to-find CD-ROM called *The Utah Centennial History Suite*.) William C. Darrah, who edited the diaries, also wrote *Powell of the Colorado*, the first biography of Powell. Darrah was a tireless researcher, and all later writers on Powell are in his debt.

Powell is the subject of two other fine biographies, Wallace Stegner's *Beyond the Hundredth Meridian* and Donald Worster's *A River Running West*. All three authors focus largely on Powell's long career as a public figure, and, in my judgment, all three vastly underestimate the dangers that Powell confronted on the Green and the Colorado. After 1869, Powell played an important role on the national stage. Knowing the trajectory of his biography creates a temptation to cast his success in 1869 as all but foreordained. It was anything but. Powell's 1869 expedition launched his career, but it came within a boat length or two of drowning him before he had become even a footnote.

UHQ is the *Utah Historical Quarterly*. (Unless otherwise noted, all references to the *UHQ* are to volume 15, 1947.) *CRC* is Robert Stanton's *Colorado River Controversies*. Many titles have been shortened. They are cited in full in the bibliography.

"fragments or relics": Powell, *Exploration*, Aug. 30, p. 130.
"Fearful Disaster": *Chicago Tribune*, July 3, 1869.

CHAPTER ONE: The Challenge

"foggy ideas": UHQ, Marston, ed., "The Lost Journal of John Colton Sumner,"
    May 24, 1869, p. 175.
"It would be impossible": Bloomington [Ill.] Pantagraph, May 24, 1869, quoted in
    Elmo Scott Watson, The Professor Goes West, p. 28.
"peaceful and unbroken": Langford, The Discovery of Yellowstone Park, p. 33.
"The rocks h-e-a-p": Powell, Exploration, June 1, pp. 16–17.
One of the crew: The river historian Dock Marston says flatly that "of the entire
    crew, there were none who had experienced any rough water—either on the
    ocean or fast water rivers." (See "Was Powell First?" Patagonia Roadrunner, 1969,
    p. 16.) But note that Sumner wrote: "[Bradley] was something of a geologist
    and, in my eyes far more important, he had been raised in the Maine codfish-
    ery school, and was a good boatman." (See CRC, p. 174.) Bradley makes no
    direct mention in his journal of previous boating experience, although he
    makes occasional remarks that might reflect time at sea. On June 14, 1869, for
    instance, he wrote, "Below each rapid there is a heavy sea and one would actu-
    ally fancy himself in a gale at sea if he could not see the land so near him."
    (See UHQ, Darrah, ed., "George Y. Bradley's Journal," p. 38.) Sumner also
    wrote, "Bradley and I were the only experienced boatmen." (See CRC, p.
    175.) But Hawkins wrote: "We were all green at the business, Bradley was the
    only one that had any experience. But he acknowledged afterwards that this
    was a little rough." (See Bass, Adventures, p. 20.) Drifter Smith, a boatman and
    river historian, guesses that Bradley did not have significant experience fishing in
    the Atlantic (since he did not mention it in his journal) but may have had some
    minor boating experience that he mentioned or played up in talking to Powell,
    in the hope it would help convince Powell to spring him from the army.
seen white water. See the interview with Marston, conducted by Haymond and
    Hoffman, p. 121. Worster writes that Powell (and his wife and brother) "descended
    the cliff walls to examine the rapids closely." (See River Running West, p. 151.)
"all experienced": Bell, New Tracks, reprinted in UHQ, p. 21.
"green at the business": Bass, Adventures, p. 20.
only one life jacket: Jack Sumner, CRC, p. 134.
"To get an idea": Nash, Big Drops, p. 1. Nash is both a historian and an accom-
    plished river runner, and his small book is one of the best single volumes on
    America's wildest rivers. Perhaps the best history of Western boatmen is
    Lavender's River Runners of the Grand Canyon.
"By modern standards": Meloy, Raven's Exile, p. 12.
only 250 people: "Dock" Marston, a celebrated river runner and Grand Canyon histo-

rian, compiled the best known list of early river runners. By Marston's tally, the total did not reach 100 until 1949. In connection with a study of injuries in the Grand Canyon, the physician and writer Tom Myers compiled his own, more recent list. Myers contends that Marston undercounted. Part of the dispute is technical— if, for example, one river runner made six Grand Canyon trips, Marston counted that as one trip and Myers as six—and part is personal. Myers believes that Marston, a colorful and controversial figure, trimmed the count in order to ensure himself a place among the "First One Hundred." The figure cited in the text is from Tom Myers, "River Runners and the Numbers Game," pp. 22–23.

CHAPTER TWO: The Crew

"By day disgusting": Samuel Bowles, quoted in Ambrose, Nothing Like It in the World, p. 219.

"the completion of the greatest": The New York Times, May 4, 1869. The Times jumped the gun in its excitement, declaring its delight a week early.

"joyous exultation": Lieutenant William Wheeler of the Thirteenth New York Battery, quoted in Hess, The Union Soldier in Battle, p. 92.

"give their time": Chicago Tribune, May 29, 1869.

"Should it be necessary": For a discussion of the contract, see Tikalsky, "Historical Controversy, Science, and John Wesley Powell," pp. 409–10; and Anderson, "First Through the Canyon," p. 394.

"The object is": Chicago Tribune, May 19, 1869.

He could not wash: On the river, this task fell to Hawkins, the cook.

"eminently a magnetic": Merrill, "John Wesley Powell," p. 332.

He was a democrat: Powell shared Jefferson's devotion to small farmers and small institutions generally. "The building of great industrial operations does not daze my vision," he wrote. "I love the cradle more than the bank counter. The cottage home is more beautiful to me than the palace." (See Powell, "Institutions for the Arid Lands," p. 116.) For a brief, cogent attempt to place Powell's views in a historical context, see Aton, Inventing John Wesley Powell.

"a Renaissance man": Reisner, Cadillac Desert, p. 24.

"as single-minded as a buzz saw": Stegner, Beyond the Hundredth Meridian, p. 20.

"He wasn't all saint": A. H. Thompson letter to Frederick Dellenbaugh, Nov. 9, 1902, quoted in Anderson, "Fact or Fiction."

"After three years' ": Chicago Tribune, May 29, 1869.

"The region last": Pyne, How the Canyon Became Grand, p. 41.

"silent, moody": Powell, "The Cañons of the Colorado," Scribner's Monthly, v. 9 (1875), p. 295.

"*about as worthless*": Sumner, *CRC*, p. 211.

"*Trout fishing can*": *Colorado Miner*, July 25, 1867, p. 4.

"*amateurs like myself*": Powell, *Exploration,* p. ix.

"*Mountains, hills, rocks*": Powell, *Truth and Error*, p. 281.

"*In our evening*": Sumner, *CRC*, p. 168.

Sumner had been raised: CRC, pp.164–5.

"*Jack Rabbit*": Sumner's story is included in Cohig, "History of Grand County, Colorado," pp. 57–8

The party believed: The first woman whose name we know who reached the summit of Pikes Peak was Julia Archibald Holmes, in 1858.

"*a very beautiful*": Bloomington [Ill.] *Daily Pantagraph,* June 7, 1909.

"*the homeliest man*": Worster, *River Running West*, p. 548.

"*could ride all day*": Lincoln, "John Wesley Powell" (Part III: "The Professor"), p. 90.

"*a quiet, pensive*": Powell, "Cañons," *Scribner's*, p. 296.

$1.50 a day: Hawkins, *CRC*, p. 142.

"*as tough as a badger*": Sumner, *CRC*, p. 174.

"*the River Styx*": Bradley, *UHQ*, Aug.11, p. 62.

the sixth grade: Darrah, "The Powell Colorado River Expedition," *UHQ*, p. 17.

"*The Major as usual*": Bradley, *UHQ*, June 11, p. 37.

Emily Dickinson's poetry: Christopher Benfey, "The Mystery of Emily Dickinson," *New York Review of Books,* April 8, 1999, p. 39.

"*whoever dares venture*": Bowles, *The Switzerland of America*, p. 82.

"*Professor Powell is well-educated*": Ibid., p. 85.

"*Is any other nation*": Ibid., p. 83.

"*The wonder*": Ibid., p. 85.

"*a bull-dozing way*": Hawkins, *CRC*, p. 157.

"*to make barometrical*": Powell's contract with Dunn, Sumner, and O. G. Howland can be found in the Frederick S. Dellenbaugh collection at the University of Arizona, Special Collections.

"*Yours till death*": See, for instance, Lucille Griffith, ed., *Yours till death: Civil War Letters of John W. Cotton.*

CHAPTER THREE: The Launch

"*swift, glossy river*": Sumner, May 24, reprinted in Marston, ed., "The Lost Journal of John Colton Sumner," *UHQ*, v. 37, no. 2 (Spring 1969), p. 175. (This "lost journal" covers May 24, 1869, to June 28, 1869. All later references to Sumner's diary between these dates are from this volume; all the entries were written in 1869.)

"almost without effort": Bradley, *UHQ*, May 24, p. 31.

"swollen, mad and seeming eager": Powell, *Chicago Tribune*, May 24, 1869. Reprinted in *UHQ*, p.74.

"Raining hard": Sumner, *UHQ*, May 25, p. 176, and May 27, p. 176.

"We knew nothing: Bass, *Adventures*, p. 19.

"Kitty's crew": Sumner, *UHQ*, May 24, p. 175.

"neither gee nor haw": Sumner, CRC, p. 175.

"In trying to avoid": Powell, *Exploration*, May 24, p. 9.

"very good water: Bass, *Adventures,* p. 20.

"The hunters managed": Powell, letter to *Chicago Tribune*, May 24, 1869. (See *UHQ*, p. 79.)

"If we fail": Sumner, *UHQ*, May 24, p. 175.

"After supper": Powell, *Exploration*, June 8, p. 23.

Did Sumner spin: Powell, "The Cañons of the Colorado," *Scribner's*, p. 295.

"It is late": Powell, *Exploration*, June 8, p. 23. It is unclear just when this conversation took place. Powell places it on the night of June 8, but we know from Bradley and Sumner that this is incorrect. The Sumner and Bradley journals rule out June 7 and June 9 as well.

The men slept: Sumner, *CRC*, p. 211.

They were Whitehalls: Bass, *Adventures*, p. 19.

"Survival of the Fastest": Stephens, "Survival of the Fastest: Evolution of the Whitehall," *Wooden Boat*, p. 49.

"Maneuverability was not": Stephens, "Survival," pp. 50–1.

But their home base: Author interview with Brad Dimock, October 25, 1999.

There were four boats: Powell, *Exploration*, May 24, p. 8. The four-foot figure is from Marston, "With Powell on the Colorado," p. 70.

seven thousand pounds: Sumner, *UHQ*, June 2, p. 178.

"the utmost care": Powell, *Exploration*, May 24, p. 9.

"stanch and strong": Powell, "Cañons," *Scribner's*, p. 295. (In *Exploration*, Powell described the boats as "stanch and firm." See May 24, p. 8.)

The Emma Dean *carried:* Powell, *Exploration*, May 24, pp. 8–9.

"The boats were ordered": Sumner, *CRC*, p. 175.

"We feel quite proud": Powell, *Chicago Tribune*, May 24, 1869.

"we make a pretty show": Sumner, *UHQ*, May 24, p. 175.

There he would stand: This description is from Powell, *Exploration*, p. 91: "I stand on deck, supporting myself with a strap, fastened on either side to the gunwale."

"I saw they were": Bass, *Adventures*, p. 20.

"we better unload": Ibid.

"a couple of hours": Sumner, *UHQ*, May 24, p. 175.

"rather slim rations": Ibid.

"exchanged tough stories": Ibid.

Somehow the taciturn: Bradley, in turn, did not know that Powell and Sumner kept journals. (See Bradley, *UHQ*, June 13, p. 37.)

We have two: The Rhodes (Hawkins) accounts are in Bass, *Adventures* and *CRC*. The other accounts can be found in *UHQ*, v. 15, 1947.

"desolate enough to suit": Sumner's diary entry in Darrah, ed., "J. C. Sumner's Journal" (July 6–Aug. 31, 1869), *UHQ*, v. 15 (1947), Aug. 4, p. 118.

"whirled it around": Sumner, *CRC*, p. 191.

"we broke many oars": Sumner, *CRC*, p. 184.

But Powell's notes: Not all of the diary has been found. The diary at the Natural History Museum covers the dates from July 2 to Aug. 28, 1869.

"the difficulty of writing": Frederick Dellenbaugh, quoted in Marston, *Brand Book II*, p. 66.

"Wrote until 10:00": Powell diary entry in Darrah, ed., "Major Powell's Journal" (July 2–Aug. 28, 1869), *UHQ*, v. 15 (1947), July 5, p. 125.

In fact, it was dictated: Gilbert, "John Wesley Powell" (Part V: "The Investigator"), p. 289. In the interest of brevity, I use such expressions as "On Aug. 5, Powell wrote . . . ," but the reader should bear in mind Powell's actual style of composition.

"I decided to publish": Powell, preface to *The Exploration of the Colorado River and Its Canyon*. (This is the 1895 edition of Powell's 1875 text.)

the river diary contains: Powell's diary covers two months of the three-month expedition. A full diary (if one exists) might therefore be some 4,500 words long, rather than 3,000.

"wet, chilled, and tired": Powell, *Exploration*, May 25, p.10.

"villainous bacon": Sumner, *UHQ*, May 25, p. 176.

"tall lifting and tugging": Bradley, *UHQ*, May 25, p. 31.

"the largest and most difficult": Bradley, *UHQ*, May 26, p. 31.

"It cannot be navigated": Sumner, *UHQ*, May 26, p. 176.

"No injury done": Bradley, *UHQ*, May 26, p. 31.

"Wrapping is": Nash, *Big Drops*, pp. 159–60.

"placid stream": Powell, *Exploration*, May 26, p. 10.

"The river winds": Bradley, *UHQ*, May 27, p. 32.

"Barren desolation": Powell, *Exploration*, May 24, p. 9.

"Dark shadows": Ibid.

"Country worthless": Sumner, *UHQ*, May 27, p.176.

"It is the grandest": Bradley, *UHQ*, May 27, p. 31.

"*yet it glides*": Powell, *Exploration*, May 26, p. 11. (Powell's dates were often unreliable. Sumner and Bradley agreed on May 27.)

"*We name it Flaming Gorge*": Powell, *Exploration*, May 26, p. 11.

CHAPTER FOUR: *Ashley Falls*

"*tired and hungry*": Bradley, *UHQ*, May 28, p. 32.

"*[We] are ready*": Powell, *Exploration*, May 27, pp. 11–12.

"*Tramped around*" : Sumner, *UHQ*, May 27, p. 176.

"*Still in camp*": Sumner, *UHQ*, May 28, p. 176.

"*We are ready*": Powell, *Exploration*, May 30, p. 13.

"*We enter the narrow*": Ibid., p. 14.

"*We took off boots*": Bradley, *UHQ*, May 30, pp. 32–3.

"*Here we have our first*": Powell, *Exploration*, May 30, p. 14.

"*watching the finny tribe*": Sumner, *UHQ*, May 30, p. 177.

"*Goodman saw one elk*": Ibid.

"*The amphitheater seems*": Powell, *Exploration*, May 30, p. 15.

"*to examine some rocks*": Sumner, *UHQ*, May 31, p. 177.

"*a bad rapid*": Ibid.

"*Had supper*": This paragraph is from Sumner, *UHQ*, May 31, p. 178, and Powell, *Exploration*, May 31, p. 15.

"*As the twilight deepens*": Powell, *Exploration*, May 31, pp. 15–16.

"*Was there ever such*": Limerick, *Desert Passages*, p. 109.

"*like lightning*": Bradley, *UHQ*, June 1, p. 33.

"*To-day we have an exciting*": Powell, *Exploration*, June 1, p. 16.

"*necessitates much bailing*": Ibid.

"*As the roaring*": Sumner, *UHQ*, June 1, p. 178. Sumner's guess that the rapid dropped ten feet in twenty-five works out to a gradient of about two thousand feet per mile, an impossibly high figure.

"*heap, heap high*": Powell, *Exploration*, June 1, pp. 16–17.

"*Then came the real hard work*": Sumner, *UHQ*, June 1, p. 178.

"*huge bowlders*": Bradley, *UHQ*, June 1, p. 33.

Some of the men thought: Powell thought the date read either 1835 or 1855. (See *Exploration*, June 2, p. 17.) Sumner said the date was 1825. (See *UHQ*, June 2, p. 178.)

"*The word 'Ashley'* ": Powell, *Exploration*, June 2, p. 17.

A host of men: Webb, *If We Had a Boat*, p. 21.

"*They went about the blank*": DeVoto, *Across the Wide Missouri*, p. 159.

"*A tough hombre*": Ibid., p. 128.

The fourth group: In his diary, Ashley wrote that he was with seven men. (See the diary entry for Wednesday, April 22, 1825.) In the narrative he compiled later, he referred to six others. (See Dale, ed., *The Ashley-Smith Explorations*, p. 138.)

upside-down umbrella: Don Bragg, a boatman turned filmmaker and river historian, pointed out the resemblance in an interview on June 1, 2000.

"One of our boats": This is from Beckwourth's autobiography, *The Life and Adventures of James P. Beckwourth*, pp. 57–8.

"To be a liar": Morgan, *Jedediah Smith and the Opening of the West* (Delmont R. Oswald quoted Morgan in his Introduction to Beckwourth's autobiography. See p. xi.)

Remarkably, his account: Beckwourth published his *Life and Adventures* in 1856. In 1904, the explorer and writer Frederick Dellenbaugh, in his *Romance of the Colorado River*, retold Beckwourth's story as if it were strict history. (See *Romance*, pp. 110–12.)

"Exosted with the fatiegue": Ashley diary, May 13, 1825. The diary can be found at the Missouri Historical Society, in St. Louis, and on the Internet.

"just as from all appearance": Ashley diary, May 14, 1825.

"destitute of game": Ibid., May 17, 1825.

"Capt. W. L. Manly": Manly, *Death Valley in '49*, p. 79.

"We put a great many 'ifs'": Ibid., pp. 73–4.

"all our worldly goods": Ibid., p. 76.

"It was not a heavy": Ibid.

Instead, the stick caught: In similar fashion, beginning boaters today often jam an oar in shallow rocks. They are unlikely to get thrown overboard, but they may well break an oar.

"huge rocks as large": Ibid., p. 80.

At about this point: Manly reports that the inscription "Ashley, 1824" was "painted in large black letters." (See *Death Valley*, p. 80.) According to Sumner, "Ashley, 1825" was "scratched on evidently by some trapper's knife." (See Sumner, *UHQ*, June 2, p. 178.)

"The current was so strong": Manly, *Death Valley*, p. 81.

"afoot and alone": Ibid.

"the stream was so swift": Ibid., pp. 85–6.

Exactly how many: Marston, "Was Powell First?" pp. 9–10.

"When I told Chief Walker": Manly, *Death Valley*, pp. 93–4. "[Wakara] undoubtedly saved our little band from a watery grave," Manly went on, "for without his advice we had gone on and on, far into the great Colorado cañon, from which escape would have been impossible and securing food another impossibility, while destruction by hostile Indians was among the strong probabilities of the case. So in a threefold way I have for these more than forty years credited the lives of myself and comrades to the thoughtful interest and humane consideration of old Chief Walker." (See *Death Valley*, p. 99.)

"*the damndest scoundrel*": Burns, *Civil War*, p. 74.

"*[We] were, I believe*": Nevins, *Frémont: The West's Greatest Adventurer*, pp. 116–17.

"*the party was left stripped*": Ibid.

Frémont's reasoning, apparently: Nevins, *Frémont: Pathmarker of the West*, v. 2, p. 351.

the heroic scout Alexis Godey: The fullest account of Frémont's 1848 expedition is Richmond's superlative *Trail to Disaster*. Richmond rediscovered all but one of Frémont's camps, many of them identified by the telltale stumps of trees—now thirty feet above the ground—that the men cut for firewood. Roberts provides a short, excellent account of the ordeal in *A Newer World*.

"*entirely satisfactory*": Nevins, *Frémont: Adventurer*, p. 119.

"*No trappers have been found*": UHQ, Morgan, "Introduction," v. 15, 1947, p. 5.

CHAPTER FIVE: *Paradise*

"*had dinner and smoked*": Sumner, UHQ, June 2, p. 178.

"*Beautiful river*": Sumner, UHQ, June 2, p. 178.

"*It was,*" he wrote, "*like sparking*": Ibid.

"*Spread our blankets*": Ibid., p. 179.

"*the sky is clear*": Bradley, UHQ, June 2, p. 34.

"*up to a pine grove park*": Powell, *Exploration*, June 3, p. 18.

"*they didn't bring in enough*": Bradley, UHQ, June 3, p. 35.

"*queer mongrel*": Sumner, UHQ, June 3, p. 179.

"*splendid ride*": Sumner, UHQ, June 4, p. 179.

"*only few rappids*": Bradley, UHQ, June 4, p. 35.

"*a most disgusting looking stream*": Sumner, UHQ, June 4, p. 179.

"*4000 head of oxen*": Sumner, UHQ, June 7, p. 181.

"*Everything [was] as lovely*": Sumner, UHQ, June 5, p. 180.

On the way back: Bradley, UHQ, June 5, p. 35.

"*no easy task*": Sumner, UHQ, June 6, p. 180.

"*I feel quite weary*": Bradley, UHQ, June 6, p. 35.

"*a mountain drinking*": Webb, *If We Had a Boat*, p. 58.

"*We have now reached*": Bradley, UHQ, June 6, p. 35.

"*When we return*": Powell, *Exploration*, June 7, p. 21.

"*a dark portal*": Ibid.

CHAPTER SIX: *Disaster*

"*As we passed along*": Dale, ed., *The Ashley-Smith Explorations*, p. 145.

"*Each canyon of the Green*": Zwinger, *Run, River, Run,* p. 155. Zwinger is a lively, learned writer whose interests range across history and natural history. She is also the author of the excellent *Downcanyon*, about a river trip through the Grand Canyon.

"sometimes the water descends": Powell, *Exploration*, June 9, p. 24.

"If I can see": Powell, *Exploration*, June 8, p. 22.

"On the ocean": Litton, "The Dory Idea," p. 16.

"The water of an ocean": Powell, *Exploration*, June 8, p. 22.

Since water cannot squeeze: More precisely, the rate of flow at a given point is the product of the river's cross section and its velocity. If the area of the cross section decreases, the velocity must increase proportionally.

On returning to the river: Bradley, *UHQ*, June 8, p. 36.

"the idea of diving": Sumner, *UHQ*, June 20, p. 184.

"as hard a day's work": Sumner, *UHQ*, June 8, p. 181.

"the wildest rapid": Bradley, *UHQ*, June 8, p. 36.

"a terrible rapid": Sumner, *UHQ*, June 8, p. 181.

"And now the three men": This account is from Powell, *Exploration,* June 9, p. 24, and Oramel Howland's letter to the *Rocky Mountain News*, June 19, 1869. Howland's letter is reprinted in *UHQ*, v. 15, 1947, pp. 95–9.

"Our position on the bar": Howland, *Rocky Mountain News*, June 19, 1869. (See *UHQ*, p. 97.)

"Right skillfully": Powell, *Exploration*, June 9, p. 25.

"with the speed": Sumner, *CRC*, p. 177.

"certain destruction": Ibid., p. 177.

"We are as glad": Powell, *Exploration*, June 9, p. 25.

"The trapper," he wrote: Sumner, *UHQ*, June 8, p. 181.

"Had he stayed": Ibid.

"nothing could have saved": Ibid.

"so much water aboard": Howland, *Rocky Mountain News*, June 19, 1869. (See *UHQ*, p. 96.)

"I walk along": Powell, *Exploration*, June 9, p. 24.

"The scouting boat": Sumner, *UHQ*, June 8, p. 181.

"About one o'clock": Howland, *Rocky Mountain News*, June 19, 1869. (See *UHQ*, p. 96.)

"As soon as Howland": Sumner, *CRC*, pp. 177–8.

Gone was about: This tally is from Sumner, *UHQ*, June 8, p. 181, and Bradley, *UHQ*, June 8, p. 36.

The Howland brothers and Goodman: Bradley, *UHQ*, June 8, p. 36.

Because clothing dragged: Howland, *Rocky Mountain News*, June 19, 1869 (see *UHQ*, p. 97), and Bradley, *UHQ*, June 8, p. 36.

"But, then," Powell asked: Powell, *Exploration*, June 9, p. 25.

"We are rather low spirrited": Bradley, *UHQ*, June 8, p. 36.

Powell gave the credit: Powell, *Exploration*, June 10, p. 26.

In his published: Powell wrote in *Exploration*, p. 9, that Hawkins and Hall crewed the *Maid of the Cañon* and Walter Powell and Bradley crewed the *Kitty Clyde's Sister*. Sumner has it the other way around (see *UHQ*, May 24, p. 175), and so does Andy Hall (see Hall's letter dated Sept. 10, 1869, in *UHQ*, v. 16–17, 1949, p. 507).

"Away they went": Sumner, *UHQ*, June 9, p. 182.

"great risk": Bradley, *UHQ*, June 9, p. 36.

an untapped ten-gallon cask: Powell referred to "Howland's whiskey" (see his letter of June 29, 1869, in *UHQ*, p. 88) and wrote that the men recovered a three-gallon cask of liquor (*Exploration*, June 10, p. 26). Sumner put it at ten gallons. (See *UHQ*, June 9, p. 182.) On this issue Sumner is the more reliable witness.

"an hour's floundering": Sumner, *UHQ*, June 9, p. 182.

"The boys set up": Powell, *Exploration*, June 10, p. 26.

CHAPTER SEVEN: Shiloh

"War was a proving": Author interview, April 19, 2000. Powell rose quickly, and so did many other capable and ambitious men. In 1861, the permanent army numbered only sixteen thousand men; a year later, Northern and Southern forces together totaled nearly one million.

"People in those days": Catton, *Reflections on the Civil War*, p. 161.

"reaching prompt decision": Davis, "Biographical Memoir of John Wesley Powell," p. 14.

"a history of the war": Powell, *Exploration*, p. 208.

"as if a thousand battles": Ibid., p. 174.

"interprets it as a fragment": Powell, "Esthetology," p. 14.

For thirty years: Darrah, *Powell*, p. 319.

At the war's outset: Some did. William Tecumseh Sherman predicted that the war would bring hundreds of thousands of deaths and was ridiculed and called insane. Sam Houston made a similarly accurate and gruesome prediction of what war would mean. Houston had as little impact on his Southern audiences as Sherman did on his Northern ones.

"what I didn't know": Frank and Reaves, *Seeing the Elephant: Raw Recruits at the Battle of Shiloh*, p. 161.

"They were convinced": Moorehead, *Gallipoli*, p. 112.

"Do you think": This remark and the observation about three out of four soldiers being untried in combat are from my interview with Stacy Allen, April 19, 2000.

"the most confused": Marshall's comment is from his foreword to Wiley Sword's *Shiloh: Bloody April*, p. vii. A library has been written on the Civil War. The total is well

over fifty thousand books, about one for every day since the war's end. At times, the meticulous reconstructions of the various battles impose a seeming order on what were in truth scenes of hellish disarray. This was especially true at Shiloh.

"just a disorganized": Burns, *The Civil War*, p. 267.

Shiloh Meeting House: Logsdon, ed., *Eyewitnesses at the Battle of Shiloh*, p. 2n.

The name, from the Hebrew: See, for example, McDonough, *Shiloh*, p. 4, or Burns, *The Civil War*, p. 115. Stacy Allen, the Shiloh historian, is dubious about the accuracy of the "Place of Peace" translation.

"I would fight them": McDonough, *Shiloh*, p. 81.

"offer battle to the invaders": Ibid., p. 78.

"My God," cried General William T. Sherman: Ibid., p. 115.

"awakened by the rattle": Letter from Powell to Colonel Cornelius Cadle, May 15, 1896. This letter is kept at Shiloh National Military Park. Stacy Allen, the park historian, provided me with a copy.

What the factory: A total of ninety Americans were killed and wounded at the Battle of Lexington and Concord. The bloodiest battle of the Revolutionary War, for the Americans, was Camden. American casualties numbered one thousand.

A single cannon: Frank and Reaves, *Seeing the Elephant*, p. 104.

"There is some places": Hess, *The Union Soldier in Battle*, p. 49.

By far the greatest: Adams, *Doctors in Blue*, p. 113. This admirable book, by far the best in its field, is essential reading for anyone with an interest in Civil War medicine.

"a beautiful, inhumane": Author interview, April 19, 2000.

Muskets were notoriously: McPherson, *Battle Cry of Freedom*, p. 473.

"might fire at you all day": McWhiney and Jamieson, *Attack and Die*, p. 29.

At five hundred yards: Frank and Reaves, *Seeing the Elephant*, p. 44, n4.

But weapons changed: A revisionist school of historians downplays the significance of the "rifle revolution" and seeks a different explanation for the Civil War's bloodiness. For a powerful statement of the conventional view, see McWhiney and Jamieson, *Attack and Die*. Griffith, *Battle Tactics of the Civil War* is an equally forceful presentation of the revisionist case.

"It was not war": McWhiney and Jamieson, *Attack and Die*, p. 3.

"We struck the Confederate": Ibid., p. 128.

"It doesn't take any": Author interview, April 19, 2000.

"This was the first": Shelby Foote, *The Civil War: Fort Sumter to Perryville*, p. 338.

CHAPTER EIGHT: The Hornets' Nest

Southern troops made at least: The fighting at the Hornets' Nest raises a number of tangled questions. First is the number of Confederate assaults. The count ranges anywhere from seven to fourteen, depending on the historian. Second, many writers describe the Hornets' Nest as the site of the fiercest fighting at Shiloh. Third, many accounts of the battle describe the Confederates making near-suicidal charges across Duncan Field. Stacy Allen, who has studied the issue in detail, disputes nearly all the conventional wisdom. He argues, convincingly, that there were seven confirmed assaults and possibly two more; that fighting elsewhere at Shiloh was more deadly than in the Hornets' Nest; and that most of the assaults on the Hornets' Nest were through dense thickets, not across open fields.

"For God's sake": Daniel, *Shiloh*, p. 211.

As the barrage continued: This account is from McDonough, *Shiloh*, pp.146–49 and Daniel, *Shiloh*, pp. 210–13. Allen survived and went on to become governor of Louisiana.

"I am lying so close": Logsdon, ed., *Eyewitnesses*, p. 29.

"Men fell around us": Ibid., p. 32.

The roar of the guns: McDonough, *Shiloh*, p. 143.

A minié ball plowed: Darrah, *Powell*, p. 57, and Powell's letter to Cadle, cited above, pp. 4–5. The source for Powell's version of the story (cited in the footnote in the text) is Baker's "Major John Wesley Powell."

"The scene upon the boat": Frank and Reaves, *Seeing the Elephant*, pp. 157–8.

"Now, now," Powell whispered: Darrah, *Powell*, p. 58.

Fifteen thousand by some accounts: Keegan, *The Mask of Command*, p. 166.

"Most of them would": Burns, *The Civil War*, p. 115.

At one point, thirty: Keegan, *Mask of Command*, p. 165.

"It aren't me": Logsdon, *Eyewitnesses*, p. 41.

"a cannonball struck him": Frank and Reaves, *Seeing the Elephant*, p. 105.

"leaning back against a tree": Ibid., p. 107.

within the flames: Many memoirs contain horrified descriptions of fires at Shiloh. One detailed account is in Ambrose Bierce's "What I Learned of Shiloh."

Corpses littered the battlefield: Stacy Allen, "Shiloh! The Second Day," pp. 10–11.

"The sight was more unendurable": McDonough, *Shiloh*, p. 188.

"a lump of sugar": McPherson, *Battle Cry of Freedom*, p. 413.

"I could have crossed": See, for example, Sword, *Shiloh*, p. 429, or Burns, *Civil War*, p. 121.

"piles of dead": McPherson, *Battle Cry of Freedom*, p. 413.

"Scarcely a tree": Sword, *Shiloh*, p. 440.

"Skulls and toes": Frank and Reaves, *Seeing the Elephant*, p. 123.

"A tramp": Castel, ed., "The War Album of Henry Dwight, Part III," p. 36.

"There's been a lot": Author interview, April 19, 2000.

In two spring days: Union losses at Shiloh were 1,754 men killed, 8,408 wounded, and 2,885 captured, for a total of 13,047. Confederate losses were 1,723 killed, 8,012 wounded, and 959 missing, for a total of 10,694. The overall casualty rate was 24 percent. The figures are cited in Foote, *The Civil War: Fort Sumter to Perryville*, p. 350. Most of the men listed as missing were captured by the enemy.

With a death toll: It is *not* the case, though the claim often appears in print, that more Americans died in the Civil War than in all the nation's other wars combined. From the American Revolution through the Gulf War but excluding the Civil War, 638,560 Americans died in combat. In the Civil War alone, 558,052 Americans lost their lives.

"The sorriest sights": Linderman, *Embattled Courage*, pp. 130–1.

three out of four: National Museum of Civil War Medicine.

A surgeon could remove: National Museum of Health and Medicine (formerly the Army Medical Museum). This fine museum, on the grounds of the Walter Reed Hospital in Washington, D.C., contains a wealth of material on the state of Civil War medicine.

The man who amputated: Darrah, *Powell*, p. 58. Worster disagrees with Darrah and writes that Medcalfe was indeed a physician. (See Worster, *River Running West*, p. 580 n13.)

Infection rates were so high: Frank and Reaves, *Seeing the Elephant*, p. 152.

It was 1868 before: Adams, *Doctors in Blue*, p. 50.

"we were playing Soldiers": Frank and Reaves, *Seeing the Elephant*, p. 120.

a soldier was twice as likely: Linderman, *Embattled Courage*, p. 115.

four hundred *women:* Frank and Reaves, *Seeing the Elephant*, p. 20.

"mouldy crackers": Ibid., p. 42.

It was common: Catton, *Reflections*, p. 43.

"A man risked": Ibid., p. 178.

a regiment's musicians: Adams, *Doctors in Blue*, p. 68.

At Fort Donelson: Ibid., p. 80.

"You may imagine": This quotation and the description in the previous sentence are from Frank and Reaves, p. 156.

"There stood the surgeons": Adams, *Doctors in Blue*, p. 118.

"However bad the wound": Ibid., p. 132.

"We operated in": Ibid., p. 125.

"Wes, you are a maimed man": Darrah, *Powell*, pp. 58–72.

CHAPTER NINE: *Hell's Half Mile*

"The red sand-stone": Bradley, *UHQ*, June 9, p. 36.

"After taking a good": Sumner, *UHQ*, June 9, p. 182

"Some packed freight": Ibid.

"Major and brother": Bradley, UHQ, June 9, p. 36.

"Can't rely on anything": Ibid.

"the shadow of a pang": Bell, New Tracks. Reprinted in UHQ, p. 26.

"The sensation on the first": Dellenbaugh, Romance, p. 323.

"Basically, Powell and his men": Author interview, October 25, 1999.

"We knew intimately": Author interview, June 7, 2000.

"The river in this cañon": Bradley, UHQ, June 10, pp. 36–7.

enormous fallen boulders: Powell, Exploration, June 10, p. 26. Powell's observation is
    dated June 10 but he may have been describing events of June 11. Here and
    in many other places, Sumner's journal and Bradley's are in agreement and
    Powell's reckoning of dates is less dependable.

"poor Kitty's Sister": Sumner, UHQ, June 10, p. 182.

"Rapids and portages": Ibid., p. 182.

"Have been working": Bradley, UHQ, June 11, p. 37.

"When night comes": Powell, Exploration, June 11, p. 26. As above, it is possible that
    Powell is off by a day or two.

"Camped . . . under": Sumner, UHQ, June 11, p. 182.

Powell "as usual [had] chosen": Bradley, UHQ, June 11, p. 37.

"only more of it": Bradley, UHQ, June 12, p. 37.

"Still there is no": Ibid.

"My eye is very black": Ibid.

"Everything is wet": Powell, Exploration, June 13, p. 27. This entry should have been
    dated June 12.

"We discover an iron": Powell, Exploration, June 10, p. 26.

"We have plenty": Bradley, UHQ, June 9, p. 36.

"for [I] don't think": Bradley, UHQ, June 13, p. 37.

"mythology": John Wesley Powell, "Mythologic Philosophy," II, p. 63.

"Our rations are getting very sour": Bradley, UHQ, June 13, p. 38.

"as ever without game": Ibid.

"There is nothing": Sumner, UHQ, June 13, p. 182.

"suddenly wheel[ed] around": Walter Powell, Chicago Evening Journal, July 3, 1869.
    (See UHQ, p. 91.)

"The prospects of success": Bradley, UHQ, June 12, p. 37.

"five times as bad": Bradley, UHQ, June 15, p. 39.

"On the east side": Powell, Exploration, June 15, p. 27.

"My boat was sunk": Bradley, UHQ, June 15, p. 39.

"Here we have three": Powell, Exploration, June 15, pp. 27–8.

"The main current": Zwinger, Run, River, Run, p. 164.

"We make a portage": Powell, *Exploration,* June 15, p. 28.

"Lining a boat is horrible": Author interview, January 6, 2000. Ghiglieri is the author of the fine book *Canyon,* which describes the Colorado from the point of view of a professional boatman.

"The boat at times": "Journal of W. C. Powell, April 12, 1871–December 7, 1872," *UHQ,* v. 16–17, 1949, p. 329.

Gary drowned: Fadiman, "Under Water," pp. 64–5, in David Quammen, ed., *The Best American Science and Nature Writing, 2000.* Fadiman's essay originally appeared in *The New Yorker,* Aug. 23, 1999.

The rising water: Webb, *Grand Canyon,* p. 102.

"Hell's Half Mile": Powell, *Exploration,* June 15, p. 28. The name was not given until the second expedition down the Green and the Colorado, in 1871.

Parsifal to memory: Nash, *Big Drops,* p. 58.

"Then we commence": Powell, *Exploration,* June 16, p. 28.

"We are letting": Ibid.

CHAPTER TEN: *Fire*

While letting the Maid": Sumner, *UHQ,* June 16, p. 183.

"We gave up": Powell, "Cañons," *Scribner's,* p. 304.

"Rejoicing that we had not": Howland, *Rocky Mountain News,* July 1, 1869. (See *UHQ,* p. 101.)

"We hope and expect": Bradley, *UHQ,* June 16, p. 39.

"cliffs and crags": Powell, *Exploration,* June 16, p. 29.

"Never before": Ibid.

"Ran many little rappids": Bradley, *UHQ,* June 17, p. 39.

"the freight boats": Sumner, *UHQ,* June 16, p. 183. It appears that this entry should be dated June 17.

"to keep her afloat": Ibid.

"in a nice little cove": Bass, *Adventures,* p. 22.

"One of the crew": Howland, *Rocky Mountain News,* July 1, 1869. (See *UHQ,* p. 100.)

"Our plates are gone": Powell, *Exploration,* June 16, p. 30.

"One gold pan": Howland, *Rocky Mountain News,* July 1, 1869. (See *UHQ,* p. 100.)

"Had supper": Sumner, *UHQ,* June 16, p. 184.

"Had a splendid ride": Sumner, *UHQ,* June 18, p. 184; Bradley, *UHQ,* June 18, p. 40.

This railroad comparison: See Powell, *Exploration,* June 1, p. 16; Walter Powell, letter to *Chicago Evening Journal,* July 3, 1869. (See *UHQ,* p. 92.)

"those in her said": Howland, *Rocky Mountain News,* June 19, 1869. (See *UHQ,* p. 95.)

In the narrowest stretches: Kieffer measured the speed of Grand Canyon rapids and found a top figure of thirty-three feet per second, in Hermit Rapid. (See "Hydraulics and Geomorphology," p. 338.) (Thirty-three feet per second is twenty-two miles per hour.) In Kieffer's "1983 Hydraulic Jump," she reports that at one point the rapid reached a speed of fourteen meters per second (at a flow of fifty thousand cubic feet per second). See p. 399. (Fourteen meters per second is thirty-one miles per hour.)

ten times faster: Kieffer, "Hydraulics and Geomorphology," p. 337.

"optic flow": Curtis Rist, "Roll Over, Newton," *Discover,* April 2001, p. 49. According to Rist, even the twenty-inch height difference between an SUV and an ordinary car changes a driver's perception of speed. An SUV driver going sixty miles an hour feels as if he is only driving at forty.

"I believe I would have": Nash, *Big Drops,* p. 56.

"I pulled the bow oars": Dellenbaugh, *Romance,* p. 328.

Nathaniel "Than" Galloway: Lavender, *River Runners,* pp. 36–8.

"every way built": Powell, *Exploration,* p. 8.

"With that flat bottom": Author interview, Sept. 11, 2000.

"calamitous": Dimock, "Buzz Holmstrom and the Evolution of Rowing Style," p. 21.

It was not until: Ibid.

"Opposite the mouth": Sumner, *UHQ,* June 18, p.184.

"Bradley was much provoked": Ibid.

"This has been a chapter": Powell, *Exploration,* June 17, p. 30. This entry should be dated June 18.

"A calm, smooth stream": Howland, *Rocky Mountain News,* June 19, 1869. (See *UHQ,* p. 98.)

"I predict that the river": Bradley, *UHQ,* June 18, p. 40.

"the worst by far": Ibid.

CHAPTER ELEVEN: *The First Milestone*

"When we have rowed": Powell, *Exploration,* June 19, p. 34.

"Mother has but one": Bradley, *UHQ,* June 20, p. 41.

"a narrow, dangerous": Sumner, *UHQ,* June 21, p. 184.

"The walls are high": Powell, *Exploration,* June 21, p. 35.

"All this volume": Ibid.

"The cañon is much narrower": Ibid.

"great alarm": Ibid., p. 36.

"Now, by passing": Powell, *Exploration,* June 21, p. 37.

"and soon had a score": Sumner, *UHQ,* June 21, p. 185.

"Mr. Howland": Sumner granted Howland another honorific "Mr." on June 26, for no clear reason.

"a splendid run": Sumner, *UHQ*, June 22, p. 185.

"One of the boats": Howland, *Rocky Mountain News*, July 1, 1869. (See *UHQ*, p. 102.)

"They hunted*"*: Bradley, *UHQ*, June 22, p. 41.

"Into the middle": Powell, *Exploration*, June 22, p. 37.

"the waters waltz": Powell, *Exploration*, June 21, p. 35.

"dancing over ...waves": Sumner, *UHQ*, June 22, p. 186.

"looking like drowned rats": Ibid., p.185.

"a splendid park": Ibid., p. 186.

"We came out into": Bradley, *UHQ*, June 22, p. 41.

little to admire: "The lover of nature, whose perceptions have been trained in the Alps, in Italy, Germany, or New England, in the Appalachians or Cordilleras, in Scotland or Colorado," wrote the geologist Clarence Dutton, "would enter this strange region with a shock." Dutton diagnosed the ailment. "Great innovations, whether in art or in literature, in science or in nature, seldom take the world by storm. They must be understood before they can be estimated." See *Tertiary History*, pp. 141–2.

"rattlesnakes crawl": Powell, *Exploration*, p. 175.

"The whole country": Sumner, *UHQ*, June 22, p. 186.

"Wonderful to relate": Bradley, *UHQ*, June 23, p. 41.

"He stopped to take a look": Bass, *Adventures*, p. 21.

"The men are on tiptoe": Bradley, *UHQ*, June 23, p. 42.

"our cuisine*"*: Powell, *Exploration*, June 20, p. 35.

"Have spread the rations": Bradley, *UHQ*, June 22, p. 41.

"I am so used to climbing": Ibid., June 24, p. 42.

"The river about four miles": Ibid.

"a muddy looking rainbow": Howland, *Rocky Mountain News*, July 1, 1869. (See *UHQ*, p. 103.)

"The park is below": Powell, *Exploration*, June 24, p. 38.

"a gloomy chasm": Ibid., p. 39.

"one vast contiguity of waste": "Uintah-Ouray Indian Reservation," *Utah History Encyclopedia*.

"splitting [it] for a distance": Powell, *Exploration*, June 24, p. 38.

"The river seems to go": Howland, *Rocky Mountain News*, July 1, 1869. (See *UHQ*, pp. 103–4.)

"We enter Split Mountain Cañon": Powell, *Exploration*, June 25, p. 39.

"One of the men sick": Sumner, *UHQ*, June 25, p.186.

"Spent two hours": Sumner, *UHQ*, June 26, p. 186.

"all hands pulled": Ibid.

"Our sentinel": Ibid.

"very poor at this season": Bradley, *UHQ*, June 26, p. 42.

"down a river": Sumner, *UHQ*, June 27, p. 187.

these geese were scrawny: Oramel Howland, *Rocky Mountain News*, July 1, 1869. (See *UHQ*, p. 104.)

"We have had a hard": Bradley, *UHQ*, June 27, p. 42.

"We must be very near": Ibid., p. 43.

CHAPTER TWELVE: *Hoax*

"a splendid meadow": Bradley, *UHQ*, July 1, p. 44.

"not much of a road": Sumner, *UHQ*, June 28, p. 188.

"Hope to receive": Bradley, *UHQ*, June 28, p. 43.

"We had the greatest ride": Hall, *UHQ*, v.16–17, 1949, pp. 506–7.

"I have written this": Sumner, *UHQ*, June 28, p. 189.

"I send manuscript journal": Marston quotes Powell's letter in his introduction to "The Lost Journal of John Colton Sumner," *UHQ*, v. 37, 1969, p. 174.

"The party has reached": Powell letter to Richard Edwards, June 29, 1869. (See *UHQ*, p. 86.)

"saw the party perish": reprinted in Watson, *The Professor Goes West*, p. 43.

Theodore Hook: Webb, *If We Had a Boat*, p. 57.

"the completeness of the disaster": Dellenbaugh, *Canyon Voyage*, p. 25. Dellenbaugh gives Hook's name as Theodore. The *Rocky Mountain News* of July 6, 1869, refers to "H. M. Hook," as does Marston in "The Lost Journal of John Colton Sumner," *UHQ*, p. 173.

On July 2, the Omaha Republican: July 2 and 3, 1869. Reprinted in the *New York Times*, July 9, 1869.

"The John Sumner of the expedition": *Chicago Tribune*, July 8, 1869.

So much for "Sumner": *Rocky Mountain News*, July 6, 1869. Reprinted in the *New York Times*, July 13, 1869.

The tiny Appleton [Wisconsin] Post: July 8, 1869.

"Mr. Risdon is an honest": *Rocky Mountain News*, July 6, 1869. Reprinted in the *New York Times*, July 13, 1869.

"the fate of Major Powell's": *Chicago Tribune*, July 3, 1869. Reprinted in the *New York Times*, July 6, 1869.

Powell's mother: *Chicago Evening Journal*, July 3, 1869. Reprinted in the *New York Times*, July 6, 1869.

"To one at all acquainted": *Detroit Post*, July 3, 1869. Reprinted in the *New York Times*, July 7, 1869.

A few days later, Emma: *Detroit Tribune*, July 5, 1869. Reprinted in the *New York Times*, July 7, 1869.

"a liar or a crazy man": Bloomington [Ill.] *Pantagraph*, July 6, 1869.

The Detroit Tribune *declared*: *Detroit Tribune*, July 5, 1869. Reprinted in the *New York Times*, July 7, 1869.

"Risdon ought to be hung": *Rocky Mountain News*, July 6, 1869. Reprinted in the *New York Times*, July 13, 1869.

"Camp at Mouth of Uintah": Powell, June 29, 1869. (See *UHQ*, p. 86.)

Less than a month: *Springfield* [Illinois] *Journal*, July 10, 1869. Reprinted in the *New York Times*, July 13, 1869.

CHAPTER THIRTEEN: *Last Taste of Civilization*

"It is a toilsome walk": Powell, *Exploration*, July 1, p. 42.

"At noon met Hall": Powell, *UHQ*, July 2, p. 125.

If Powell kept a diary: Stanton, whose *CRC* is strongly critical of Powell, argues that Powell did not keep a diary in May and June. His evidence is that Powell referred to each night's camp by number, and his first entry after leaving the agency and returning to the river is labeled "Camp No. 1." (See *CRC*, p. 120.)

From this point on: Powell's river diary has no entries for the dates between July 6 and July 20.

"It is rather pleasant": Powell, *Exploration*, July 1, p. 42.

"We feel lonesome": Bradley, *UHQ*, July 2, p. 44.

"One of the men says": Ibid.

"I did nothing today": Worster, *River Running West*, p. 146, quoting Rhodes Allen.

"We shall have a poor 4": Bradley, *UHQ*, July 3, p. 44. The reference to duck hunting and fishing is from Sumner, *CRC*, p. 183.

"Three successive 4ths: Bradley, *UHQ*, July 4, p. 45.

"seen danger enough": Powell, *Exploration*, July 5, p. 43.

"one of those that lost": Bradley, *UHQ*, July 5, p. 45.

"one of the wrecked party": Howland, *Rocky Mountain News*, July 1, 1869. (See *UHQ*, p. 105.)

"Goodman, having had": Sumner, *CRC*, p. 183.

"As our boats are rather": Powell, *Exploration*, July 5, p. 43.

"The major is from near Blumington": Hall, *UHQ*, v. 16–17, 1949, p. 506.

"The boats seem to be": Powell, *UHQ*, pp. 86–7.

"i think that we ar now": Hall, *UHQ*, v. 17, 1949, p. 507.

"So far," he noted simply: Sumner, *UHQ*, June 28, p. 188.

Nor, despite their hopes: Bradley, *UHQ*, Aug. 30, p. 72.

"letters from home": Bradley, *UHQ*, June 28, p. 43.

"[Powell] has got 300 lbs.": Bradley, *UHQ*, July 5, p. 45.

"Major Powell was gone five": Sumner, *CRC*, p. 183.

"Our rations . . . have been wet": Powell, *UHQ*, p. 87.

"They're much more difficult": Author interview, Oct. 25, 1999.

Powell's problem would have been: Author interview with Brad Dimock, September 11, 2000.

"perched on a sack": Sumner, *UHQ*, June 26, p. 86. Michael Ghiglieri called my attention to the relevance of the Sumner remark to the sweep-oar debate. Author interview, March 19, 2001.

"They're definitely not": Author interview, Oct. 25, 1999.

When boatmen of their skill: Michael Ghiglieri rejects this argument. "Modern boatmen are spoiled by having experience—and conditioned responses—only with top-notch and more appropriate boats," he insists. "When you go from a modern, lightweight dory to one of Powell's boats, it's like going from a sports car to a Flintstone car, where you have to kick your feet. If you *only* knew Whitehall haulers, you'd figure out how to row them." (Author interview, Mar. 19, 2001.)

an all-star team: Brad Dimock, Regan Dale, and Martin Litton, among others, are in the pro-sweep camp; the anti-sweep side includes boatmen Drifter Smith, Michael Ghiglieri, Pat Reilly, and the historian Martin Anderson.

See the notes: The most relevant entry in any contemporary account was Bradley's journal entry on Aug. 28. Bradley had gotten in trouble in a rapid but found a way to rescue himself. "By putting an oar first on one side then on the other I could swing her around and guide her very well," he wrote, and the remark seems hard to reconcile with the boat having had a sweep oar. (See Bradley, *UHQ*, Aug. 28, p. 71.) But Powell described the same episode, although he did not write his account until 1875: "Bradley seizes the great scull oar, places it in the stern rowlock, and pulls with all his power." (See Powell, *Exploration*, Aug. 28, p. 101.)

Which man had it right? When we come to Aug. 28, we will see that Powell's account of the day's events is suspect in other regards, but there is no denying that we have to make a judgment call here. A fuller discussion of this issue can be found in the notes for Aug. 28, in Chapter Twenty-five.

"It would have been a tremendous": Author interview with Brad Dimock, Oct. 25, 1999.

"Was exceedingly glad": Bradley, *UHQ*, July 6, p. 45.

"After 7 days of weary": Sumner, *UHQ*, July 6, p. 113.

"intention to plant some corn": Powell, *Exploration*, July 6, p. 45.

"We had read that 'stolen'": Bradley, *UHQ*, July 6, p. 45.

"The Professor, Dunn and Hall": Sumner, *UHQ*, July 6, p. 113.

"Such a gang of sick": Sumner, *CRC*, p. 184.

"and we tumble around": Powell, *Exploration*, July 6, p. 45.

"didn't think potato tops": Sumner, *CRC*, p. 184.

"we shan't eat any": Bradley, *UHQ*, July 6, p. 46.

"with no more current": Sumner, *UHQ*, July 7, p. 113.

"I . . . put my name": Bradley, *UHQ*, July 7, p. 46. Bradley's stone has vanished.

CHAPTER FOURTEEN: *Trapped*

"The walls are almost": Powell, *Exploration*, July 8, p. 46.

"We start up a gulch": Powell, *Exploration*, June 18, p. 33. The date is wrong because, as discussed in the text, Powell places the adventure at Echo Rock.

"Climbed the mountain": Bradley, *UHQ*, July 8, p. 46. "We have named the long peninsular rock on the other side Echo Rock," Powell wrote. "Desiring to climb it, Bradley and I take the little boat and pull up stream as far as possible, for it cannot be climbed directly opposite." (See *Exploration*, p. 33.) Then he continued with the story as told here. Echo Rock was far upstream, and Sumner, Bradley, and Howland each described it; none of them mentioned a word about a climbing mishap. Powell's reference to Echo Rock must have been a slip of memory or, perhaps, a literary flourish intended to provide a suitably grand backdrop for a monumentally dramatic scene.

It is possible to read Bradley's account as not quite confirming Powell's. Perhaps Powell merely needed assistance ("In one place Major . . . couldn't get up"), not rescue. But this reading seems strained. The story of the drawers seems to indicate that Powell was in a desperate plight, as do the words "he got up safe."

"wild and desolate": Bradley, *UHQ*, July 8, p. 46.

"It is a wild": Ibid., p. 47.

"I should run it": Bradley, *UHQ*, July 8, p. 47.

"If you're the leader": Author interview, Jan. 6, 2000.

"We need only a few": Bradley, *UHQ*, July 9, p. 47.

"the wildest day's run": Ibid., p. 47.

Powell and Oramel Howland climbed: Powell, *Exploration*, July 10, p. 47.

"a long, slow, anxious": Ibid., p. 48.

"On the way back": Dellenbaugh, *Romance*, p. 281.

"a voice like a crosscut": Bradley, *UHQ*, July 10, p. 48.

"bad" but runnable: Ibid.

"rowled over and over": Bradley, *UHQ*, July 11, p. 48. Michael Ghiglieri cites this episode as additional evidence that the *Emma Dean* did not have a sweep oar. With one pair of oars lost, Ghiglieri argues, the man without oars would certainly have manned the sweep and not have sat idly by. Author interview, Feb. 26, 2001.

"I soon find that swimming": Powell, *Exploration*, July 11, p. 48.

"We broke many oars": Sumner, *CRC*, p. 184.

"The guns and barometer": Powell, *Exploration*, July 11, p. 49.

The last claim: "Dock" Marston, who had little use for Powell, discussed this incident and others in "With Powell on the Colorado," pp. 65–76.

"Major had to leave": Bradley, *UHQ*, July 11, p. 48.

"Sunday again": Ibid.

"we came suddenly upon": Bradley, *UHQ*, July 12, p. 48.

"To us who are below": Powell, *Exploration*, July 12, p. 49.

"except a glorious ducking": Bradley, *UHQ*, July 12, pp. 48–9. Like Powell's story of Bradley rescuing him with his drawers, Powell's account of Bradley's swamping is hard to evaluate. Bradley does describe a "big wave . . . knocking me over the side so that I held the boat by one hand," but does not say that his foot was pinned beneath his seat. Does the discrepancy between Bradley's brief account of his "glorious ducking" and Powell's more elaborate version reflect Bradley's modesty or Powell's tendency to embellish a good yarn?

CHAPTER FIFTEEN: "Hurra! Hurra! Hurra!"

"About three o'clock": Powell, *Exploration*, July 12, p. 50.

Dunn was a strong: Hawkins, *CRC*, p. 158.

"a short portage": Powell, *Exploration*, July 12, p. 50.

"We camp on a sand": Ibid.

"The wind continues to blow": Ibid.

"Pulled out early": Sumner, *UHQ*, July 13, p. 114.

"We glide along": Powell, *Exploration*, July 13, p. 50.

"a barren parched dessert": Bradley, *UHQ*, July 13, p. 49.

"Plains, and hills, and cliffs": Powell, *Exploration*, July 13, p. 51.

"as desolate a country": Sumner, *UHQ*, July 14, p. 114.

"We have run so many": Bradley, *UHQ*, July 13, p. 49.

"This is the place": Powell, *Exploration*, July 13, p. 51.

"As we get farther": Bradley, *UHQ*, July 14, p. 50.

"Arrowheads are scattered": Powell, *Exploration*, July 14, p. 52.

"A sad, bitter business": Sumner, *CRC*, p. 185.

"We go around a great": Powell, *Exploration*, July 15, p. 53.

"River very crooked": Sumner, *UHQ*, July 15, p. 114.

"The sun was so hot": Bradley, *UHQ*, July 15, p. 50.

"We are all in fine": Powell, *Exploration*, July 15, p. 54.

"The landscape everywhere": Ibid., July 17, p. 55. This journal entry, which also describes reaching the junction with the Grand, should have been dated July 16.

"strange, weird, grand": Ibid.

"What strange anomaly": Meloy, *Raven's Exile*, p. 9.

"Hurra! Hurra! Hurra!" Bradley, *UHQ*, July 16, p. 50.

"calm and wide": Ibid.

"It is possible we are allured": Ibid., p. 51.

Denis Julien: Zwinger, *Run, River, Run*, p. 251.

"not timber enough": Sumner, *UHQ*, July 16, p. 115.

"when the summer-sun comes": Powell, *Exploration*, p. 4.

CHAPTER SIXTEEN: *Outmatched*

the bigger of the two: Sumner judged the Green to be 70 or 80 yards wide and the Grand 125. Sumner, *UHQ*, July 16, p. 114.

"Our rations, . . . we find": Powell, *Exploration*, July 18, p. 57.

"about 500 pounds of flour": Sumner, *CRC*, p. 186. (Bradley put the total at six hundred pounds. See *UHQ*, July 17, p. 51.)

"seemed as if commissioned": Bradley, *UHQ*, July 17, p. 51.

"From that date there": Sumner, *CRC*, p. 200.

"Howland would sometimes get": Sumner, *CRC*, p. 201.

"got things somewhat": Sumner, *CRC*, p. 189.

"They are quite decent": Bradley, *UHQ*, July 16, p. 51.

"While we are eating": Powell, *Exploration*, July 19, p. 59.

the nearest pie: Powell, *Exploration*, July 19, p. 59.

"for we cannot think": Bradley, *UHQ*, July 17, p. 51.

"Sunday again": Bradley, *UHQ*, July 18, p. 51.

"We must not conceive": Powell, *Exploration*, July 17, p. 55.

"Climbed 'Cave Cliff' ": Drifter Smith called my attention to Powell's repeated references to "Bradey" rather than "Bradley." Martin Anderson, in turn, had pointed out the misspelling to Smith. The transcript of Powell's river diary in the *UHQ* silently corrects this error, but the handwritten original shows "Bradey" for "Bradley." In his published accounts of the expedition, Powell did spell Bradley's name correctly. Bradley never complained in his diary that Powell *called* him "Bradey," so presumably Powell knew his name.

"Then we try the rocks": Powell, *Exploration*, July 19, p. 57.

"we determine to attempt": Ibid., p. 58.

"Away to the west": Ibid.

"The scenery from the top": Bradley, *UHQ*, July 20, p. 53.

Powell, inclined by nature: Powell built his journal account from his river diary, which was markedly richer than Bradley's diary but far more restrained than Powell's journal account would be. Powell's diary entry for July 20, 1869, read as follows: "Climbed 'Cave Cliff' with Bradey. Summit of cliffs full of caves, hence name. Pinnacles in the red sandstone. The terraces, the monuments of the stages of erosion. Found a cool spring in gulch on our way up. One cave 75 paces long, dome, skylight at each end connected by fissure 6 or 8 inches wide, from 10 to 40 ft. wide, 5½ ft. high."

"We start this morning": Powell, *Exploration*, July 21, p. 60.

"Rapids commenced": Bradley, *UHQ*, July 21, p. 53.

"Two very hard portages": Powell, *Exploration*, July 21, p. 60.

"We are thrown into": Ibid.

"Have made two portages": Bradley, *UHQ*, July 21, p. 53.

"So I conclude the Colorado": Ibid.

"We come at once": Powell, *Exploration*, July 23, p. 61.

"All the way rapid": Bradley, *UHQ*, July 23, p. 53.

"We camp tonight": Ibid.

"Our way . . . is through'": Powell, *Exploration*, July 23, p. 62.

"It was only a few": Stanton, *Down the Colorado*, p. 55.

"We have as yet found": Bradley, *UHQ*, July 23, p. 54.

Emery showed the flickering: The information on the Kolbs's film is from an advertising handout called "See This at Grand Canyon, Arizona" distributed to would-be filmgoers. See Babbitt's *Grand Canyon: An Anthology* for a short essay on the Kolbs and for many of their still photographs.

"We always thought we needed": Kolb, *Grand Canyon*, p. 138.

"Fisher's raft flipped": American Whitewater Association, "Year End Whitewater Fatality Report, 1997."

"The conclusion to which": Powell, *Exploration*, July 23, p. 62.

"How will it be": Ibid.

"They speculate over the serious": Ibid.

"Let it come": Bradley, *UHQ*, July 23, p. 54.

"Among these rocks": Powell, *Exploration*, July 24, p. 62.

"Had to take everything": Bradley, *UHQ*, July 24, p. 54.

"Kitty's Sister had another": Sumner, *UHQ*, July 24, p. 115.

"They tell me [it]": Bradley, *UHQ*, July 24, p. 54.

"¾ blue marble": Sumner, *UHQ*, July 24, p. 116.

"God gone mad": Houston, "Cataract," p. 123, reprinted in *Prize Stories 1999: The O. Henry Awards.*

"Driftwood 30 ft. high": Sumner, *UHQ,* July 24, p. 116.

"The waves are rolling": Powell, *Exploration,* July 24, p. 63.

CHAPTER SEVENTEEN: *Flash Flood*

"As soon as we had": Kolb, *Grand* Canyon, p. 139.

"The Emma Dean *is caught"*: Powell, *Exploration,* July 25, p. 63. Nash's excellent *Big Drops* has a chapter on the history of attempts to run the Colorado through Cataract Canyon.

Forty years would pass: Nathaniel Galloway made the first successful run, in 1909. See Nash, *Big Drops,* p. 87.

The white-water passages: The story, which deserves every bit of its acclaim, is "Cataract." Houston discussed her story with me in a telephone interview on Oct. 18, 2000.

"I was worried about": Houston, "Cataract," *O. Henry Prize Stories,* p. 135.

"not of smooth concrete": Staveley, *Broken Waters Sing,* p. 173.

"jagged rocks, like the bared": Kolb, *Grand Canyon,* p. 149.

"bottomless, thrashing abysses": Staveley, *Broken Waters,* p. 174.

"In the instant": Ibid., p. 176.

"still one foaming torrent": Bradley, *UHQ,* July 25, p. 55.

"I have a barometer": Powell, *Exploration,* July 26, p. 64.

"People die in chubascos": Childs, *The Secret Knowledge of Water,* p. 201.

"The waters that fall": Powell, *Exploration,* Aug. 15, p. 86.

"I have stood in the middle": Abbey, *Desert Solitaire,* p. 150.

"Canyons are basically nets": Childs, *Secret Knowledge,* p. 239.

Two people were killed: Arizona Daily Sun, "Canyon Trails Will Need Major Rebuilding Effort," July 17, 1999. *Arizona Republic,* "Bright Angel Cleanup Begins," July 19, 1999.

"An entire oak tree": Childs, *Secret Knowledge,* p. 244.

"The flood bounded": Ibid., p. 246.

"I find a thousand": Powell, *Exploration,* July 26, p. 65.

"Another day wasted": Bradley, *UHQ,* July 26, p. 55.

"5 of the men": Sumner, *UHQ,* July 26, p. 116.

"very bad": Bradley, *UHQ,* July 27, p. 55.

"In the present reduced state": Ibid.

"a Godsend": Sumner, *UHQ,* July 27, p. 116.

"But fresh meat": Powell, *Exploration,* July 27, p. 66.

"The walls suddenly close in": Powell, *Exploration*, July 28, pp. 66–7.

"Now that it is past": Ibid., p. 67.

CHAPTER EIGHTEEN: *To the Taj Mahal*

"The untransacted *destiny"*: Smith, *Virgin Land*, p. 37.

"The greatest gash on the earth's": Townshend, *A Tenderfoot in Colorado*, pp. 41–2.

"Is it a trout-stream?": Powell, *Exploration*, July 28, p. 67.

"The water is about as filthy": Sumner, *UHQ*, July 28, p. 116.

"stinks bad enough": Sumner, *CRC*, p. 187.

Hawkins would later dispute: Hawkins, *CRC*, p. 154.

"Major named the new": Bradley, *UHQ*, July 29, p. 56.

He traveled week after: Michael Ghiglieri called my attention to Bradley's forbearance. Author interview Mar. 19, 2001.

Ran 20 miles: Bradley, *UHQ*, July 29, p. 56.

"How they contrived": Sumner, *UHQ*, July 29, p. 117.

"as destitute of vegetation": Sumner, *UHQ*, July 30, p. 117.

"Here I stand": Powell, *Exploration*, July 29, p. 69.

"The curves are gentle": Powell, *Exploration*, July 31, p. 70.

"almost hissing hot": Bradley, *UHQ*, July 31, p. 57.

"The remainder of the afternoon": Powell, *Exploration*, July 31, p. 70.

"They seem to have no fear": Bradley, *UHQ*, Aug. 1, p. 57.

"[The sheep] are very good": Ibid.

"And this," Powell marveled: Powell, *Exploration*, Aug. 1, p. 71.

rock-shock: Harrington, *Dance of the Continents*, p. 151.

The Howland brothers and Bill Dunn: Dellenbaugh described finding the Howland and Dunn inscriptions in *Canyon Voyage*, p. 141.

"Old Shady": The lyrics to "Old Shady" can be found with the Civil War Song Sheets at the Rare Book and Special Collections Division of the Library of Congress, or at the Library of Congress website.

"a point of commanding view": Powell, *Exploration*, Aug. 2, p. 71.

"In the same camp": Bradley, *UHQ*, Aug. 2, p. 57.

"We sleep again": Powell, *Exploration*, Aug. 2, p. 71.

"Major has been taking": Bradley, *UHQ*, Aug. 2, p. 57.

"If we could get game": Ibid.

"Major has agreed": Ibid.

"easy run": Bradley, *UHQ*, Aug. 3, p. 58.

"Where the water is still": Ibid.

Old photographs: Meloy, *Raven's Exile*, pp. 171–2.

"Past these towering monuments": Powell, *Exploration,* Aug. 3, p. 72.

"Many huge Mts.": Powell, *UHQ,* Aug. 3, p. 128.

"excepting daily duckings": Sumner, *CRC,* p. 189.

"The dark red rock": Fradkin, *A River No More,* p. 194.

The two sides agreed: Brower, *Let the Mountains Talk, Let the Rivers Run,* pp. 26–7. Brower, one of the towering figures in the environmental movement, deserved much of the credit for keeping dams out of the Grand Canyon. But he castigated himself for having given in to the Glen Canyon compromise, which he saw as the great mistake of his life. "Glen Canyon died in 1963," he wrote in the foreword to *The Place No One Knew,* "and I was partly responsible for its needless death. So were you. Neither you nor I, nor anyone else, knew it well enough to insist that at all costs it should endure."

"To grasp the nature": Abbey, *Desert Solitaire,* p. 189.

"Probably no man-made artifact": Blaustein, *The Hidden Canyon,* p. 25.

But the Bureau of Reclamation: Powell's legacy is so confused, in part, because of a tendency to forget that he was a man firmly of the nineteenth century. We tend today to lump together the words "preservationist" and "conservationist," but, as the historian Patricia Limerick notes, the two terms had sharply different meanings in the 1800s. Powell was a conservationist, which meant, in his day, that he favored conserving scarce resources, water chief among them. He favored reclaiming the land, not preserving it unspoiled like a colossal museum exhibit. Powell talked happily of "conquered rivers" and looked to a future where rivers would be tamed by man "as wild beasts have been domesticated for his use." The alternative to dams and reservoirs, Powell wrote repeatedly, was that rivers would "run to waste," spilling their treasure uselessly into the sea. See Limerick, *Desert Passages,* pp. 169–72, and Aton, *Inventing John Wesley Powell.* For a typical expression of Powell's views, see his essay on the Johnstown flood, "The Lesson of Conemaugh." For a book-length attempt to place Powell's views in their historical context, see Worster's *River Running West.*

"through a dark calm cañon": Bradley, *UHQ,* July 16, p. 51.

"Today the walls grow": Powell, *Exploration,* Aug. 4, p. 73.

This narrow spot: Rusho, *Powell's Canyon Voyage,* p. 24.

"a perfect tornado": Bradley, *UHQ,* Aug. 4, p. 58.

"With some feeling of anxiety": Powell, *Exploration,* Aug. 5, p. 73.

CHAPTER NINETEEN: Grand Canyon

"We were in no condition": Sumner, *CRC,* p. 190.

John D. Lee: The information in this paragraph is from Juanita Brooks's path-breaking *Mountain Meadows Massacre.* (See p. 187.) A summary of what is known about each of Lee's nineteen wives, and a bar graph showing the years they were married to Lee, is in Brooks, *John Doyle Lee*, pp. 378–84.

"It is desolate enough": Sumner, *UHQ*, Aug. 4, p. 118.

the governor of Arizona Territory: Egan, *Lasso the Wind*, p. 125.

"Just below our camp": Bradley, *UHQ*, Aug. 4, p. 59.

"I said yesterday": Bradley, *UHQ*, Aug. 5, p. 59.

"long and difficult": Sumner, *UHQ*, Aug. 5, p. 118.

"Very hard work": Ibid.

"Am very tired tonight": Bradley, *UHQ*, Aug. 5, p. 59.

"Made 3 portages": Sumner, *UHQ*, Aug. 6, p. 118.

"Three times today": Bradley, *UHQ*, Aug. 6, p. 59.

Powell described the procedure: Powell, *Exploration*, Aug. 6, p. 74.

"We succeeded in landing": Bradley, *UHQ*, Aug. 6, p. 59.

Robert Brewster Stanton, a railroad engineer: This account is based on Stanton's *Down the Colorado*, pp. 30–90.

"my heart sank": Stanton, *Down*, p. 36.

"I, with Reynolds": Stanton, *Down*, p. 34.

"It would be a great": Ibid., p. 46.

"disaster every day": Reynolds, "In the Whirlpools of the Grand Cañon of the Colorado," p. 29. Frail as the boats were, it was their instability rather than their fragility that proved disastrous. (See P. T. Reilly, "How Deadly?" *UHQ*, p. 254.)

"The matter of supplies": Stanton, *Down*, p. 50.

"Such work as we had": Ibid., p. 56.

"Ghastly suggestion": Ibid., p. 61.

It looked like nothing more: Writing independently, P.T. Reilly and Michael Ghiglieri, both of them Grand Canyon boatmen and white-water historians, described this scenario as their best guess. McDonald himself apparently pinned the blame on an underwater current. "Just as we turned, in what seemed to be smooth waves," McDonald wrote, "a heavy wave came up out of a whirl on upper side of boat & instantly upset boat, throwing us both into river away from the boat." Stanton endorsed McDonald's view but Reilly dismissed it contemptuously. "Since the three deaths which ensued from the use of this equipment [the boats] did not result from that which Stanton called defective, it is interesting to see him grope for explanations and advance such things as 'up-shoots' and 'boiling fountains.'" (See Reilly, "How Deadly?" pp. 254–5; Ghiglieri, *Canyon*, p. 158; Stanton, *Down*, pp. 78–9.)

"They are all right": Stanton, *Down*, p. 86.

"I then realized fully": Ibid., p. 87.

"death's canyon": Ibid., p. 88.

"something like a large": Ibid., p. 89.

"at a place where it seems": Powell, *Exploration*, Aug. 6, p. 74.

"The line of totality": "The Recent Total Eclipse of the Sun," *Nature*, Nov. 4, 1869.

"Tomorrow is the eclipse": Bradley, *UHQ*, Aug. 6, p. 59.

"but clouds come on": Powell, *Exploration*, Aug. 7, p. 75.

"At last we lose our way": Ibid.

"Daylight comes, after": Powell, *Exploration*, Aug. 8, p. 75.

"Constant banging against rocks": Bradley, *UHQ*, Aug. 7, p. 59.

CHAPTER TWENTY: *Time's Abyss*

"Though it is Sunday": Bradley, *UHQ*, Aug. 8, p. 60.

"Pulled out early": Sumner, *UHQ*, Aug. 8, p. 118.

"It is with very great labor": Powell, *Exploration*, Aug. 8, p. 75.

"We begin to be a ragged": Bradley, *UHQ*, Aug. 8, p. 60.

"I had a pair": Hawkins, *CRC*, p. 146.

"The limestone is coming up": Bradley, *UHQ*, Aug. 8, p. 60.

"And now, the scenery": Powell, *Exploration*, Aug. 9, pp. 75–6.

"There is marble enough": Sumner, *CRC*, pp. 191–2.

"a rose-red city": Steve Jones, *Darwin's Ghost*, p. 193.

To study the age of rocks: William Jennings Bryan was fond of a similar formulation, to the effect that he was interested in the rock of ages rather than the age of rocks.

One hundred generations: Lucchitta, *Canyon Maker: A Geological History of the Colorado River*, p. 5.

"It has taken several years": Powell, *Exploration*, pp. 20–1.

"If the Eiffel Tower": Twain, *The Damned Human Race*, p. 170.

almost exact contemporaries: Powell was born in 1834, Mark Twain in 1835.

"In the Grand Canyon": Powell, *The Exploration of the Colorado River and Its Canyons*, p. 390. (This passage is from the 1895 edition of Powell's journal.)

"library of the gods": Powell, *Exploration*, p. 193.

"so regular and beautiful": Ibid., p. 174.

"High up in the North": Eicher cites this story in his excellent *Geologic Time*. The tale itself is from Hendrick Van Loon's *Story of Mankind*.

The fixed and solid-seeming: Thomas Huxley made this comparison in the second of his *Six Lectures to Working Men* in 1863. See *On Our Knowledge of the Causes of the Phenomena of Organic Nature*.

"continents would crawl": McPhee, *Annals of the Former World*, p. 170.

The life span of a mountain: Hamblin, *Earth's Dynamic Systems*, p. 275.

"not by an extravagant": Powell, *Exploration*, p. 162.

"Water-drops have worn": Troilus and Cressida, act 3, scene 2.

At one site in Wyoming: Wallace, *The Bonehunters' Revenge*, p. 266.

During a brief stop: Ibid., pp. 52–3.

The oldest rocks: Price, *An Introduction to Grand Canyon Geology*, p. 23. The oldest rock in the Grand Canyon is Vishnu Schist. The oldest exposed rocks in the world, on the shores of Canada's Great Slave Lake and in Greenland, are four billion years old.

On a timeline: Eicher, *Geologic Time*, p. 18.

"a dog's discovery of the Moon": Dellenbaugh, *Romance*, p 36.

"From the top they could": Hammond and Rey, *Narratives of the Coronado Expedition.* Reprinted in Babbitt, ed., *Grand Canyon: An Anthology*, p. 12.

"Standing there or rather lying": Langford, *The Discovery of Yellowstone Park*, p. 32.

four and six million years old: Webb, *Grand Canyon*, p. 144.

Five hundred thousand tons of sand: L. Greer Price, *Grand Canyon*, p. 8. The numbers were far higher during floods. In 1948, scientists at the U.S. Geological Survey determined that in a particular 24-hour span the river carried some 9.5 million tons of sand, rock, and silt past Lee's Ferry. See Carothers and Dolan, "Dam Changes on the Colorado River," p. 78.

"The most emphatic lesson": Davis, "The Lessons of the Colorado Canyon," p. 346.

"the Great Unconformity": At Blacktail Canyon, as Ann Zwinger notes in *Downcanyon*, a hiker can span the missing 1.2 billion years of the Great Unconformity with her hands. (See Zwinger, p. 103.)

"Beds hundreds of feet": Powell, *Exploration*, p. 208.

"The finest workers": Thoreau, *A Week on the Concord and Merrimak River.* Thoreau makes the remark in "Wednesday."

Come the spring thaw: The stonecutting analogy and the remark about the river's intermittent work habits are from Hamblin, *Earth's Dynamic Systems*, p. 245.

"The river was the saw": Powell, *Exploration*, pp. 152–3.

And the canyon is so wide: Webb, *Grand Canyon*, p. 144.

"The history of the rocks there": Cadell, "The Colorado River of the West," p. 456.

The Spanish conquistadors: Babbitt, *Grand Canyon: An Anthology*, p. 11.

"It looks about large enough": Clarence E. Dutton, *Tertiary History of the Grand Canyon District*, p. 89.

CHAPTER TWENTY-ONE: *The Great Unknown*

"but we can't get down": Bradley, *UHQ*, Aug. 8, p. 60.

"We are interested now": Ibid.

"in a cave": Bradley, *UHQ*, Aug. 8, p. 61.

"Hard at work": Sumner, *UHQ*, Aug. 9, p. 118.

"this series of heavy": Bradley, *UHQ*, Aug. 9, p. 61.

"The river turns sharply": Powell, *Exploration*, Aug. 9, p. 76.

"the unending barrenness": Bradley, *UHQ*, Aug. 9, p. 61.

"the prettiest sight": Sumner, *CRC*, p. 191.

"The white water over the blue": Sumner, *UHQ*, Aug. 9, p. 118.

"The water sweeps rapidly": Powell, *Exploration*, p. 75. This quotation from Powell is dated Aug.8, which cannot be correct, since Bradley and Sumner agree that the expedition only reached Vasey's Paradise, *upstream* of Redwall Cavern, on Aug. 9.

photographs of the same rapid: Compare photos of Lava Falls from 1872 and 1968, in Stephens and Shoemaker, *In the Footsteps of John Wesley Powell*, pp. 272–3. Stephens and Shoemaker reshot photographs originally taken in 1871 to 1872. See also Webb's *Grand Canyon, A Century of Change: Rephotography of the 1889–1890 Stanton Expedition*.

By the time Powell: It is impossible to know exactly what rapids Powell faced in 1869. Some Grand Canyon rapids have changed since his day (Soap Creek was unrunnable, for instance; today's much-feared Crystal was not much more than a riffle; MNA did not exist). Floods, debris flows, rock falls, and the level of the river can all transform the rapids.

"A creek comes in": Powell, *Exploration*, Aug. 14, p. 82.

Powell was the first: Webb, *Grand Canyon*, p. 142. Perhaps Bradley deserves at least a share of the credit. In his journal on Aug. 18, he wrote, "Rapids very numerous and very large. A great many lateral cañons come in almost as large as the one in which the river runs and they sweep down immense quantities of huge rocks which at places literally dam up the river, making the worst kind of a rapid because you can see rocks rising all over them with no channel in which to run them."

"As the rain increased": Stanton, *Down*, pp. 90–1.

a 280-ton boulder: Webb, *Grand Canyon*, p. 138.

"The parents' bedroom": John McPhee, "Los Angeles Against the Mountains," *The Control of Nature*, p. 185.

Debris flows are rare: Webb, *Grand Canyon*, p. 128.

nowhere else in North America: Childs, *Secret Knowledge*, p. 189.

"Debris flows move entire": Webb, *Grand Canyon*, p. 144.

depth is unchanging: Ibid.

"We run them all": Bradley, *UHQ*, Aug. 10, p. 61.

"It is a lothesome": Ibid.

"as disgusting a stream": Sumner, *UHQ*, Aug. 10, p. 119.

"slime and salt": Sumner, *CRC*, pp. 192–3.

a painful forty-eight degrees: Stevens et al., *Fateful Journey,* p. 37. For a discussion of the impact of this cold water on fish in the Colorado, see Carothers and Dolan, "Dam Changes on the Colorado River," p. 79.

mouthfuls of water: Stevens et al., *Fateful Journey,* p. 36.

"We are sorry": Bradley, *UHQ,* Aug. 10, p. 62.

"The ascent is made": Powell, *Exploration,* Aug. 11, p. 78.

"As we were on the edge": Sumner, *CRC,* p. 193.

"tumbling down": Powell, *UHQ,* Aug. 11, p. 129.

"Have you never seen": Brewer, "John Wesley Powell," *American Journal of Science* 14, 1902.

"filthy with dust": Bradley, *UHQ,* Aug. 11, p. 62.

Modern boatmen wear: Mark Thatcher invented Tevas, which now account for more than $50 million in annual sales. He sold a total of two hundred pair in his first year in business.

"I have given away": Bradley, *UHQ,* Aug. 11, p. 62.

"If this is a specimen": Ibid.

"There remains nothing": Bradley, *UHQ,* Aug. 12, p. 63.

"I am surprised": Ibid.

"Take obs. Capt. climbed Mt.": Powell, *UHQ,* Aug. 12, p. 129.

"We are now ready": Powell, *Exploration,* Aug. 13, p. 80.

CHAPTER TWENTY-TWO: *Sockdolager*

At eight in the morning: Sumner, *UHQ,* Aug. 13, p. 119. Bradley wrote that it was nine o'clock.

"With some eagerness": Powell, *Exploration,* Aug. 13, p. 80.

"The rapids are almost": Bradley, *UHQ,* Aug. 13, p. 63.

"the worst rapid": Ibid.

The asbestos was valuable: Zwinger, *Downcanyon,* p. 128.

Hance abandoned mining: Ghiglieri, *Canyon,* p. 143.

"about 1 mile long": Sumner, *UHQ,* Aug. 13, p. 119.

"i was 10 days With out": Stanton, *CRC,* p. 10.

"i see the hardes time": Ibid.

White's story was widely: A few consider it merely unlikely rather than impossible, partly on the grounds that the alternative—that White walked around the Grand Canyon and entered the Colorado somewhere near the canyon's downstream boundary—seems equally preposterous. For typical modern views, see Stevens et al., *Fateful Journey,* p. 122, and Litton, "Introduction" to Blaustein, *Hidden Canyon,* pp. 11–12. The debunker who made the most detailed anti-White argument was Stanton, the engineer who had set out on the ill-fated 1889 Grand Canyon expedition and who became the second, after Powell, to follow the Colorado through the Grand Canyon. Stanton, a man for whom the word "dogged" might have been coined, assembled all the documentary evidence and concluded that White's story simply did not hold together. Then he found, to his astonishment, that the hero of the tale, presumed long dead, was alive and well in Colorado. In 1907, Stanton hurried off to interview the seventy-year-old White. Stanton emerged from his interview convinced that White had been sincere but deeply mistaken about where he had drifted. (See *CRC*, pp. 3–93.)

"the water was so smooth": Stanton, *CRC*, p. 58.

White was simply lost: This is Stanton's conclusion, which has won general acceptance. (See *CRC*, p. 35.)

Powell claimed that the idea: In the first paragraph of the preface to his *Exploration*, for example, Powell wrote that "the result of the summer's study [of 1867] was to kindle a desire to explore the cañons of the Grand, Green, and Colorado Rivers, and the next summer I organized an expedition with the intention of penetrating still further into that cañon country." On p. 7 of the same book, he wrote again of how "a desire to explore the Grand Cañon itself grew upon me," and described how "a small party was organized for this purpose" and "boats were built in Chicago" for the expedition.

Sumner made the same: Sumner wrote that Powell had asked him to join an exploring party to the Bad Lands of Dakota. "I declined the proposition and fired back at him the counter-proposition—the exploration of the Colorado River of the West, from the junction of the Green and Grand rivers to the Gulf of California." According to Sumner, Powell initially rejected the idea as foolhardy and impossible and only agreed to it after long and heated debate. "The idea was certainly not his own," Sumner claimed. (See *CRC*, pp. 170–1.)

"The absence of any distinct": Stanton, *CRC*, p. 22.

Whether or not Powell: Bradley, nearly always reliable as far as one can tell, mentioned White's story, noted that he didn't believe it, and added that "Major has seen the man." (See Bradley, *UHQ*, Aug. 10, p. 62.) Powell never mentioned such a meeting.

"At the lower end": Sumner, *UHQ*, Aug. 13, p. 119.

"Heretofore, hard rocks": Powell, *Exploration*, Aug. 14, p. 81.

"Major has just come in": Bradley, *UHQ*, Aug. 13, p. 63.

"We can see but a little": Powell, *Exploration*, Aug. 14, p. 81.

"The granite gorge seemed": Kolb, *Grand Canyon*, p. 211.

"At the very introduction": Powell, *Exploration*, Aug. 14, pp. 81–2.

"The sound grows louder": Powell, *Exploration*, Aug. 14, p. 82.

"*Fall 30 ft. probably*": Powell, *UHQ*, Aug. 14, p. 129.

"*We must run the rapid*": Powell, *Exploration*, Aug. 14, p. 82.

This was Sockdolager: The rapid was named in 1871.

"*Just as the dinner hour*": Dellenbaugh, *Canyon Voyage*, p. 226.

"*We finally encountered*": Sumner, *CRC*, pp. 195–6. Sumner's estimate of a thirty-foot drop matched the estimate Powell gave in his river diary.

"*Thank God we are still*": Lavender, *River Runners*, p. 39.

By portaging and lining every: I owe this observation to Drifter Smith.

"*a perfect hell*": Sumner, *UHQ*, Aug. 10, p. 119.

"*I decided to run it*": Sumner, *CRC*, p. 196.

"*We step into our boats*": Powell, *Exploration*, Aug. 14, p. 82.

"*The Emma Dean had not made*": Sumner, *CRC*, p. 196.

"*Still, on we speed*": Powell, *Exploration*, Aug. 14, pp. 82–3. Bradley wrote: "The little boat being too small for such a frightful sea filled soon after starting and swung around head up river almost unmanageable but on she went and by the good cool sense of those on board she was kept right side up through the whole of it (more than half a mile)." (See *UHQ*, Aug. 14, p. 63.)

He only regretted: Sumner, *CRC*, p. 197.

"*I have been in a cavalry*": Ibid.

"*liquid predator*": Ghiglieri, *Canyon*, p. 3.

not *a continuous string:* The two biggest rapids in the Grand Canyon created by rockfalls are MNA, at Mile 26.6, and Sinyala, at Mile 153.3. (See Webb, *Grand Canyon*, p. 128.)

Rapids form only 10: Webb, *Grand Canyon*, p. 144.

The effect is to exaggerate: William Calvin makes this point in *The River That Flows Uphill*, p. 47.

"*This is big water*": Author interview, July 10, 1999.

"*In high water there are a lot*": Sumner, *CRC*, pp. 187–8.

"*The Emma Dean was caught*": Sumner, *CRC*, p. 188.

In the 1983 flood: Author interview with Clair Quist, Aug. 18, 2000.

they range from three feet across: Leopold, "The Rapids and the Pools—Grand Canyon," p. 140.

"*At thirty-two thousand cubic feet per second*": Stevens, "The 67 Elephant Theory," p. 25.

"*The important thing to remember*": Author interview, July 10, 1999.

"*You can say, 'Oh,'*": Author interview, Oct. 25, 1999.

"*Down in these grand*": Powell, *Exploration*, Aug. 14, p. 83.

CHAPTER TWENTY-THREE: Fight

"The walls now, are more": Powell, *Exploration*, Aug. 14, p. 83.

"This is emphatically the wildest": Bradley, *UHQ*, Aug. 10, p. 62.

"a rapid that cannot": Sumner, *UHQ*, Aug. 10, p. 119.

Bradley yearned to take a chance: Bradley, *UHQ*, Aug. 14, p. 64.

"tucked around [the cliff]": Ibid.

"Rowed into the eddy": Ibid.

Somewhere in the chaos, Oramel: Bradley, *UHQ*, Aug. 15, p. 64.

"not easy to describe": Powell, *Exploration*, *UHQ*, Aug. 15, p. 85.

"to moisten a postage stamp": Sumner, *CRC*, p. 198.

"The river is very deep": Powell, *Exploration*, Aug. 15, p. 85.

"clear as crystal": Bradley, *UHQ*, Aug. 15, p. 64.

Powell named it Silver Creek: Powell, *UHQ*, Aug. 15, p. 129.

"Stretched our weary": Sumner, *UHQ*, Aug. 15, p. 120.

"so terrific it seems": Sumner, *CRC*, p. 198.

"rapids, daily duckings": Sumner, *CRC*, p. 198.

"considerable of a task": Ibid.

"They have come to think": Bradley, *UHQ*, Aug. 16, p. 65.

"rotten flour mixed": Sumner, *CRC*, p. 199.

"Our rations are still spoiling": Powell, *Exploration*, Aug. 17, p. 88.

"To add to our troubles": Sumner, *CRC*, pp. 193–4.

As if to mark their misery: This account is based on recollections by Hawkins and Sumner. Hawkins told the story in two similar but not identical versions, both reprinted in *CRC* (pp. 138–53 and pp. 154–63, especially pp. 154–60). Sumner's account can also be found in *CRC* (pp. 167–218, especially pp. 201–3). The various versions corroborate one another in places, contradict one another in other places, and leave a host of questions unresolved. But if particular details are debatable, the overall contour of the story line seems fairly certain. For more on the reliability of the Hawkins and Sumner accounts in Stanton, see the long note in the Epilogue.

"At noon one day": Hawkins, *CRC*, p. 147.

"Dunn told him a bird": Ibid.

Dunn "really should have": Sumner, *CRC*, p. 201.

no great loss: Hawkins, *CRC*, pp. 154–5.

"he would have to come and get": Hawkins, *CRC*, p. 156.

On August 16: August 16 is an educated guess, not a hard fact. On Aug. 15 Bradley wrote, "Howland had the misfortune to lose his notes and map of the river from Little Colorado down to this point." (See Bradley, *UHQ*, Aug. 15, p. 64.) Sumner described

Powell rebuking Howland. He did not specify a date but wrote, "Then, after having had a spat with Howland in the forenoon, Major Powell at the noonday camp informed Dunn that he could leave the camp immediately or pay him fifty dollars a month for rations as long as he was with the outfit." (See *CRC*, p. 201.)

"he couldn't come any damned military": Worster, *River Running West*, p. 180.

Then Powell turned to Dunn: The account in this paragraph and the next is from Hawkins, *CRC*, pp. 159–60.

"pulled out again for more": Sumner, *CRC*, p. 199.

"There was not much talk": Ibid.

"This part of the canyon": Ibid.

"We had to move on": Sumner, *CRC*, p. 199.

"We must make all haste": Powell, *Exploration*, Aug. 17, p. 88.

CHAPTER TWENTY-FOUR: *Misery*

"Although very anxious": Powell, *Exploration*, Aug. 17, pp. 88–9.

"Have been thoroughly drenched": Ibid., p. 89.

"very bad": Bradley, *UHQ*, Aug. 17, p. 65.

"some vast spout": Powell, *Exploration*, Aug. 15, p. 86.

"It is especially cold": Powell, *Exploration*, Aug. 17, p. 89.

"very numerous and very large": Bradley, *UHQ*, Aug. 18, p. 69.

"Major Powell . . . was a nuisance": Sumner, *CRC*, p. 195.

"I climb so high": Powell, *Exploration*, Aug. 18, p. 89.

"This P.M. we have had": Bradley, *UHQ*, Aug. 18, p. 66.

"Still it rains": Powell, *Exploration*, Aug. 18, p. 89.

Now they drank rainwater: Bradley, *UHQ*, Aug. 18, p. 66.

"only a labyrinth": Powell, *Exploration*, Aug. 18, p. 89.

"Still we are in our granite prison": Powell, *Exploration*, Aug. 19, p. 89.

"The waves were frightful": Bradley, *UHQ*, Aug. 18, p. 66.

"the whirlpools below caught us": Bradley, *UHQ*, Aug. 19, p. 67.

"It seems a long time": Powell, *Exploration*, Aug. 19, p. 90.

Remarkably, they lost nothing: Bradley's account and Powell's disagreed on whether the oars were lost. Bradley wrote, "Fortunately nothing was lost but a pair of oars." (See Bradley, *UHQ*, Aug. 19, p. 67.) Powell wrote, "The oars, which fortunately have floated along in company with us, are gathered up." (See Powell, *Exploration*, Aug. 19, p. 90.) I have chosen to follow Bradley, generally considered the most reliable diarist on the expedition.

"a ceaseless grind": Sumner, *CRC*, p. 195.

"everything was as smooth": Sumner, *CRC*, p. 202. Sumner's version of the narrowly

averted gunfight in camp differed in many ways from Hawkins's version (both accounts centered on Walter Powell, but Sumner made no mention of a wrestling match between Walter and Hawkins. Instead, Sumner wrote, "Walter Powell tried a bluff and was immediately called to settle, as there was a pretty little sand bar just about long enough for Colt's forty-fours." As Sumner told it, Walter refused to duel and the incident blew over). See Sumner, *CRC*, p. 203.

"We must be getting near": Bradley, *UHQ*, Aug. 20, p. 67.

"first for dashing wildness": Bradley, *UHQ*, Aug. 21, p. 67.

"a perfect hell": Sumner, *UHQ*, Aug. 21, p. 91.

"The excitement is so great": Powell, *Exploration*, Aug. 21, p. 91.

"Just here," Powell rejoiced: Ibid.

"Ten miles in less": Ibid.

"I feel more unwell": Bradley, *UHQ*, Aug. 21, p. 67.

"We wheel about": Powell, *Exploration*, Aug. 21, p. 91.

"What it means I don't know": Bradley, *UHQ*, Aug. 22, p. 68.

"Ran the granite up and down": Sumner, *UHQ*, Aug. 21, p. 121.

"a part of our flour": Powell may have dated this entry incorrectly. "We have made but little over seven miles today," he wrote, "and a part of our flour has been soaked in the river again." Bradley and Sumner each recorded Aug. 22's advance as just over eleven miles.

"Camped on the south side": Sumner, *UHQ*, Aug. 23, p. 121.

"This P.M. we got out": Bradley, *UHQ*, Aug. 23, p. 68.

"all marble": Sumner, *UHQ*, Aug. 24, p. 121.

"We cannot now be very far": Bradley, *UHQ*, Aug. 24, pp. 68–9.

more than 120 miles: Bradley, *UHQ*, Aug. 24, p. 68.

"It is curious how anxious": Powell, *Exploration*, Aug. 24, p. 94.

Before the boat could make: Bradley, *UHQ*, Aug. 25, p. 69.

"What a conflict": Powell, *Exploration*, Aug. 25, p. 95. It is worth noting that Powell's enthusiasm here was *not* an after-the-fact addition. His river diary betrays the same fascination with the lava as does his 1875 account, though the language is less polished.

Lava Falls: The rapid is formed by a debris flow, like nearly all Grand Canyon rapids, and not by the remains of a lava dam.

"Thirty five miles today": Powell, *Exploration*, Aug. 25, p. 95.

"We commenced our last": Bradley, *UHQ*, Aug. 25, p. 69.

"What a kettle": Powell, *Exploration*, Aug. 26, p. 96.

"What a supper": Ibid.

"A few days like this": Powell, *Exploration*, Aug. 26, p. 96.

"Now and then the river": Powell, *Exploration*, Aug. 27, p. 96.

"About nine o'clock we come": Ibid.

CHAPTER TWENTY-FIVE: *Separation Rapid*

"The water dashes": Bradley, *UHQ*, Aug. 27, p. 69.

"To run it," Powell concluded: Powell, *Exploration*, Aug. 27, p. 97.

Up they climbed: Powell added a long story here about finding himself trapped on the cliff, unable to advance or retreat, "suspended 400 feet above the river, into which I should fall if my footing fails." He called for help, and his men (unnamed) managed a complicated rescue. No mention of the incident appears in Powell's river diary, or in Sumner's or Bradley's. (Powell's only reference to climbing in his Aug. 27 diary was "Spent afternoon in exploration." See Powell, *UHQ*, Aug. 27, p. 131, and Powell, *Exploration*, Aug. 27, p. 97.) Some modern white-water historians insist there is no spot on the canyon walls that fits Powell's description. See Anderson, "John Wesley Powell's *Explorations* . . . Fact, Fiction, or Fantasy?" p. 378, or Marston, "Separation Marks," p. 7.

Only five days' food: This paragraph is drawn from Powell's first account of his expedition, in Bell's *New Tracks*, 1870. Marston, in "Separation Marks," takes Powell to task for writing, in 1870, that the cliffs were eight hundred feet high but then writing, in 1875 (in his account of being stranded while climbing), that they were four hundred feet tall. See Marston, p. 7.

"We appeared to be up": Sumner, *CRC*, pp. 202–3.

"I did what I could": Sumner, *CRC*, p. 203.

"[Howland] had fully made": Ibid.

"We have only subsistence": Bradley, *UHQ*, Aug. 27, p. 70.

"There is discontent in camp": Ibid.

"We have another short": Powell, *Exploration*, Aug. 27, p. 98. The description of Powell's thoughts on Aug. 27 comes from his long, detailed account on pp. 96–9.

"At one time, I almost": Powell, *Exploration*, Aug. 27, p. 99.

Powell woke Walter: Hawkins told a different version, in which he played a decisive role. In this account, Powell came to Hawkins and said, "Well, Billy, we have concluded to abandon the river for the present." Hawkins asked if Powell would sell his boat to him and Hall, so they could continue downstream. "Then the Major said, 'Well, Billy, if I have one man that will stay with me I will continue my journey or be drowned in the attempt.'" (See Hawkins, *CRC*, pp. 150–1.)

"I shall be one to try": Bradley, *UHQ*, Aug. 27, p. 70.

"At last daylight comes": Powell, *Exploration*, Aug. 28, p. 99.

"came to the determination": Bradley, *UHQ*, Aug. 28, p. 70.

"Abandoned the small boat": Sumner, *UHQ*, Aug. 28, p. 122.

Left on shore, too: Powell, *Exploration*, Aug. 28, p. 99, and Sumner, *CRC*, p. 205.

"With great labor": Bradley, *UHQ*, Aug. 28, p. 70. Powell mentions a 25- or 30-foot boulder (*Exploration*, p. 99), as does Sumner (*CRC*, p. 204).

Once there, they snubbed: Sumner, *CRC*, p. 204.

In the haste: Darrah, "The Powell Colorado River Expedition of 1869: Introduction," *UHQ*, p. 16.

"It is rather a solemn": This paragraph and the previous one are based on Powell, *Exploration*, Aug. 27, p. 100.

"Three men refused to go": Bradley, *UHQ*, Aug. 28, p. 70. Hall saw the separation in much the same light. In a letter to his brother in September 1869, he wrote, "Just before we came out of the canyon three of the men left us on the head of rapids. They were afraid to run it so they left us in a bad place." (See *UHQ*, v. 16–17, 1949, p. 507.)

The Kitty Clyde's Sister, *with Powell:* "My old boat left, I go on board of the *Maid of the Cañon*," Powell wrote, but Andy Hall wrote that Powell joined him and Hawkins in the *Kitty Clyde's Sister*. (See Powell, *Exploration*, Aug. 28, p. 100, and Hall, *UHQ*, v. 16–17, 1949, p. 507.) Hall's account is the better bet: Powell wrote several years after the fact, and on other occasions he had made mistakes in matching boats and crews. Hall's account was written at the time and was fairly detailed.

"Dashed out into the boiling": Bradley, *UHQ*, Aug. 28, p. 70.

"too large to do anything": Ibid.

Somehow—it happened so fast: This account is based on Bradley, *UHQ*, Aug. 28, p. 70, and Powell, *Exploration*, Aug. 28, p. 100.

not even an oar: Bradley, *UHQ*, Aug. 28, p. 70.

It had taken perhaps: Powell, *Exploration*, Aug. 28, p. 100. Powell wrote that "although it looked bad from above, we have passed many places that were worse." In 1870, in Bell's *New Tracks*, he wrote that "we really found it less dangerous than a hundred we had run above." Bradley and Sumner emphatically disagreed. "We had never such a rapid before," Bradley wrote. (See *UHQ*, Aug. 28, p. 70.) Sumner wrote later that "if I remember rightly, Major Powell states it was not as bad as it looked, and that we had run worse. I flatly dispute that statement." (See Sumner, *CRC*, p. 205.)

"The men that were left": Powell, in Bell, *New Tracks*. Reprinted in *UHQ*, p. 25. Powell wrote the account reprinted in Bell's book in 1870. By 1875, he had made a curious change in his description, writing, "We are behind a curve in the cañon, and cannot see up to where we left them, and so we wait until their coming seems hopeless, and push on." (See Powell, *Exploration*, p. 101.) This description of "a curve in the canyon" that prevented the two groups from seeing one another is

incorrect; the Colorado runs straight for two miles beyond Separation Canyon.

Earlier in the trip: Author interview, Michael Ghiglieri, February 26, 2001.

"The last thing we saw": Sumner, *CRC*, p. 205.

"Ran 10 more rapids"; Sumner, *UHQ*, Aug. 28, p. 122.

Bradley stayed in his boat: This account is based on Bradley, *UHQ*, Aug. 28, p. 71.

Now two frantic scenes: Bradley's account here is confusing. He wrote that "the water roared so furiously that I could not make them hear and they could not see me," but a few sentences later he wrote that "after what seemed like half an hour and just as they were uniting the two ropes, the boat gave a furious shoot out into the stream." (See Bradley, *UHQ*, Aug. 28, p. 71.)

"On I went": Bradley, *UHQ*, Aug. 28, p. 71. As remarked in the notes to Chapter Thirteen, this passage figures in the controversy about whether the boats had a sweep oar at the stern. Bradley's description of "putting an oar first on one side then on the other" would seem to speak *against* the existence of sweep oars, but Powell's 1875 account described "Bradley seiz[ing] the great scull oar." (See Powell, *Exploration*, Aug. 28, p. 100.) In other instances where Bradley and Powell disagree on matters of fact, most historians have gone with Bradley.

Much of the confusion stems from Powell's flimflammery with dates. Powell mounted a second expedition through the Grand Canyon in 1871, this time with photographers and artists, and the boats on this second expedition definitely did have steering oars. The problem is that Powell lifted several incidents from the 1871 trip and inserted them into his 1869 narrative. Similarly, the text of his 1869 adventure was illustrated with drawings of his 1871 boats.

Making matters worse, much of the eyewitness testimony was recorded decades after the fact and is hard to evaluate. In 1907, for example, Robert B. Stanton tracked down Billy Hawkins and convinced him to write a short account of the 1869 expedition. At one point, declared Hawkins, "the high waves . . . were over fifteen feet in height, but Hall had the boat under such headway that I could manage it with my steering oar." (See Hawkins, *CRC*, p. 152.) But then what do we make of a second account of the 1869 trip that Hawkins wrote in 1919, shortly before his death and fifty years after the events it describes? On the first day on the river, Hawkins wrote, "I was steering the boat with one oar behind." (See Bass, *Adventures*, p. 20.) This seems unequivocal, but, as Brad Dimock noted, Powell's boats did *not* have steering oars on the first day of the trip. Michael Ghiglieri suggests that on the trip's first day, when the river was swift but there were no rapids to speak of, Hawkins may have left the rowing to Hall and taken an oar to the stern to use as a makeshift rudder. (Author interview, Mar. 19, 2001.) And, steering oars aside, Hawkins's account contained glaring mistakes of chronology and geography. (He put

Island Park upstream of Disaster Falls, for instance, but the expedition reached Disaster Falls two weeks *before* Island Park.) In the end, we are left dangling.

Some skeptics have taken a different approach. Drifter Smith notes that the 1869 expedition became a race for survival and wonders if the men would have gambled their precious time on carving steering oars. Smith notes the problem of raw materials as well. The men had trouble finding wood that would do for an ordinary oar. A piece of wood suitable for a sweep oar would have been longer and presumably even harder to find.

"The boat is fairly turned": Powell, *Exploration*, Aug. 28, pp. 101–2.

"But he is in a whirlpool": Powell, *Exploration*, Aug. 28, p. 102.

"A wave rolls over us": Ibid.

"Major says nothing ever gave": Bradley, *UHQ*, Aug. 28, p. 71.

"A No. 1 of the trip": Ibid.

"Boys left us": Powell, *UHQ*, Aug. 28, p. 131. The diary entry, as printed in the *UHQ*, differs in a few minor ways from the handwritten original. Darrah, the editor, added periods and capital letters, changed the phrase "camp on left bank" to "Make camp on left bank," and silently corrected the spelling of Bradley's name.

CHAPTER TWENTY-SIX: *Deliverance*

"the first Sunday": Bradley, *UHQ*, Aug. 29, p. 71.

Everyone was exhausted: Ibid., p. 72.

"for we have no time": Ibid.

"At twelve o'clock": Powell, *Exploration*, Aug. 29, p. 102.

"We came out": Sumner, *UHQ*, Aug. 29, p. 122.

"It was a strange and delightful": Sumner, *CRC*, p. 206.

Rowing easily now: The rest of the paragraph is drawn from Sumner, *CRC*, p. 206.

"The relief from danger": Powell, *Exploration*, Aug. 29, pp. 102–3.

"Ever before us has been": Powell, *Exploration*, Aug. 29, p. 103.

"The river rolls by": Ibid.

"nothing but smooth water": Sumner, *UHQ*, Aug. 30, p. 122.

By noon, they had covered: Bradley, *UHQ*, Aug. 29 and 30, p. 72.

Sumner pulled out a spyglass: Sumner, *CRC*, p. 207.

Three men and a boy: Bradley, *UHQ*, Aug. 30, p. 72.

"for any fragments": Powell, *Exploration*, Aug. 30, p. 104.

"We could hardly credit": Bradley, *UHQ*, Aug. 30, p. 72.

"we laid our dignified": Sumner, *UHQ*, Aug. 30, p. 123.

"We talked and ate": Sumner, *UHQ*, Aug. 31, p. 123.

"Rapids ran 414": Ibid.

"I find myself penniless": Ibid., p. 124.

The final tally: Powell, *Exploration*, Aug. 30, p. 104.

CHAPTER TWENTY-SEVEN: *The Vanishing*

"when Proff Powell left us": Sumner's critique of Powell's leadership appeared in the *Denver Post* in October 1902, and was reprinted in Stegner, "Jack Sumner and John Wesley Powell," pp. 62–6.

In his old age: "[Powell] gave me $60 and Hall $60, and said that he would send us a government voucher for the rest," Hawkins grumbled in 1907. (For a detailed account of the money Hawkins felt he was owed, see Hawkins, *CRC*, pp. 142–4.)

At St. George, Powell asked: Darrah, *Powell*, p. 144n.

"Three of the Powell Expedition Killed": Powell journal, *UHQ*, p. 141.

by some accounts, raped: Sumner, *CRC*, p. 209.

"I have known O. G. Howland personally": Darrah, *Powell*, p. 145.

At nine the same evening: Powell, *UHQ*, p. 145.

"Major Powell, of the Powell Expedition": Ibid.

"with a loyalty": Brooks, "Jacob Hamblin," p. 316.

"I tell the Indians": Powell, *Exploration*, p. 129.

"We will be friends": Powell, *Exploration*, pp. 129–30.

Here was confirmation, then: Hamblin heard a slightly different version of the tale from one of the Indians after the official meeting broke up. In this account, the Shivwits had given the three men food and sent them on their way. Soon after, "an Indian from the east side of the Colorado" appeared, carrying news of a group of miners who had killed an Indian woman in a drunken brawl. Furious, the Shivwits had set out after the Howlands and Dunn and killed them in the mistaken belief that they were the "miners." (See Powell, *Exploration*, pp. 130–1.)

"That night I slept": Powell, *Exploration*, p. 131.

He never sought to punish: The killings are sometimes attributed to one To-ab, based on a hearsay account written decades later by Anthony Ivins, a prominent Mormon and onetime mayor of St. George, Utah. To read Ivins's account is to realize that the case against To-ab is virtually nonexistent. (See Ivins, "A Mystery of the Grand Canyon Solved." Ivins's essay can also be found in *Pioneer Stories*, edited by Preston Nibley.)

An armed posse arrived: Author interview with Scott Thybony, Feb. 27, 2001.

"the red-bellies would surely": Sumner, *CRC*, pp. 205–6.

"From one with a listless": Sumner, *CRC*, pp. 208–9.

"I am positive I saw": Sumner, *CRC*, p. 209.

Then Wesley Larsen: This account is based on Larsen, "The 'Letter,' or Were the Powell Men Really Killed by Indians?" pp. 12–19; and on Ghiglieri and Myers, *Over the Edge: Death in Grand Canyon*; on two unpublished essays on Larsen that Scott Thybony kindly shared with me; and on interviews with Ghiglieri and Thybony.

blood would rise "to their knees": Scott Thybony, "Hurricane Cliffs," unpublished manuscript. Thybony is a writer and historian with a wide knowledge of the Southwest in general and the Grand Canyon in particular (his books include *Official Guide to Hiking the Grand Canyon* and *Burntwater*). For a research project of his own, he investigated Larsen's story and the mystery of the Dunn-Howland killings.

Three theories, then: For still another theory, see Dobyns and Euler, "The Dunn-Howland Killings: Additional Insights," pp. 87–95. The authors contend that Dunn and the Howlands were killed by Indians. According to Dobyns and Euler, the motive for the killings was neither revenge nor robbery but anti-white hostility stemming from the still unresolved "Walapai War."

They should have made it: Author interview with Scott Thybony, Feb. 27, 2001.

Scratched into the rock: Belshaw, "The Dunn-Howland Killings," p. 416.

If the inscription is a hoax: Harvey Butchart, author of *Grand Canyon Treks* and the great authority on hiking the Grand Canyon, doubts that the inscriptions are authentic, on the grounds that "the three men would be trying to get to St. George without any touristy detours. The view from the top wouldn't help them find waterholes or the best route." Private communication.

Ten years ago: Author interview with Scott Thybony, Feb. 27, 2001.

The bodies of the Howland: In 1907, Hawkins wrote, "Some years afterwards I, with a party of some others, buried their bones in the Shewits Mountains, below Kanab Wash." (See Hawkins, *CRC*, p. 162.) Hawkins never added any details beyond those in that single sentence, and no one has been able to confirm his claim.

EPILOGUE:

One white-water historian: Webb, *If We Had a Boat*, p. 84.

His death went unnoticed: Personal communication, Margaret Motes, Historical Society of Old Newbury, Cushing House Museum, Newburyport, Mass.

The irrepressible Andy Hall: Hall's story is drawn from Woody and Schwartz, *Globe, Arizona*, pp. 65–77.

"May 24th 1869," he wrote: Stegner, "Jack Sumner and John Wesley Powell," p. 64.

both men kept: Their critiques of Powell appear in Robert Stanton's *Colorado River Controversies*, a book with a curious history. Stanton was the engineer who had hoped to build a railroad through the canyon and who saw three of his companions drown in the Colorado's rapids. He became the second man to lead an expedition through the Grand Canyon. In 1891, he ran into a grizzled prospector who asked if he could spare any tobacco. The man turned out to be Jack Sumner. Stanton and Sumner began to chat, Stanton referred to Powell's famous *Exploration of the Colorado*, and Sumner grumbled that "there's lots in that book besides the truth."

Stanton had read Powell's *Exploration* "as I would the Gospel of St. John, with an almost worshipful reverence" (*CRC*, p. 107), but later he grew disillusioned with his hero's attempt "to glorify his work and belittle mine." (*CRC*, p. 110). Spurred by Sumner's remark, Stanton embarked on a quest to correct the record. Or, perhaps, to puncture Powell's reputation. Whatever his motivation, Stanton set out after Powell like Ahab after Moby Dick. Until his death in 1922, Stanton worked tirelessly to gather information on anything and everything to do with the Colorado, and especially with Powell. (It was Stanton who unearthed Powell's river diary.) At his death, he left a huge, unwieldy manuscript that was eventually edited and published as *Colorado River Controversies*.

Stanton documented, in parallel columns, instances where Powell's river diary and his 1875 journal differed. He pinned down multiple cases of events that took place on Powell's second expedition that Powell transplanted to the first. He proved in agonizing detail that Powell had indeed worn a life jacket but never mentioned it.

And he asked Hawkins and Sumner, by this time old men, endless questions about their 1869 expedition. In 1907, in independent accounts written for Stanton, the two men relived that landmark journey. In 1919, only months before his death, Hawkins produced another short recollection of the 1869 expedition, which made its way into print in William W. Bass's *Adventures in the Canyons of the Colorado*. (Stanton reprinted this account, minus its first five pages, in his own book.) In these late accounts, details ignored or glossed over in the river diaries took center stage. Where Bradley had remarked merely that the men were "uneasy and discontented and anxious to move on," for example, Sumner and Hawkins told tales of mutiny and screaming arguments and a near gunfight.

Powell partisans brush these stories aside as the ramblings of bitter old men, unable to accept Powell's fame and their own anonymity. Undoubtedly, the sour-grapes theory has some merit. More than that, many of the stories Hawkins and Sumner told were plainly exaggerated or distorted by the passage of many decades. Hawkins, in particular, indisputably confused some dates and misplaced some incidents. If he were a witness in a lawsuit, a cagey lawyer could tie him in knots. But if one's purpose is not to win an argument but to try to unravel a tangled tale, it might be a mistake to discount Hawkins or Sumner.

Hawkins and Sumner wrote independently, and in a surprising number of instances their stories agree. Both described Powell's fury at Dunn for soaking his watch, for instance, although the two differed on the details. Hawkins recalled that Powell had first insisted that Dunn either pay for the watch or leave the party, and then relented and said that Dunn could pay $30 for the watch when the trip was over. (See *CRC*, pp. 156–7.) In Sumner's version, Powell demanded that Dunn leave camp immediately or pay $50 a month for food as long as he stayed with the expedition, and then Sumner intervened on Dunn's behalf. (See *CRC*, pp. 202–3.) Often the same details feature in both men's recollections, and more often than that, the same broad themes come through. In particular, both highlight the bitter feuding that grew to a crescendo at Separation Rapid. (See Stegner's "Jack Sumner and John Wesley Powell" for a fervent defense of Powell.)

One great cause: See Sumner's letter to the *Denver Post* in October 1902, in reply to an article reporting Powell's death. Sumner's letter is reprinted (and discussed) in Stegner, "Jack Sumner and John Wesley Powell."

"I can say one thing truthfully": Hawkins, *CRC*, p. 146.

"Without dwelling at length": Sumner's obituary appeared in the *Rocky Mountain News*, July 10, 1907.

One sister or another: Worster, *River Running West*, p. 339.

"At one time, perhaps for two": Worster, *River Running West*, p. 609n.

The Colorado is "the world's": Kieffer, "Hydraulics and Geomorphology," p. 333, in Beus and Morales, eds., *Grand Canyon Geology*.

"direct from scenes": *Chicago Republican*, Sept. 21, 1869.

The crew was new: The new boats were slightly modified versions of the 1869 ones. The original boats had two bulkhead sections, one fore and one aft. The new boats had a third covered section in the middle, an improvement that made it less likely the boats would fill with water. (Powell lashed a chair to the middle deck so that he could scan downriver for trouble ahead.) The new boats also had a long sweep oar at the stern, a great advantage for maneuverability.

Instead of feisty: Since the first expedition, Powell had sharpened his sense of how to appeal to the public. This time the crew members included a professional photographer and a talented amateur artist. The best-known early photographs of the Grand Canyon date from the second expedition. (The first-ever photographs of the Grand Canyon were taken by Timothy O'Sullivan, with the Wheeler expedition in 1871.) See Pyne's *How the Canyon Became Grand* for a sophisticated history of attempts to depict the Grand Canyon in words and pictures.

The second expedition endured: As remarked in the footnote in the text, many of the men on the second expedition felt that Powell had robbed them of the credit they deserved. Even Frederick Dellenbaugh, who became the youngest member of Powell's crew, at seventeen, and was smitten with a case of hero worship that lasted throughout his long life, ventured a criticism of Powell on this point. "It has always seemed to me," he wrote, "that the men of the second party, who made the same journey, who mapped and explored the river and much of the country roundabout, doing a large amount of difficult wok in the scientific line, should have been accorded some recognition." (See Dellenbaugh, *Romance*, preface, p. vi. For a defense of Powell on this point, see Worster, *River Running West*, pp. 256–7.)

"Everybody felt like praising God": Fowler, ed., "Jack Hiller's Diary," p. 142.

Even if every river: See Powell's *Report on the Lands of the Arid Region*. Powell also made his case in several short, vivid essays. See "The Irrigable Lands of the Arid Region," "The Non-Irrigable Lands of the Arid Region,"and "Institutions for the Arid Lands."

"In readiness to receive": Gilpin, *Mission of the North American People*, pp. 50–1.

"I was lucky": *New York Herald Tribune*, Aug. 18, 1889.

"Many years have passed": John Wesley Powell, *The Exploration of the Colorado River and its Canyons*, preface.

The Washington Post *ranked*: Sept. 25, 1902.

The New York Times *made do*: Sept. 24, 1902.

"a moral giant": McGee, "In Memory of John Wesley Powell," pp. 788–90.

"Major Powell, throughout his life": Meadows, *John Wesley Powell*, p. 17.

BIBLIOGRAPHY

Anyone writing about the Grand Canyon comes to rely on Earle Spamer's superlative *Bibliography of the Grand Canyon and the Lower Colorado River, 1540–1980.* It is available in a free, updated, searchable version on the Internet. See http://www.grandcanyon.org/biblio/bibliography.htm.

Abbey, Edward. *Desert Solitaire.* New York: Ballantine, 1968.

Adams, George Worthington. *Doctors in Blue: The Medical History of the Union Army in the Civil War.* Dayton, Ohio: Morningside, 1985.

Allen, Stacy D. "Shiloh: The Campaign and First Day's Battle." *Blue and Gray Magazine,* Feb. 1997.

———. "Shiloh: The Second Day's Battle and Aftermath." *Blue and Gray Magazine,* April 1997.

Ambrose, Stephen. *Nothing Like It in the World: The Men Who Built the Transcontinental Railroad.* New York: Simon & Schuster, 2000.

Anderson, Martin J. "Artist in the Wilderness: Frederick Dellenbaugh's Grand Canyon Adventure." *Journal of Arizona History,* v. 28 (Spring 1987), pp. 47–68.

———. "First Through the Canyon: Powell's Lucky Voyage in 1869." *Journal of Arizona History*, v. 20 (Winter 1979), pp. 391–408.

———. "John Wesley Powell's "Exploration of the Colorado River: Fact, Fiction, or Fantasy?" *Journal of Arizona History*, v. 24 (Winter 1983), pp. 363–80.

Ashley, William. "Ashley's 1825 Diary." William H. Ashley Papers, Missouri Historical Society, St. Louis.

Aton, James M. "Inventing John Wesley Powell: The Major, His Admirers and Cash-register Dams in the Colorado River Basin." Distinguished Faculty Lecture No. 9, Southern Utah State College, Dec. 1, 1988.

———. *John Wesley Powell*. Boise State University Western Writers Series No. 114, 1994.

Babbitt, Bruce, ed. *Grand Canyon: An Anthology*. Flagstaff, Ariz.: Northland Press, 1978.

Bain, David Howard. *Empire Express: Building the First Transcontinental Railroad*. New York: Viking, 1999.

Baker, Marcus. "Major J. W. Powell: Personal Reminiscences of One of His Staff." *Open Court*, v. 17 (1903), pp. 348–51.

Barber, Lynn. *The Heyday of Natural History*. Garden City, N.Y.: Doubleday, 1980.

Barnard, Frederick A. P. "Inaugural Discourse." *Proceedings at the Inauguration of Frederick A. P. Barnard as President of Columbia College*. New York: Hurd and Houghton, 1865.

Barry, Patricia. *Surgeons at Georgetown: Surgical and Medical Education in the Nation's Capital, 1849–1969*. Franklin, Tenn: Hillsborough, 2001.

Bass, William Wallace. *Adventures in the Canyons of the Colorado*. Grand Canyon, Ariz.: 1920.

Beer, Bill. *We Swam the Grand Canyon: The True Story of a Cheap Vacation That Got a Little Out of Hand*. Seattle: Mountaineers, 1988.

Belknap, Buzz. *Belknap's Waterproof Grand Canyon River Guide*. Evergreen, Col.: Westwater Books, 1989.

Belshaw, Michael. "The Dunn-Howland Killings: A Reconstruction." *Journal of Arizona History*, v. 20 (Winter 1979), pp. 409–22.

Beus, Stanley S., and Michael Morales, eds. *Grand Canyon Geology*. New York: Oxford University Press, 1990.

Blaustein, John. *The Hidden Canyon*. San Francisco: Chronicle, 1999.

Bonner, Thomas D., ed. *The Life and Adventures of James P. Beckwourth*. New York: Alfred A. Knopf, 1931.

Boorstin, Daniel J. *The Americans: The National Experience*. New York: Random House, 1965.

Boslough, John. "Rationing a River." *Science 81* (June 1981), pp. 26–37.

Bowles, Samuel. *Across the Continent*. New York: Hurd & Houghton, 1865.

———. *The Switzerland of America*. New York: American News, 1869.

Bradley, George Young. "George Y. Bradley's Journal." Utah Historical Quarterly, v. 15 (1947), pp. 31–72.

Brewer, William H. "John Wesley Powell." American Journal of Science, v. 14 (1902), pp. 377–82.

Brodie, Fawn M. No Man Knows My History: The Life of Joseph Smith. New York: Knopf, 1945.

Brooks, Juanita. "Jacob Hamblin: Apostle to the Lamanites." Pacific Spectator, v. 2 (Summer 1948), pp. 315–30.

———. John Doyle Lee: Zealot—Pioneer Builder—Scapegoat. Glendale, Cal.: Clark, 1962.

———. The Mountain Meadows Massacre. Norman, Okla.: University of Oklahoma Press, 1962. (Originally published 1950.)

Brower, David. Let the Mountains Talk, Let the Rivers Run. New York: HarperCollins, 1995.

Bruce, Robert V. The Launching of Modern American Science 1846–1876. Ithaca, N.Y.: Cornell University Press, 1987.

Bulger, Harold. "First Man Through the Grand Canyon." Bulletin of the Missouri Historical Society (July 1961), pp. 321–31.

Butchart, Harvey. Grand Canyon Treks. Bishop, Cal.: Spotted Dog Press, 1997.

Cadell, H. M. "The Colorado River of the West." The Scottish Geographical Magazine. v. 3 (1887), pp. 441–60.

Calvin, William H. The River That Flows Uphill: A Journey from the Big Bang to the Big Brain. New York: Macmillan, 1986.

Carothers, Steven W., and Robert Dolan. "Dam Changes on the Colorado River." Natural History, v. 91, no. 1 (1982), pp. 75–83.

Castel, Albert, ed. "The War Album of Henry Dwight" (Part III: Shiloh), Civil War Times Illustrated (May 1980), pp. 32–6.

Catton, Bruce. The Civil War. New York: Doubleday, 1960.

———. Reflections on the Civil War. New York: Doubleday, 1981.

Childs, Craig. The Secret Knowledge of Water: Discovering the Essence of the American Desert. Seattle: Sasquatch Books, 2000.

Cohig, Ruth Cowdery. History of Grand County, Colorado (M.A. thesis, University of Denver, 1922).

Collier, Michael. An Introduction to Grand Canyon Geology. Grand Canyon, Ariz: Grand Canyon Natural History Association, 1980.

Crumbo, Kim. A River Runner's Guide to the History of the Grand Canyon. Boulder, Col.: Johnson Books, 1994.

Dale, Harrison Clifford, ed. The Ashley-Smith Explorations and the Discovery of a Central Route to the Pacific, 1822–1829, with the Original Journals. Cleveland: Arthur Clark, 1918.

Dana, James D. "Presidential Address." *Proceedings of the American Association for the Advancement of Science.* v. 9 (1854), pp. 1–36.

Daniels, George H. *Science in American Society: A Social History.* New York: Knopf, 1971.

Dark, Larry, ed. *Prize Stories 1999: The O. Henry Awards.* New York: Random House, 1999. (Anthology includes Pam Houston, "Cataract.")

Darrah, William C. "John Wesley Powell and an Understanding of the West." *Utah Historical Quarterly,* v. 37 (1969), pp. 146–51.

Darrah, William C., ed. (Darrah not only edited the 1869 expedition journals but also wrote short essays on each man's life. See "Major John Wesley Powell 1834-1902," pp. 19–21; "George Young Bradley 1836–1885," pp. 29–30; "Walter Henry Powell 1842–1915," pp. 89-90; "The Howland Brothers and William Dunn," pp. 93-5; "Hawkins, Hall, and Goodman," pp. 106-8; "John C. Sumner 1840–1907," pp. 109–12.) *UHQ,* v. 15 (1947).

———. "The Powell Colorado River Expedition of 1869." *Utah Historical Quarterly,* v. 15, pp. 9–18.

———. *Powell of the Colorado.* Princeton: Princeton University Press, 1951.

Davis, W. M. "Biographical Memoir of John Wesley Powell." *National Academy of Sciences,* v. 9 (Feb. 1915), pp. 11–83.

———. "The Lessons of the Colorado Canyon." *Bulletin of the American Geographical Society,* v. XLI, no. 6 (1909), pp. 345–54.

Dellenbaugh, Frederick S., *A Canyon Voyage: The Narrative of the Second Powell Expedition.* Tucson: University of Arizona Press, 1996. (Originally published in 1908.)

———. *The Romance of the Colorado River.* Chicago: Rio Grande Press, 1902.

———. *Across the Wide Missouri.* Boston: Houghton Mifflin, 1947.

DeVoto, Bernard. "Geopolitics with the Dew on It." *Harper's,* March 1944, pp. 313–23.

———. *The Course of Empire.* Boston: Houghton Mifflin, 1952.

Dimock, Brad. "Buzz Holmstrom and the Evolution of Rowing Style." *The Confluence,* v. 5, Issue 2 (Summer 1998), pp. 21–3.

———. "What's It Like to Row Those Old Boats?" *The Waiting List,* v. 3, no. 4 (Dec. 1999), p. 31.

Dimock, Brad, Vince Welch, and Cort Conley. *The Doing of the Thing: The Brief Brilliant Whitewater Career of Buzz Holmstrom.* Flagstaff: Fretwater Press, 1998.

Dobyns, Henry F., and Robert C. Euler. "The Dunn-Howland Killings: Additional Insights." *Journal of Arizona History,* v. 20 (1980), pp. 87–95.

Dolan, Robert, Alan Howard, and Arthur Gallenson. "Man's Impact on the Colorado River in the Grand Canyon." *American Scientist,* v. 62 (1974), pp. 393–401.

Duffield, John T. *The Anthropology of Evolutionism and the Bible.* Princeton: Press Printing Establishment, 1878.

Egan, Timothy. *Lasso the Wind: Away to the New West.* New York: Alfred A. Knopf, 1998.

Eiseley, Loren. *The Firmament of Time.* New York: Atheneum, 1971.

Emmons, David M. *Garden in the Grasslands: Boomer Literature of the Central Great Plains.* Lincoln: University of Nebraska Press, 1971.

Euler, Robert C., and Frank Tikalsky, eds., *The Grand Canyon: Intimate Views.* Tucson: University of Arizona Press, 1992.

Fowler, Don D. *"Photographed All the Best Scenery": Jack Hillers' Diary of the Powell Expeditions, 1871–1875.* Salt Lake City: University of Utah Press, 1972.

———. *The Romance of the Colorado River.* Chicago: Rio Grande Press, 1902.

———. *The Western Photographs of John K. Hillers: Myself in the Water.* Washington, D.C.: Smithsonian Institution Press, 1989.

Fradkin, Philip L. *A River No More: The Colorado River and the West.* New York: Alfred A. Knopf, 1981.

Frazier, Ian. *Great Plains.* New York: Farrar, Straus and Giroux, 1989.

Gabriel, Ralph Henry. *The Course of American Democratic Thought: An Intellectual History Since 1815.* New York: Ronald, 1940.

Ghiglieri, Michael P. *Canyon.* Tucson: University of Arizona Press, 1992.

Ghiglieri, Michael P., and Thomas M. Myers. *Over the Edge: Death in Grand Canyon.* Flagstaff: Puma, 2001.

Gilbert, G. K. "John Wesley Powell." *Science,* Oct. 10, 1902, pp. 561–7.

———. "John Wesley Powell: V. The Investigator." *Open Court,* v. 17 (1903), pp. 228–39, 281–90.

Gilpin, William. *Mission of the North American People.* Philadelphia: Lippincott, 1873.

Glaab, Charles N. "Visions of Metropolis: William Gilpin and Theories of City Growth in the American West." *Wisconsin Magazine of History,* v. 35 (Autumn, 1961), pp. 21–31.

Goetzmann, William H. *Exploration and Empire: The Explorer and the Scientist in the Winning of the American West.* New York: Vintage, 1966.

———. *New Lands, New Men: America and the Second Great Age of Discovery.* New York: Viking, 1986.

Gould, Stephen Jay. *Time's Arrow, Time's Cycle: Myth and Metaphor in the Discovery of Geological Time.* Cambridge, Mass.: Harvard University Press, 1987.

Graf, William L. "Rapids in Canyon Rivers." *Journal of Geology,* v. 87 (1979), pp. 533–51.

Griffith, Paddy. "The Myth of the Rifle Revolution in the Civil War." *North & South,* v. 1, no. 5 (1998).

Hall, Andrew. "Three Letters by Andrew Hall." *Utah Historical Quarterly,* v. 15, (1947) pp. 505–8.

Hamblin, W. Kenneth. *Earth's Dynamic Systems*. New York: Macmillan, 1992.

Harrington, John W. *Dance of the Continents: Adventures with Rocks and Time*. Boston: Houghton Mifflin, 1983.

Hess, Karl Jr. "John Wesley Powell and the Unmaking of the West." *Environmental History*, v. 2, no. 1 (Jan. 1997), pp. 7–28.

Hitchcock, Edward, *Religion of Geology and its Connected Sciences*. Boston: Phillips, 1854.

Howland, O. G. "Letters of O. G. Howland to the *Rocky Mountain News*." *Utah Historical Quarterly*, v. 15 (1947), pp. 95–105.

Hume, James D. "An Understanding of Geologic Time." *Journal of Geological Education*, v. 26, 1978, pp. 141–3.

Hunt, Charles B. "Geologic History of the Grand Canyon." In *The Colorado River Region and John Wesley Powell*. Geological Survey Professional Paper 669. Washington, D.C.: 1969.

———. "Grand Canyon and the Colorado River, Their Geologic History." In R. S. Babcock et al., *Geology of the Grand Canyon*. Flagstaff: Museum of Northern Arizona, 1974.

Huxley, Thomas. *On Our Knowledge of the Causes of the Phenomena of Organic Nature*. London: R. Hardwicke, 1863.

Ivins, Anthony. "A Mystery of the Grand Canyon Solved." *Improvement Era*, v. 27, no. 11 (Sept. 1924), pp. 1017–25.

Jaffe, Mark. *The Gilded Dinosaur: The Fossil War Between E. D. Cope and O. C. Marsh and the Rise of American Science*. New York: Crown, 2000.

James, Preston E. *All Possible Worlds: A History of Geographical Ideas*. New York: Bobbs-Merrill, 1972.

Jones, Steve. *Darwin's Ghost*. New York: Random House, 1999.

Keegan, John. *The Mask of Command*. New York: Viking, 1987.

Kieffer, Susan Werner. "Hydraulics and Geomorphology of the Colorado River in the Grand Canyon." (In Beus and Morales, *Grand Canyon Geology*.)

———. "The 1983 Hydraulic Jump in Crystal Rapid: Implications for River-Running and Geomorphic Evolution in the Grand Canyon." *Journal of Geology*, v. 93, no. 4 (July 1985), pp. 385–406.

Kolb, Ellsworth L. *Through the Grand Canyon from Wyoming to Mexico*. New York: Macmillan, 1914.

Langford, Nathaniel Pitt. *The Discovery of Yellowstone Park*. Lincoln: University of Nebraska Press, 1972. (Originally published in 1905.)

Larsen, Wesley P. "The 'Letter': Were the Powell Men Really Killed by Indians?" *Canyon Legacy*, no. 17 (1993), pp. 12–19.

Lavender, David. *River Runners of the Grand Canyon*. Grand Canyon, Ariz.: Grand Canyon Natural History Association, 1985.

Leopold, Luna B. "The Rapids and the Pools—Grand Canyon." In *The Colorado River Region and John Wesley Powell*. Geological Survey Professional Paper 669. Washington, D.C.: GPO, 1969.

Leopold, Luna B., and W. B. Langbein. "River Meanders." *Scientific American* (June) 1966, pp. 60–70.

Limerick, Patricia Nelson. *Desert Passages: Encounters with the American Deserts*. Albuquerque: University of New Mexico, 1985.

———. *The Legacy of Conquest: The Unbroken Past of the American West*. New York: Norton, 1987.

Lincoln, Mrs. M. D. "John Wesley Powell," Part 1: "Boyhood and Youth." *Open Court*, v. 16 (1902), pp. 702–15. Part 2: "The Soldier." Ibid, v. 17 (1903), pp. 14–25. Part 3: "The Professor." Ibid., pp. 86–94.

Litton, Martin. "The Dory Idea." *Oar and Paddle* (May–June, 1974), pp. 14–18.

Loewenberg, Bert James. "Darwinism Comes to America, 1859-1900." *Mississippi Valley Historical Review*, v. 28 (Dec. 1941), pp. 339-68.

Logsdon, David R., ed. *Eyewitnesses at the Battle of Shiloh*. Nashville: Kettle Mills, 1994.

Long, John. *The Liquid Locomotive: Legendary Whitewater River Stories*. Helena, Mont: Falcon, 1999.

Lucchitta, Ivo. *Canyon Maker: A Geological History of the Colorado River*. Flagstaff: Museum of Northern Arizona, 1988.

Malin, James C. *The Grassland of North America: Prolegomena to its History*. Lawrence, Kan., 1947.

Manning, Richard. *Grassland: The History, Biology, Politics, and Promise of the American Prairie*. New York: Viking, 1995.

Marston, O. Dock. "For Water-Level Rails Along the Colorado River." *Colorado Magazine*, v. 46, no. 4 (Fall 1969), pp. 287–303.

Marston, O. Dock, ed. "Interview with Dock Marston by Jay M. Haymond and John F. Hoffman." Utah State Historical Society Oral History Program, May 28, 1976. (See also P. T. Reilly's comments on Marston's interview.)

———. "The Lost Journal of John Colton Sumner." *Utah Historical Quarterly*, v. 37 (1969), pp. 173–189.

———. "River Runners: Fast Water Navigation." *Utah Historical Quarterly*, v. 28, no. 3 (July 1960), pp. 291–308.

———. "Separation Marks: Notes on 'the worst rapid' in the Grand Canyon." *Journal of Arizona History* (Spring 1976), pp. 1–20.

———. "Was Powell First?" *Patagonia Roadrunner*, May 1969. (This is a transcript of a talk to the Tenth Annual Arizona Historical Convention, delivered May 3, 1969.)

————. "Who Named the Grand Canyon?" *Pacific Historian*, v. 12, no. 3 (Summer 1968), pp. 4–8.

————. "With Powell on the Colorado." *Brand Book II* (The Westerners, San Diego Corral), 1971, pp. 65–76.

McCairen, Patricia C. *Canyon Solitude: A Woman's Solo River Journey Through Grand Canyon*. Seattle: Seal Press, 1998.

McCully, Patrick. *Silenced Rivers: The Ecology and Politics of Large Dams*. London: Zed Books, 1996.

McDonough, James Lee. *Shiloh—in Hell Before Night*. Knoxville: University of Tennessee Press, 1977.

McGee, W. J. et al. "In Memory of John Wesley Powell." *Science*, v. 16 (1902), pp. 782–90.

McKee, Edwin D. "Stratified Rocks of the Grand Canyon." In *The Colorado River Region and John Wesley Powell*. Geological Survey Professional Paper 669. Washington, D.C.: GPO, 1969.

McPhee, John. *Annals of the Former World*. New York: Farrar, Straus and Giroux, 1998.

————. *The Control of Nature*. New York: Farrar, Straus and Giroux, 1989.

McPherson, James M. *Battle Cry of Freedom*. New York: Oxford University Press, 1988.

————. *Drawn with the Sword: Reflections on the American Civil War*. New York: Oxford University Press, 1996.

————. *For Cause and Comrades: Why Men Fought in the Civil War*. New York: Oxford University Press, 1997.

McWhiney, Grady, and Perry D. Jamieson. *Attack and Die: Civil War Military Tactics and the Southern Heritage*. Tuscaloosa: University of Alabama Press, 1982.

————. "No Myth! The Rifle Revolution." *North & South*, v. 1, no. 5. (1998), pp. 22–30.

Meadows, Paul. *John Wesley Powell: Frontiersman of Science*. Lincoln: University of Nebraska Press, new series, no. 10 (1952).

Meloy, Ellen. *Raven's Exile: A Season on the Green River*. New York: Henry Holt, 1994.

Merrill, George P. "John Wesley Powell." *American Geologist*, v. 31, no. 6 (June 1903), pp. 327–33.

Moody, Tom. "Rocks, Rapids and the Hydraulic Jump." *The News*, v. 6, no. 3, (1993). (*The News* is now *Boatman's Quarterly Review*.)

Morgan, Dale L. "Introduction" to "The Powell Colorado River Expedition of 1869." *Utah Historical Quarterly*, v. 15 (1947), pp. 1–8.

Morris, Lindsey Gardner. "John Wesley Powell: Scientist and Educator." *Illinois State University Journal*. v. 31, no. 3 (Feb. 1969), pp. 3–48.

Morris, Ralph C. "The Notion of a Great American Desert East of the Rockies." *Mississippi Valley Historical Review*, v. 13, no. 2 (Sept. 1926), pp. 190–200.

Myers, Tom. "River Runners and the Numbers Game." *Boatman's Quarterly Review*, v. 10, no. 1, 1997, pp. 22–3.

Nash, Roderick. *The Big Drops: Ten Legendary Rapids of the American West*. Boulder, Col.: Johnson Books, 1989.

Nevins, Allan. *Frémont: Pathmarker of the West*. New York: Longmans, Green, 1955.

———. *Frémont: The West's Greatest Adventurer*. New York: Harper, 1928.

Newberry, J. S. "Presidential Address." *Proceedings of the American Association for the Advancement of Science* (1867), pp. 1–15.

Numbers, Ronald L. *Darwinism Comes to America*. Cambridge, Mass.: Harvard University Press, 1998.

O'Connor, Cameron, and John Lazenby, eds. *First Descents: In Search of Wild Rivers*. Birmingham, Ala.: Menasha Ridge Press, 1989.

O'Reilly, Sean, James O'Reilly, and Larry Habegger. *Grand Canyon: True Tales of Life Below the Rim*. San Francisco: Travelers' Tales, 1999.

Persons, Stow, ed. *Evolutionary Thought in America*. New York: George Braziller, 1956.

Peterson, Levi S. *Juanita Brooks: Mormon Woman Historian*. Salt Lake City: University of Utah Press, 1988.

Porter, Eliot. *The Place No One Knew: Glen Canyon on the Colorado*. San Francisco: Sierra Club, 1963.

Powell, J. W. "The Cañons of the Colorado." *Scribner's Monthly*, v. 9 (1875), pp. 293–310, 394–409, 523–37.

———. "Certitudes and Illusions." *Science* (Feb. 21, 1896). pp. 426–33.

———. "Esthetology, or the Science of Activities Designed to Give Pleasure." *American Anthropologist*, v. 1, no. 1 (Jan. 1899), pp. 1–40.

———. "Evolution of Music from Dance to Symphony." *Proceedings of the American Association for the Advancement of Science* (July, 1890).

———. *Exploration of the Colorado River of the West and Its Tributaries. Explored in 1869, 1870, 1871, and 1872, Under the Direction of the Secretary of the Smithsonian Institution*. Washington, D.C.: Government Printing Office, 1875. (Reprinted in slightly revised form as *The Exploration of the Colorado River and Its Canyons*, 1895.)

———. "Institutions for the Arid Lands." *Century*, v. 40 (May 1890), pp. 111–6.

———. "The Irrigable Lands of the Arid Region." *Century*, v. 39 (April 1890), pp. 766–76.

———. "The Lessons of Folklore." *American Anthropologist*, v. 2, no. 1 (Jan. 1900), pp. 1–36.

———. "Letters of Major J. W. Powell to the *Chicago Tribune*." *Utah Historical Quarterly*, v. 15, pp. 73–88.

———. "Major Powell's Journal." *Utah Historical Quarterly,* v. 15 (1947), pp. 125-31.

———. "Memorial Address for James Dwight Dana." *Science* (Feb. 7, 1896), pp. 181–5.

———. "The Non-Irrigable Lands of the Arid Region." *Century,* v. 39 (April 1890), pp. 915–22.

———. "Relation of Primitive Peoples to Environment, Illustrated by American Examples." *Smithsonian Institution Annual Report, 1895.* Government Printing Office, 1896.

———. *Report on the Lands of the Arid Lands of the United States.* Washington, D.C.: Government Printing Office, 1879.

———. "The Subject of Consciousness." *Science* (June 5, 1896), pp. 845–7.

Powell, Walter Clement. "Journal of W. C. Powell." *Utah Historical Quarterly,* v. 16 (1949), pp. 257–489.

Powell, Walter Henry. "Letter of W. H. Powell to the *Chicago Evening Journal,*" v. 15 (1947), pp. 90–2.

Preuss, Charles. *Exploring with Frémont.* Norman, Okla.: University of Oklahoma Press, 1958.

Price, L. Greer. *Grand Canyon: The Story Behind the Scenery.* Las Vegas, Nev.: KC Publications, 1991.

———. *An Introduction to Grand Canyon Geology.* Grand Canyon, Ariz.: Grand Canyon Association, 1999.

Pyne, Stephen J. *How the Canyon Became Grand.* New York: Viking, 1998.

Quammen, David, ed. *The Best American Science and Nature Writing 2000.* Boston: Houghton Mifflin, 2000. (Anthology includes Anne Fadiman, "Under Water.")

Rabbitt, Mary C. "John Wesley Powell: Pioneer Statesman of Federal Science." In *The Colorado River Region and John Wesley Powell.* Geological Survey Professional Paper 669. Washington, D.C.: GPO, 1969.

Redfern, Ron. *Corridors of Time: 1,700,000,000 Years of Earth at Grand Canyon.* New York: Times Books, 1980.

Reilly, P.T. "How Deadly Is Big Red?" *Utah Historical Quarterly,* v. 37, no. 2 (Spring 1969), pp. 244–60.

Reisner, Marc. *Cadillac Desert: The American West and Its Disappearing Water.* New York: Viking, 1986.

Reynolds, Ethan Allen. "In the Whirlpools of the Grand Cañon of the Colorado." *Cosmopolitan* (Nov. 1889), pp. 25–34.

Richmond, Patricia Joy. *Trail to Disaster.* Denver: Colorado Historical Society, 1989.

Roberts, David. *A Newer World: Kit Carson, John C. Frémont, and the Claiming of the American West.* New York: Simon & Schuster, 2000.

Ross, John R. "Man Over Nature: Origins of the Conservation Movement." *American Studies,* v. 16, no. 1 (1975), pp. 49–62.

Rusho, W. L. *Powell's Canyon Voyage*. Palmer Lake, Col.: Filter Press, 1969.

Sadler, Christa, ed. *There's This River: Grand Canyon Boatman Stories*. Flagstaff: Red Lake Books, 1994.

Sibley, George. "A Tale of Two Rivers: The Desert Empire and the Mountain." *High Country News*, Nov. 10, 1997.

Simmons, George C., and David L. Gaskill. *River Runners' Guide to the Canyons of the Green and Colorado Rivers With Emphasis on Geologic Features*, v. 3. Flagstaff: Northland Press, 1969.

Smith, Gusse Thomas. "The Unconquerable Colorado." *Arizona Highways* (Feb. 1947), pp. 34–9.

Smith, Henry Nash. "Rain Follows the Plow: The Notion of Increased Rainfall for the Great Plains, 1844–1880." *Huntington Library Quarterly* v. 10, no. 2 (Feb., 1947), pp. 169–93.

———. *Virgin Land: The American West as Symbol and Myth*. Cambridge, Mass.: Harvard University Press, 1950.

Smith, J. Lawrence. "Presidential Address." *Proceedings of the American Association for the Advancement of Science* (1874), pp. 1–26.

Solomon, Ben. *Kayaking on the Edge*. Birmingham, Ala: Menasha Ridge Press, 1999.

Stanton, Robert B. *Colorado River Controversies*. New York: Dodd, Mead, 1932.

———. *Down the Colorado*. Norman, Okla.: University of Oklahoma Press, 1965.

Staveley, Gaylord. *Broken Waters Sing: Rediscovering Two Great Rivers of the West*. Boston: Little, Brown, 1971.

Stegner, Page. *Grand Canyon: The Great Abyss*. New York: HarperCollins, 1995.

Stegner, Wallace. "Jack Sumner and John Wesley Powell." *Colorado Magazine*, v. 26 (1949), pp. 61–9.

———. *Beyond the Hundredth Meridian: John Wesley Powell and the Second Opening of the West*. Boston: Houghton Mifflin, 1954.

Stephens, Robert W. "Survival of the Fastest: Evolution of the Whitehall." *Wooden Boat* (Sept/Oct. 1995), pp. 48–55.

Stevens, Larry. "A Boatman's Lessons." *Plateau*, v. 53, no. 3 (1981), pp. 24–8.

———. *The Colorado River in Grand Canyon: A Guide*. Flagstaff: Red Lake Books, 1983.

———. "The 67 Elephant Theory or Learning to Boat Big Water Hydraulics." *River Runner* (Jan.-Feb. 1985), pp. 24-5.

Stevens, Lawrence E., Thomas M. Myers, and Christopher C. Becker. *Fateful Journey: Injury and Death on Colorado River Trips in Grand Canyon*. Flagstaff: Red Lake Books, 1999.

Still, Bayrd, ed. *The West: Contemporary Records of America's Expansion Across the Continent, 1607–1890*. New York: Capricorn, 1961.

Sullivan, Walter. *Landprints*. New York: Times Books, 1984.

Sumner, Jack. "John C. Sumner's Journal." *Utah Historical Quarterly*, v. 15 (1947), pp. 113–24.

Sword, Wiley. *Shiloh: Bloody April*. Dayton, Ohio: Morningside, 1974.

Teal, Louise. *Breaking into the Current: Boatwomen of the Grand Canyon*. Tucson: University of Arizona Press, 1994.

Thybony, Scott. *Official Guide to Hiking the Grand Canyon*. Grand Canyon, Ariz.: Grand Canyon Natural History Association, 1994.

Tikalsky, Frank D. "Historical Controversy, Science and John Wesley Powell." *Journal of Arizona History*, v. 23 (1982), pp. 407-22.

Townshend, R. B. *A Tenderfoot in Colorado*. Norman, Okla.: University of Oklahoma Press, 1968. (Originally published 1923.)

Twain, Mark. *Letters from the Earth*. New York: Harper & Row, 1962.

U.S. Geological Survey. *The Colorado River Region and John Wesley Powell*. Geological Survey Professional Paper 669–A.

Wallace, David Rains. *The Bonehunters' Revenge*. Boston: Houghton Mifflin, 1999.

Ward, Geoffrey C. (with Ken Burns and Ric Burns), *The Civil War: An Illustrated History*. New York: Alfred A. Knopf, 1990.

Ward, Lester. "Sketch of Professor John W. Powell." *Popular Science Monthly*, v. 20 (1882), pp. 390–7.

Watson, Elmo Scott. *The Professor Goes West*. Bloomington, Ill.: Illinois Wesleyan University Press, 1954.

Webb, Robert H. *Grand Canyon, a Century of Change: Rephotography of the 1889–1890 Stanton Expedition*. Tucson: University of Arizona Press, 1996.

Webb, Roy. *Call of the Colorado*. Moscow, Idaho: University of Idaho Press, 1994.

———. *If We Had a Boat: Green River Explorers Adventurers and Runners*. Salt Lake City: University of Utah Press, 1986.

Webb, Walter Prescott. "The American West: Perpetual Mirage." *Harper's* (May 1957). pp. 25–31.

White, Richard. *It's Your Misfortune and None of My Own: A History of the American West*. Norman, Okla.: University of Oklahoma Press, 1991.

Wilson, R. J. *Darwinism and the American Intellectual*. Homewood, Ill.: Dorsey, 1967.

Woody, Clara T., and Milton L. Schwartz. *Globe, Arizona*. Tucson: The Arizona Historical Society, 1977.

Worster, Donald. *A River Running West: The Life of John Wesley Powell*. New York: Oxford University Press, 2001.

Zwinger, Ann Haymond. *Downcanyon: A Naturalist Explores the Colorado River Through the Grand Canyon*. Tucson: University of Arizona Press, 1995.

———. *Run, River, Run: A Naturalist's Journey Down One of the Great Rivers of the American West*. Tucson: University of Arizona Press, 1975.

ACKNOWLEDGMENTS

For the past few years, I have spent my working days either burrowing deep in library subbasements or wandering through the grandest vistas on the American continent. Along the way, I have accumulated countless debts. If there is a Grand Canyon boatman I have not pestered with questions, I apologize for the oversight. Thanks to Steiner Beppu, Don Bragg, Regan Dale, Brad Dimock, Ed Hench, Zeke Lauck, Martin Litton, Scott Mosiman, Clair Quist, John Running, Bruce Simballa, Larry Stevens, and Dave Stinson. All of these patient teachers have been disabused of the notion that there is no such thing as a stupid question.

Three boatmen went miles out of their way on my behalf. Drifter Smith, who has put in nearly twenty years on the Colorado and racked up some 150 Grand Canyon trips, is a river runner, a geologist, and a student of all topics even tangentially related to the Grand Canyon. On my desk is a long, continually revised sheet of paper headed "Drifter Questions." In two years, I never stumped him. Michael Ghiglieri, another veteran boatman and an author, disagreed with my views on Powell, on writing, and on everything in between.

He fussed and fretted, and helped, at every stage of this long journey. Ben Solomon is a superb kayaker and a better-than-superb teacher. He labored mightily in an attempt to convey something of how a river looks when seen through a professional's eyes.

Stacy Allen, Jeff Hall, and Andy Trudeau helped me navigate the literature of the Civil War. Michele Missner, a researcher who is herself a precious find, tracked down endless arcane articles. Kate Headline labored so diligently in pursuit of Grand Canyon photos that museum and library curators across the nation came to recognize her voice.

Thanks to Jane Meredith Adams, a dazzling writer, for wise and thoughtful editorial counsel, and to David Smith, for careful and diligent scrutiny of the entire manuscript. Both these writers neglected their own work in favor of mine.

Rafe Sagalyn, my agent and my friend, shepherded this project along from before Day One. Hugh Van Dusen proved as superlative an editor as his reputation had led me to expect, and higher praise is hard to imagine.

Ruth and Bill Holmberg read this manuscript in its earliest incarnations. For their editorial advice, and countless other gifts, I owe more than I know how to express. For Lynn, Sam, and Ben, for bottomless reserves of inspiration, insight, and encouragement, my fervent and inadequate gratitude.

INDEX

Page numbers of illustrations appear in italics.

MAP OF THE
SOUTHWEST UNITED STATES
1869

Central Pacific Railroad

UTA
TERRIT

NEVADA

CALIFORNIA

GRAND CANYON

SEPARATION
CANYON

ARIZONA
TERRITOR

ML

7/82